Asymmetry in Plants

Asymmetry in Plants
Biology of Handedness

Edited by
Bir Bahadur
K. V. Krishnamurthy
Monoranjan Ghose
S. John Adams

CRC Press
Taylor & Francis Group
Boca Raton London New York

CRC Press is an imprint of the
Taylor & Francis Group, an **informa** business

CRC Press
Taylor & Francis Group
6000 Broken Sound Parkway NW, Suite 300
Boca Raton, FL 33487-2742

First issued in paperback 2020

© 2019 by Taylor & Francis Group, LLC
CRC Press is an imprint of Taylor & Francis Group, an Informa business

No claim to original U.S. Government works

ISBN-13: 978-1-138-58794-6 (hbk)
ISBN-13: 978-0-367-73059-8 (pbk)

Visit the Taylor & Francis Web site at
http://www.taylorandfrancis.com

and the CRC Press Web site at
http://www.crcpress.com

Contents

Foreword

Left-right asymmetries offer a fascinating opportunity to study the interplay between development, morphology, and evolution on a broad stage. What makes them so intriguing is that unlike most aspects of organismal form, common questions may be asked across a wide range of taxa (e.g., protists, fungi, plants, animals). What determines direction of asymmetry, genes, environment, or chance? How is symmetry broken during development? What is the functional significance of asymmetry? How did asymmetrical forms evolve from symmetrical ancestors?

Plants exhibit a glorious diversity of asymmetrical forms, growth patterns, and movements. The history of research on them is long and colorful, including early work by Darwin on spiral growth in climbing plants. But many important discoveries and syntheses are in out-of-the way venues, which has lessened their accessibility and impact.

In this welcome volume, Bir Bahadur and the co-editors bring together in one place a broad survey and synthesis of what is known about asymmetries in plants, with the added bonus of a chapter on asymmetries in fungi. In addition to overviews of concepts, and surveys of floral and vegetative asymmetries, the contributions cover an impressive range of phenomena including: molecular mechanisms of asymmetrical development; cellular asymmetry; spiral growth of vascular tissue; embryo rotation and seedling handedness; helical growth and movement of vines, tendrils, and the growing tips; the ever-mesmerizing spiral patterns of petals, bracts, scales, and seeds and their concordance to Fibonacci series; symmetry and asymmetry in flower petal arrays; and stylar polymorphisms including enantiostyly (left- and right-bending styles).

Settle down with a steaming mug of coffee or tea and savor these many engaging perspectives on plant asymmetries.

A. Richard Palmer, FRSC
Systematics and Evolution Group
Department of Biological Sciences
University of Alberta
Edmonton, Alberta, Canada

Preface

Symmetry is always appealing, whether it is in art, craft, and architecture or in sciences like biology and chemistry. One of the very important and obvious manifestations of pattern in living organisms or their parts is their symmetric construction. Symmetry is noticed in both their external form and internal organization. But yet asymmetry or symmetry breaking is also as common as or even more common than symmetry in nature. A very critical study of asymmetry reveals that its presence in living organisms should not be considered as an erroneous deviation from the norm but as an evolutionary and functional adaptation to the environmental conditions operating in any situation at any time. In several instances it is selected through Natural selection, provided it has selection and fitness values. In fact, without some asymmetric features the organism may not even survive. Therefore, from an evolutionary and functional perspective, asymmetry is as appealing and "beautiful" as symmetry.

All through academic history, symmetry/asymmetry has attracted the attention of animal scientists to a far greater extent than the plant scientists, although this phenomenon in plants is almost as common as in animals. Several interesting books, review articles, and research papers have been published on animal asymmetry till this date. Although the Darwin's have indicated its importance in plants as early as in the latter part of the nineteenth century and D'Arcy Thomson in 1917 had made a beautiful treatment (from a mathematical perspective) on it in his famous book "On Growth and Form", not much has been done in understanding asymmetry in plants. Kihara's 37-page review in 1992 on "Right- and left-handedness in plants: A review" and a chapter on symmetry by Sinnott in 1960 in his famous book on "Plant Morphogenesis" were perhaps the first detailed accounts on symmetry/asymmetry in plants. These accounts created some renewed interest on this subject among plant scientists in different parts of the world but this interest was not sufficiently sustained. The only book, as far as the present authors are aware of, which used the word "Symmetry" in its title was "Symmetry in Plants" by R. V. Jean and D. Barabe published in 1998. This book almost exclusively covered phyllotaxy, with a greater emphasis on mathematical aspects. Thus, there was an urgent need to reemphasize the need for renewing the interest on symmetry/asymmetry in plants.

Hence, this edited volume was conceived. This book has two major goals: one is to consolidate as much as the scattered information on the various aspects of symmetry/asymmetry in plants in one place. The second is to highlight the gaps that are still wide open in our understanding of this phenomenon in plants so that future researchers may undertake fruitful studies. With these two goals in mind, we have identified topics that needed greater and focus coverage and left out topics like phyllotaxy, where we have fairly sufficient knowledge. We have also identified competent people to write on the topics chosen, from across the world. The book is organized into 24 chapters, including one on concluding remarks. We hope that this volume would be of great interest to students, teachers, and researchers in general biology, botany, genetics, evolution, mathematics, and computation biology. We would like to apologize for any mistakes, omissions, and failures that might have crept into the book; suggestions for improvement are welcome.

We wish to express our grateful thanks to all the contributing authors for readily accepting our request and promptly sending their excellent contributions. We are also thankful to them for enduring the editorial suggestions and comments on their manuscripts and willfully improved their writing. We also acknowledge the huge support and suggestion that we received from many distinguished colleagues like Prof. Denes Nagy (Budapest, Hungary), Prof. A. R. Palmer (Edmonton, Alberta, Canada), Prof. C. P. Malik (New Delhi, India), Dr. Menno Schilthuizen (Netherlands), and several others. We are also thankful to our respective family members for bearing with us for not spending our time with them during the preparation of this book. We wish to express our appreciation for the help rendered by Dr. Chuck Crumly, senior editor, and Ms. Jennifer Blaise, senior editorial assistant/associate editor of CRC Press, Taylor and Francis Publisher, and their team including Ms. Preethi Sekar of Lumina Datamatics during the various stages of production of this book.

Bir Bahadur
K. V. Krishnamurthy
Monoranjan Ghose
S. John Adams

About the Editors

Dr. Bir Bahadur, born April 5, 1938, studied at City College, Hyderabad, for 5 years, including an Intermediate Course (Osmania University, Hyderabad, Telangana), graduated from Nizam College, Hyderabad, Telangana, and postgraduated from University College, Osmania University, both in the first division. He obtained his PhD in Plant Genetics from Osmania University. He was closely associated with late Prof. J.B.S. Haldane, F.R.S., a renowned British geneticist who encouraged him to study heterostyly and incompatibility in Indian plants, a subject first studied by Charles Darwin. He made significant contributions in several areas, especially heterostyly, incompatibility, plant genetics, mutagenesis, plant tissue culture and biotechnology, morphogenesis, application of SEM in botanical research, *plant asymmetry*, plant morphology and anatomy, and the biofuel plants Jatropha and castor. He served as Lecturer and Reader at Osmania University and as Reader and Professor at Kakatiya University, Warangal, Telangana. He also served as Head of Department; Chairman, Board of Studies; Dean, Faculty of Science; and Coordinating Officer, UGC Affairs at Kakatiya University. He has over 40 years of teaching and over 50 years of research experience. He has supervised 29 PhD students and 3 MPhil. students in both these universities and has published about 250 research papers/reviews/chapters, which are well received and cited in national and international journals, textbooks, and reference books.

He was a postdoctoral fellow at the Institute of Genetics, Hungarian Academy of Sciences, Budapest, and worked on mutagenesis and chromosome replication in *Rhizobium*. He is a recipient of the direct award from the Royal Society Bursar, London worked at Birmingham University (UK). He was conferred with the title of Honorary Research Fellow by the Birmingham University. He studied species differentiation in wild and cultivated solanums using interspecific hybridization and the enzyme-etched seeds technique in combination with scanning electron microscopy to assess the relationship among various *Solanum* species. At the invitation of the Royal Society, he visited Oxford, Leeds Reading, London Universities, including the Royal Botanic Gardens, Kew. He was invited for international conferences by the US Science Foundation at the University of Missouri, St. Louis, and the University of Texas, Houston (USA), and at the SABRO international conference at Tsukuba, Japan. He has extensively visited most countries of Eastern and Western Europe as well as Tanzania and the Middle East.

He has authored/edited ten books. One of his important books is entitled *Jatropha, Challenges for a New Energy Crop*, Vols. 1, 2, and 3 published by Springer, New York, USA, 2013, jointly edited with Dr. M. Sujatha and Dr. Nicolas Carels. These books are considered significant contributions to bioenergy in recent times. He has also edited a five-volume book on *Ethnobotant of India* jointly with Prof. T. Pulliah and Dr. K. V. Krishnamurthy from CRC Press. He was chief editor in *Proceedings of Andhra Pradesh Akademi of Sciences* (Hyderabad) and executive editor in *Journal of Palynology* (Lucknow).

He is the recipient of the Best Teacher Award by the Andhra Pradesh Government for mentoring thousands of students in his teaching career spanning over 40 years. He was honored with the Prof. Vishwamber Puri Medal of the Indian Botanical Society for his original contributions in various aspects of plant sciences. He has been honored with the Bharat Jyoti Award at New Delhi for outstanding achievements and sustained contributions in the fields of education and research. He has been listed as 1 of the 39 prominent alumni of City College, a premier institution with a long history of about 98 years as per the latest update on its website. He has been chosen for distinguished standing and has been conferred with an Honorary Appointment to the Research Board of Advisors by the Board of Directors, Governing Board of Editors, and Publications Board of the American Biographical Institute, USA. He is a fellow of over a dozen professional bodies in India and abroad, including the following: Fellow of the Linnean Society, London and Chartered Biologist and Fellow of the Institute of Biology, London.

K. V. Krishnamurthy is currently working as a consultant at R & D, Sami Labs Ltd. before joining Sami; he was an adjunct professor in the Institute of Trans-Disciplinary Health Science and Technology (IHST), Bangalore, India, and offers consultancy services in Ayurvedic Pharmacognosy. He obtained his MSc in Botany with University First rank from Madras University, Chennai, in 1966 and PhD in Developmental Plant Anatomy from the same university in 1973. After a brief stint in Government colleges in Tamil Nadu, he joined the present Bharathidasan University, Tiruchirappalli, in 1977 and became a full professor in 1989. He has more than 47 years of teaching and research experience and has guided 32 PhD scholars, more than 50 MPhil scholars, and 100 master degree holders. He has published more than 172 research papers and 20 books including *Methods in Cell Wall Cytochemistry* (CRC Press, USA) and a *Textbook of Biodiversity* (Science Publishers, USA). His major research areas include plant morphology and morphogenesis, biodiversity, wood science, cytochemistry, plant reproductive biology and ecology, waste land reclamation, tissue culture and herbal medicine, and pharmacognosy. He has operated more than 15 major research projects so far. He has been a Fulbright visiting professor at the University of Colorado, Boulder, in 1993 and has visited and given lectures in various universities in UK in 1989. His outstanding awards and recognitions include the following INSA Lecture Award 2011; Prof. A Gnanam Endowment Lecture Award, 2010; President 2007 Indian Association for Angiosperm Taxonomy; Prof. V. Puri Award 2006 by Indian Botanical Society; Rashtriya Gaurav Award 2004 by India International Friendship Society, New Delhi; Scientist of the Year Award 2001 by National Environmental Science Academy, New Delhi; Tamil Nadu State Scientist Award 1997–1998 in the field of Environmental Science; Dr. V. V. Sivarajan Gold Medal Award by Indian Association on Angiosperm Taxonomy for Field study in the year of 1997–1998; Prof. Todla Ekambaram Endowment Lecture Award, Madras University, 1997; Prof. G. D. Arekal Endowment Lecture Award, Mysore University, 1997–1998; Prof. V. V. Sivarajan Endowment Lecture Award, Calicut University, 1997; Prof. Rev. Fr. Balam Memorial Lecture Award 1997; the 1984 Prof. Hiralal Chakravorthy Award instituted by the Indian Science Congress in recognition of the significant contributions made to the Science of Botany, Madras

University as University First in MSc Botany, 1960; Dr. Pulney Andi Gold Medal awarded by Madras University as University First in MSc Botany, 1966; Dr. Todla Ekambaram Prize awarded by Madras University for standing First in MSc Plant Physiology; The Maharaja of Vizianagram Prize awarded by Presidency College, Madras, for outstanding postgraduate student in Science, 1965–1966; and Prof. Fyson Prize awarded by Presidency College, Madras, for the best plant collection and herbarium, 1965–1966. He has been a fellow of the following: Fellow of National Academy of Sciences of India (FNASc.); Fellow of the Linnaean Society, London (FLS); Fellow of Indian Association for Angiosperm Taxonomy (FIAT); Fellow of International Association of Wood Anatomists, Leiden; Fellow of the Plant Tissue culture Association of India; and Fellow of the Indian Botanical society. He has been the editor and editorial member of many journals in and outside India and has also been reviewer of research articles for many journals. He also served in various committees the major funding organization of India and several universities of India. He has been the Registrar, Director of college and curriculum Development Council, Member of Syndicate and Senate, Coordinator of the school of Life Sciences and Environmental Sciences, Head of the Department of the Plant Science, and a visiting professor in the Department of Bioinformatics in Bharathidasan University, Tiruchirappalli, before assuming the present job after retirement.

Monoranjan Ghose joined as lecturer at the Agricultural & Ecological Research Unit of the Indian Statistical Institute Kolkata, India, in 1983. Subsequently he became associate professor and full professor in 1997. He served as professor-in-charge of Biological Sciences Division and Head of the Agricultural and Ecological Research Unit of the Indian Statistical Institute, Kolkata, till retirement. He worked as the chairman of the Research Fellow Advisory Committee (RFAC) of the Biological Sciences Division. He retired from active service from the Institute on July 31, 2012.

Professor Ghose did his PhD under the guidance of the late Prof. T. A. Davis (The Fibonacci Man of India) in 1982. He has published more than 60 research papers in reputed national and international standard journals. His early works were mainly on anatomical aspects of palms, published in journals like *Phytomorphology*, *Principes*, *Acta botanica Neerlandica*, etc. His three research articles on palms were quoted by Professor P. B. Tomlinson, well-known international authority on palms, in his book "The Structural Biology of Palms" (Clarendon Press Oxford, 1990). Prof. Ghose has worked jointly with the renowned scientist, Prof. B. M. Johri, and published three papers on palms.

In addition, Prof. Ghose researched on the mangroves vegetation of the Sundarbans, India. His several PhD students worked on (i) anatomy, morphology, and palynological aspects of mangroves, (ii) eco-physiology and anatomical aspects, (iii) community structure and biomass estimation of mangroves, and (iv) mycorrhizal status of mangroves of the Sundarbans. Prof. Ghose was also actively engaged on the ecology, propagation, and utilization of rattan palms of the Northeast India, such as *Calamus* sp., *Demonorhops* sp., and *Plectocomia* sp.

Prof. Ghose attended many national and international conferences and presented papers. He is a fellow of the Indian Botanical Society and the Indian Association of Biological Sciences and a life member of many academic societies such as

International Society of Plant Morphologists, Indian Botanical Society, Indian Association of Biological Science, and Indian Science Congress Association.

S. John Adams is at present a scientist in the Department of Phyto-Pharmacognosy at Sami Labs ltd., Bangalore, a leading herbal based industry. Previously he was a research fellow at Foundation of Revitalization of Local Health and Traditions (FRLHT), Bangalore, for four years and subsequently a research associate at Department of Pharmacognosy, The Himalaya Drug Company, Bangalore, for a year. He completed his higher education in St. Joseph's College, Tiruchirappalli, and Madras Christian College, Chennai, Tamil Nadu. His PhD work is on the Ayurvedic Ashtavarga plants. He has published 22 research articles, some of which have appeared in *Plant Biology and Biotechnology* published by Springer, India and in *Ethnobotany of India* (Volumes 1–5) published by CRC: Apple Academic Press, USA. He (along with Prof. K. V. Krishnamurthy) has just prepared a book on *Methods in Plant Histochemistry*. His present interest includes analysis of plant organization from morphological and developmental perspectives.

Contributors

S. John Adams
Research and Development
Sami Labs Limited
Bengaluru, India

Thiago Magalhães Amorim
Programa de Pós-Graduação em
Ecologia e Recursos Naturais
Universidade Federal do Ceará
Fortaleza, Brazil

Bir Bahadur
Department of Botany
Kakatiya University
Warangal, India

Oleh Bodnar
Institute of Architecture
Lviv Polytechnic National University
Lviv, Ukraine

Robyn J. Burnham
Department of Ecology and
Evolutionary Biology
University of Michigan
Ann Arbor, Michigan

Rainiellen S. Carpanedo
Instituto de CiênciasNaturais
Humanas e Sociais
Universidade Federal de Mato Grosso
Sinop, Brazil

Ana Isabel D. Correia
Faculdade de Ciencias
Universidade de Lisboa
Lisboa, Portugal

Cristiane Miranda da Cruz
Instituto de CiênciasNaturais
Humanas e Sociais
Universidade Federal de Mato Grosso
Sinop, Brazil

Fernão Vistulo de Abreu
Departamento de Fisica
Universidade de Aveiro
Aveiro, Portugal

Natan Messias de Almeida
Department of Biology
Universidade Federal Rural de
   Pernambuco
Garanhuns, Brazil

Wladimir Hermínio de Almeida
Instituto de Ciências Naturais
Humanas e Sociais
Universidade Federal de Mato
   Grosso
Sinop, Brazil

Cibele Cardoso de Castro
Department of Biology
Universidade Federal Rural de
   Pernambuco
Garanhuns, Brazil

**Rozangela Cristina Alves de
Oliveira**
Instituto de CiênciasNaturais
Humanas e Sociais
Universidade Federal de Mato Grosso
Sinop, Brazil

Evandro F. dos Santos
Instituto de Ciências Naturais
Humanas e Sociais
Universidade Federal de Mato Grosso
Sinop, Brazil

Aline C. S. Dresch
Instituto de Ciências Naturais
Humanas e Sociais
Universidade Federal de Mato Grosso
Sinop, Brazil

Monoranjan Ghose
Biological Sciences Division
Agricultural and Ecological Research
  Unit
Indian Statistical Institute
Kolkata, India

Maria Helena Godinho
Departamento Ciência dos Materiais
Faculdade Ciências e Tecnologia
Universidade NOVA de Lisboa
Caparica, Portugal

Polina V. Karpunina
Faculty of Biology
Lomonosov Moscow State University
Moscow, Russia

A. V. P. Karthikeyan
P.G. and Research Department of Botany
Government Arts College
  (Autonomous)
Karur, India

Robert W. Korn
Department of Biology
Bellarmine University
Louisville, Kentucky

K. V. Krishnamurthy
Research and Development
Sami Labs Limited
Bengaluru, India

Thiruppathi Senthil Kumar
Department of Botany
School of Life Sciences
Bharathidasan University
Tiruchirappalli, India

Monique Machiner
Instituto de Ciências Naturais
Humanas e Sociais
Universidade Federal de Mato Grosso
Sinop, Brazil

T. N. Manohara
Tree Improvement and Genetics
Institute of Wood Science and
  Technology
Bangalore, India

C. Manoharachary
Department of Botany
Osmania University
Hyderabad, India

D. Nagaraju
Department of Botany
Government Degree College
Warangal, India

Riichirou Negishi
Advanced Science Research
  Laboratory
Saitama Institute of Technology
Fukaya, Japan

Maxim S. Nuraliev
Faculty of Biology
Lomonosov Moscow State
  University
Moscow, Russia

Alexei A. Oskolski
Department of Botany and Plant
  Biotechnology
University of Johannesburg
Johannesburg, South Africa

Christian Westerkamp
Department of Agronomia,
Universidade Federal do Cariri
Crato, Brazil

Lucinere P. Pinto
Instituto de Ciências Naturais
Humanas e Sociais
Universidade Federal de Mato
  Grosso
Sinop, Brazil

T. Pullaiah
Department of Botany
Sri Krishnadevaraya University
Anantapur, India

A. J. Solomon Raju
Department of Environmental Sciences
Andhra University
Visakhapatnam, India

Túlio Freitas Filgueira Sá
Department of Biology
Universidade Federal Rural de
    Pernambuco
Garanhuns, Brazil

M. A. Akbar Sha
Department of Animal Sciences
Bharathidasan University and Research
    Coordinator
National College (Autonomous)
Tiruchirappalli, India

Pedro E. S. Silva
Department of Materials Science
NOVA University
Caparica, Portugal

Andrey A. Sinjushin
Associate Professor
Genetics Dept., Biological Faculty
M.V. Lomonosov Moscow State
    University

Arlete Aparecida Soares
Departamento de Biologia
Universidade Federal do Ceará
Fortaleza, Brazil

Dmitry D. Sokoloff
Faculty of Biology
Lomonosov Moscow State University
Moscow, Russia

Bradley Spilka
Department of Ecology & Evolutionary
    Biology
University of Michigan
Ann Arbor, Michigan

N. Rama Swamy
Department of Botany
Kakatiya University
Warangal, India

K. Tennakone
Former Director
Institute of Fundamental Studies
Sri Lanka

Christian Westerkamp
Agronomia, Universidade Federal do
    Cariri
Crato, Brazil

1 The Concepts of Handedness, Asymmetry, Chirality, Spirality, and Helicity

K. V. Krishnamurthy, Bir Bahadur,
Monoranjan Ghose, and S. John Adams

CONTENTS

1.1 INTRODUCTION

A critical study of history shows that the concepts relating to the "metaphorical good and sinister" (Levin et al., 2016), now known by such terms as right-/left-handed, symmetric/asymmetric, achiral/chiral, etc., have had a four-step evolution. These concepts arose as a philosophic component; subsequently became a religious entity; later got transformed into a social reality; and finally got incorporated as scientific fact. This, then, is the story of a traditional belief that gradually became a practical reality and prevalent in human art and craft, architecture, music and dance, designs, etc. ca 5000 years ago (Coren and Porae, 1977). The philosophy of all great ancient civilizations of the world—Mesopotamian, Greek, Roman, Egyptian, Indus Valley, Aryan, Dravidian, and Chinese—strongly believed in these concepts, which got deeply rooted in the firmament of human experience and worldviews. The philosophy of these civilizations generally, but not always, equated good with right-handedness and symmetry and the sinister with the opposite, although most of these civilizations were strongly biased towards good, right-handedness and symmetry. For example, according to ancient Chinese philosophy, all things on this earth were either the proverbial *Yin* or *Yang*; *Yin* was associated with femininity, submission, blood, dark earth, right-handedness, etc., while *Yang* with masculinity, leadership, light, sun, left-handedness, etc. The Pythagorean philosophy lists ten principles, each of which consisted of pairs of opposites; right is listed on the same side as male, straight, light, darkness, evil, etc. Greek philosophers like Aristotle always associated the right with good and the left with evil and criminality. Plato was convinced that both right and left are naturally of equal strength and ability and that any significance attached to anyone of them is culturally imparted. Both are very vital for "completeness," but as Levin et al. (2016) puts it, "Too much in either direction disrupts a critical balance" and that "symmetry unchecked by asymmetry transmutates order, harmony and beauty into static, sterile and monotonous"; similarly, "asymmetry unchecked by symmetry becomes aberrant, unrestrained and chaotic." This philosophic concept continued to prevail throughout history, and is even now commonly seen in almost all philosophic discourses, works, and literature (see Giemsa, 2017).

Slowly, the philosophic tone of good and sinister (i.e., right/left and symmetry/asymmetry) was incorporated into traditional and formal religion (see Findly, 2008). Ancient Egyptians, Greeks, Romans, and Indians believed that the right hand of gods/goddesses had healing and benevolent character while their left hand inflicted wounds and injuries. Thus, the analogy to right and left hands came into religious fold, although the modern descriptive words that particularly mean/indicate right and left hands were not used.

Under social context, the strongly antileft Egyptians often depicted their enemies as left-handed. Most of the ancient societies and cultures used the right hand for performing ceremonies and for eating. An ancient Assyrian decorative bas-relief sculpture shows figures using their right hands for performing a ritual. In medieval times, right-handed warriors had a distinct advantage to sword fights; they held their shield with their left hand.

Nowhere else in the world than in the medieval Tamil country in south India did the concept of handedness became an absolute social reality. As far as the authors of this chapter are aware of, the use of the word "handedness" seems to have originated around 10th century CE in the Tamil country. The term *Valangai* (a Tamil word meaning right hand) is first mentioned in the Chola king Rajaraja I's period epigraphic inscription to denote the right wing of his army (Venkayya, 1948). Subsequently, an early 11th century inscription speaks of *Valangai* peoples' endowments to a temple (IMP. Vol. 2 p. 287, 341/1907 dated CE 1014). References to the word *Idangai* (a Tamil word meaning left hand) first appeared in CE 1072 (A.R.I.E. 1936–1937, para 27), during another Chola king Kulothungan I's period, which speaks of clashes between the left-handed and right-handed social communities. Thus, by the late 11th century CE, there is evidence of two broad, and at times hostile, social divisions of the people in the societies of Tamil country as well as in other parts of south India. However, detailed analyses of the origins and functions of these dual divisions of the society have posed difficulties (Kearns, 1876); these still remain obscure. The connotations appear to be more positional, with *Valangai* people being positioned on the right-hand side of gods, sages, Brahmins or kings (Srinivasachari, 1929). *Valangai* people were considered to be in a high position in social hierarchy, while *Idangai* were inferior. However, it was difficult to determine with certainty the castes/subcastes belonging to each of the two divisions, and sometimes this distinction was applied to wealthy *versus* poor, settled *versus* nomadic, merchants *versus* farmers, etc. depending on place and time. It is likely that the terms *Valangai* and *Idangai* that prevailed in use very strongly in pre-colonial south India (and prevailed even more strongly due to aggravation by the colonists, particularly by the British, during the colonial period) (Brimness, 1999) were selected by the British to indicate any two contrasting and conflicting entities/situations; it is also likely that British considered it fertile to search for a single, consistent and substantive property underlying the conflicts between the right- and left-hand social systems, but conflicts themselves were important to them; further, handedness seemed to them to provide a general form through which these conflicts could be expressed (Brimness, 1999). It is only after the 16th century that in British writings the word "handedness" were used, and in all probability this usage was "borrowed" from Tamil country in India, which was colonized by the British, since several British government official documents on the administration of Tamil country were deposited in British archives, many of which are in fact used by Brimness (1999) while writing his book.

The actual use of the word "handedness" in science, particularly in natural science, may be said to have come into force during the 18th century CE. The origin of the word "symmetry" is believed to be in the middle of 16th century CE either from the French word "symmétrie," from the Latin word "symmetria" or from the Greek word "summetria." The word "chirality" is stated to have been derived from the Greek χείρ (pronounced as "kheir"), which means "hand," a very familiar chiral object. This word was believed to be first used by Lord Kelvin in 1893 in the second Robert Boyle Lecture at Oxford University, which was published in 1894. We do not have information on the origins of usage of "spiral" and "helix," although these may have originated long back in history. Oster, (1969; 1974) described various types of natural and physical spirals.

1.2 HANDEDNESS

In this chapter, the structures that are situated to the right side and to the left side of a body are respectively called right-handed and left-handed, although these words are also used, as established in physics, to define right-handed (where movement of a body turns clockwise away from the observer) and to define left-handed (where movement turns counterclockwise away from the observer) (Gardner, 1990) However, in biological literature, for the latter two situations, dextral (right-handed, spiral, and helical) and sinistral (left-handed, spiral, and helical) are generally applied. However, in plants, particularly in palms where leaf arrangement (phyllotaxy) is alternate, one may be able to see the leaves arranged in three or more spirals veering upwards in clockwise or counterclockwise direction. The clockwise palm is regarded as "left-handed" and the counterclockwise, "right-handed" (Davis, 1971). This property of handedness is also said to be exhibited by two forms of structures, which are mirror images of one another (i.e., chiral structures) and which cannot be super imposable on each other (Palmer, 2005). Palmer also emphasizes that handedness is a term that implies a sense of orientation of bilateral or spiral asymmetries and that right- and left-handedness is more familiar than dextral and sinistral; however, Palmer prefers the latter two for three reasons of which only one seems to be scientifically reasonable and the other two are semantic.

1.3 ASYMMETRY

The Oxford online dictionary defines symmetry as the "quality of being made up of exactly similar parts facing one another or around an axis." In mathematical and geometrical terms, symmetry is generally defined as "invariance of the configuration of elements under a group of automorphic transformations," such as congruence, which can be obtained by composing "rotations, reflections and translations" (Prusinueurez and Lindenmayer 2004). Classical mathematical biologists like D'Arcy Thompson (1969) consider the sphere (such as in a spherical cell) as the most perfect example for symmetrical objects and all other derived shapes as asymmetric (such as in elliptical and cylindrical cells). Asymmetry, opposite to symmetry, means "the lack of equality or equivalence between parts or aspects of something," or "having parts which fail to correspond to one another in shape, size or arrangement." Although, according to Palmer (2005), asymmetry is a form that is not superimposable on its mirror image, it may not be the only point that characterizes asymmetry.

A number of terms have been used in the past to document asymmetry in a population or populations of a species. The terms used include: complete right- versus left-handed symmetry; racemic; antisymmetry; pure antisymmetry; incomplete asymmetry; complete asymmetry; weakly monostrophic; strongly monostrophic; extremely monostrophic; directional asymmetry; handed asymmetry; and biased asymmetry. Palmer, (2005) simplified the terminology and resolved the confusing issues. His recommendations are followed in this chapter and the following four terms are accepted: (i) directional asymmetry is applied to any population where the "rarer enantiomorph" does not make up significantly more than 5%; (ii) enantiomorphy is applied to any population, which is "a mixture of two enatiomorphs" with the rarer form making up "significantly more than 5%; (iii) antisymmetry is applied to any population, which is a 'mixture of two enantiomorphs' that have equal frequency in the population. In other words, it implies equality of any two quantities for which it holds in both directions." Each individual within any population or species is conspicuously a symmetric, but the direction of this asymmetry is rather random (= random asymmetry). Antisymmetry is a very important type of phenotypic variation since, barring very few exceptional cases, the direction of asymmetry is not inherited; for example, asymmetric phenotype is heritable but "right-handed" phenotype or left-handed phenotype within that cannot be inherited and hence antisymmetry lacks a genetic basis; (iv) Biased antisymmetry is applied to any population with a mixture of two enantiomorphs out of which one is significantly more frequent than the other; rarer enantiomorph is significantly less than 50% but significantly more than 5% in the population.

The expression "broken symmetry" should be explained here. This usage was long known in physics (Anderson, 1972) but was introduced into biology only recently (see Piano and Khemphues 2000; Pillitteri et al., 2016). Broken symmetry essentially means the loss of symmetry, usually under the influence of some internal or external factor/factors (see Chapter 5 in this volume).

Whenever we visualize symmetry/asymmetry, the presence of an axis, which is so characteristic of the plant or its parts, is often realized by us to manifest, "although not so much as an actual material structure but as an axis of symmetry, a geometric core of plane around which or on the two sides of which the structures are symmetrically disposed" (Sinnott, 1960). It means that for some plant structures, such as a cell, cell organelle, or even an organic molecule, there is no material axis but only an imaginary/theoretical axis. In other words, it can be said that symmetry exists at all levels of biological organization and integration (Figure 1.1), from the lowest and simplest level to the highest and most complex level. On the smallest scale, an asymmetry in the form of parity violation was found with angular distribution of electrons in the beta decay of spin-oriented ^{60}Co. Thus, symmetry is seen in both the

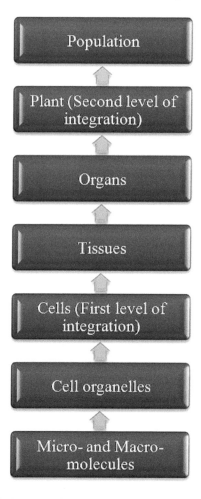

FIGURE 1.1 Levels of plant organization and integration. (From Krishnamurthy, K.V., *Growth and Development in Plants*, Scientific Publishers, Jodhpur, India, 2015.)

external form and internal structure of any organism, such as a plant, as well as in its various parts, however small the latter may be. On the largest scale, the universe with its galaxies is seen with well-defined asymmetry (Gardner, 1990; Longo, 2008; Giemsa, 2017).

Symmetry can be radial or rotational, dorsiventral or translational, bilateral or reflexional, helical/spiral, or a combination of these (Kazlacheva, 2013). In plants, some components show radial symmetry (cylindrical stems and roots), some dorsiventral symmetry (leaves, petals, etc.), others bilateral symmetry (certain flowers), and yet others helical symmetry (phyllotaxy). There can be one or two evenly placed longitudinal plants of symmetry, especially in macroscopic plant structures, or they can even be in almost infinite planes of symmetry. Radial symmetry is very common in plants and is considered by many as the basic, primitive, or conserved form of symmetry, especially by those who believed in the Telome theory and the axial theory of flowers. In bilateral symmetry, there are two planes of symmetry so that the front and back, and right and left sides of the structure are similar. In dorsiventral symmetry, there is only one plane of symmetry. Details on helical/spiral symmetry are given later in this chapter.

The four forms of symmetry can be explained as follows (see Palmer, 2004, 2005). The coordinate system, which provides positional information to the developing plant (or animal) or its various parts, consists of four axes and not three, as most people imagine (Figure 1.2). The anteroposterior and dorsiventral axes are both single axes, and they define the midplane. No single, left-right axis exists since no single physical or chemical gradient apparently extends from left to right. Rather, the left and right axes are "separate mediolateral axes which originate at and extend in opposite directions away from the midplane." Since these two mediolateral

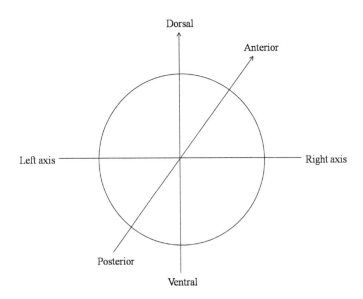

FIGURE 1.2 The four axes of a developing biological structure that provide it positional information.

axes are mirror images, "an extra symmetry breaking step must occur for one to differ from the other." This implies that bilateral symmetry is a default state once the anteroposterior and dorsiventral axes are defined; similarly, radial symmetry, according to Palmer (2004), is a default state when only one axis exists. Left-right differences, therefore, arise, according to Palmer (2004) because some kind of switch causes the mediolateral axis on one side to differ from the axis on the other side (Meinhardt, 2001), although the mechanism involved remains nuclear in most organisms. Conspicuous asymmetries have evolved independently and several times in many plants (and animals). The great varieties of morphological and structural symmetries (Ludwig, 1932; Palmer, 1996, 2004) defy simple categorization.

One can also speak of organic and inorganic symmetries (Sinnott, 1960), the latter often resembling that of the former. We do not know the degree of relationships between the two, although some believe that the latter might have given rise to the former. These relationships are discussed to some extent by Sinnott (1960) and D'Arcy Thompson (1969). However, it can be said that the molecular forces, such as surface forces (for example, surface tension), that operate in inorganic asymmetry are probably not important in determining the symmetry of larger organic bodies, such as plant organs, but certainly play an important role in organic symmetry exhibited at the subcellular, cellular, and tissue levels in plants. D'Arcy Thompson (1969) emphasized that surface tension is important in surfaces with a minimum area (not absolute minimum but relative minimum). Therefore, asymmetries at different levels of plant organization and integration (Figure 1.1) should be dealt with quite differently, at least from a mathematical and physical perspective, although some basic principles governing asymmetry at different levels of organization and integration may be identical. It should also be emphasized that there is a possibility for the infinite number of similar planes of symmetry around a theoretical or material organic axis in contrast to the very limited number of two to six planes of symmetry in inorganic symmetries, in that the latter has two distinct traits: possession of multiple parts and the helical arrangement of these parts (see later for a discussion on helicity and spirality).

Symmetry and organic form are very closely related to one another; any change in organic form is associated with asymmetry. The transition from regular symmetry to asymmetry is a very vital and cardinal event in plant organization and function. This transition can be seen at all levels of biological organization. The causes of organic symmetry/asymmetry are not fully known. In multicellular plants, the origin of symmetry (or asymmetry) can be traced back to cell division, growth, and interrelationships of cells at the meristematic regions (see Korn in this volume). Dorsiventral symmetry, unlike radial symmetry, is, in most cases, not established at the meristem itself but has its origin in changes that arise later, but it is certain that it is derived from radial symmetry (D'Arcy Thompson, 1969).

1.4   CHIRALITY

An object/system is said to be *chiral* if it is distinguishable from its mirror image and that it cannot be superimposed on its mirror images (Palmer, 2004). A chiral macroscopic object and its mirror images are called *enantiomorphs* (meaning "opposite forms"), while chiral molecules are called *enantiomers* (Palmer, 2005). The latter is widely used in chemical literature. An object or molecule that is not chiral is called *achiral* or *amphichiral*. It is surprising to note that chiral molecules in organisms exist almost exclusively as single enantiomers. In a population of species, if only one of two possible mirror-image forms occurs, and if the orientation is not specified, then the term "homochirality" is used. A mixture of two mirror-image forms with unequal or equal frequency is referred to by terms such as "heterochirality" (Palmer, 2004). The term "antimer" is used often to describe the homologous structure on the opposite side of an individual (Palmer, 2005).

Molecular chirality can be compared to the right- and left-handedness of the human, which are normally mirror images of each other; the only important difference between the two hands is in the direction that one takes to go from the thumb to the other fingers. This sense of direction is termed chiral handedness and, with reference to a molecule, whether it has a left and right orientation. Thus, two molecules can be mirror images of each other, alike in almost all aspects except for their handedness. The building blocks of proteins contain only one of the two mirror-image forms of amino acids (i.e., L-forms), and sugars of nucleic acids are only one of the D-forms (Riehl, 2011).

1.5   SPIRALITY AND HELICITY

1.5.1   SPIRALITY

In biological literature, spirality and helicity are often confused with one another, and many people including learned scientists use them interchangeably (e.g., Sinnott, 1960); this confusion must be first set right. Generally speaking, a spiral is a type of curve, which starts from "a point of origin and from that point continually diminishes in curvature as it recedes from that point"; in other words, its radius of curvature continually increases (D'Arcy Thompson, 1969) (Figure 1.3). It may also be said that spirality is the property of a curved line that extends outward from the point of

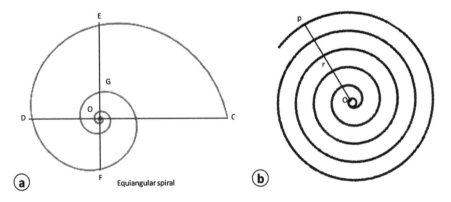

FIGURE 1.3   Two types of spirals: (a) equiangular or logarithmic spiral. C, D are the right-left axes; E, F are dorsal-ventral axes; O is the point of origin of spiral; and G is the spiral; and (b) Archimedes spiral. o is point of origin of the spiral; r is the radius of vector (straight line having its extremity in the pole and revolving about it; and p is the point traveling along the radius vector under definite conditions of velocity.

origin in a "path of continuously increasing distance." This property can apply to a single plane as in a clock spring or to a core-shaped space (Smyth, 2016). These are very wide definitions, and these include a number of different curves, but these definitions definitely exclude a curve that is made by a simple screw, a cylindrical helix. This curve neither starts from a definite point of origin nor changes its curvature as it proceeds further (D'Arcy Thompson, 1969). Helix is technically not a spiral, as it has a fixed radius (Smyth, 2016). This distinction must be definitely followed while dealing with spiral and helical structures of plants (and animals). Further discussion on helicity is found later in this chapter.

We do not lack true organic spirals, and a huge variety of these, mostly on animals, are described and beautifully illustrated in Cook (1903, 1914) and D'Arcy Thompson (1969). In the animal kingdom, examples of spirals include the horns of ruminants, molluscan shells, foraminifera shells, a lock of hairs on a human head, a staple of wool, an elephant's trunk, circling spires of a snake, coils of a cattle-fish's arm, a monkey's or chameleon's tail, etc. These spiral forms are all examples of the remarkable curve known as the equiangular or logarithmic curve or spiral (Figure 1.3a). This type of spiral continually increases in breadth and does so in a steady and unchanging ratio. "Any plane curve proceeding from a fixed point and such that the vectorial area of any sector is always a gnomon to the whole preceding figure, is called an equiangular, or logarithmic, spiral" (D'Arcy Thompson, 1969). The spiral of objects mentioned above, although they may be different from one another in outward appearance, in nature and origin, they all belong to one particular class of conformations. In the majority of cases, there is no or very little reason to consider one part of the existing spiral structure as older as or younger than another. But in the mollusc shell or horns of ruminants, part of the existing structure is older and part younger. These structures, although belonging to a living animal, are themselves not living but are formed by living cells. Thus, these structures show an equiangular spiral increase or accumulate materials rather than grow in the true sense. Hence, in a logarithmic spiral, there is always a time element.

Let us now focus our attention on plant structures that are categorized under spiral curves. D'Arcy Thompson (1969) cites the examples of the scorpioid and helicoids cymose inflorescences (Figure 1.4), respectively also called bostryx and cincinnus, and the head inflorescence as in the sunflower. In the helicoids inflorescence, the primary axis gives rise to a secondary axis at a certain definite angle; the secondary axis in turn gives rise to another axis on the same side at the same angle, and this process continues but the lengths of successive axes diminish at a constant ratio. The deflection/curvature, thus, is continuous and progressive in which no extrinsic but only intrinsic force operates. We may regard each successive axis as "forming, or defining, a gnomon to the preceding structure" (D'Arcy Thompson, 1969). In scorpioid cyme, successive secondary axes are formed on alternate sides but all at equal angles of divergence just equal to 180° as in the helicoids cyme. In both cases, according D'Arcy Thompson, there is an equiangular spiral, but botanists without understanding the mathematical significance have labeled wrongly the inflorescence as helicoid (it is truly a spiral not a helix). In the sunflower, according to D'Arcy Thompson (1969), a spiral conformation is built by many separate and successive florets and that the serially arranged portions of this composite structure are similar to

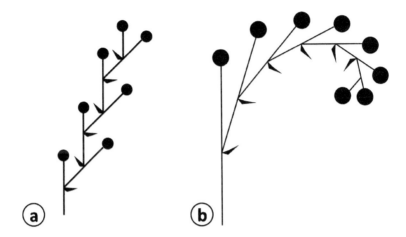

FIGURE 1.4   Diagrammatic figures of the scorpioid (a) and helicoid (b) inflorescences.

one another in form but differ in age and also in their magnitude. Thus, according to him, the head inflorescence shows an equiangular spiral with the significant presence of a time element in it. However, according to Mathai and Davis (1974), the formation of the individual florets on the sunflower head are formed one at a time on the disc. A flower primordium is differentiated on a side of the stem apex, and the subsequent florets are generated at a fast rate with a constant time interval between any two consecutive individuals. As the flower gets differentiated, the tip of the meristematic axis rotates so much so that the older florets are seen to move away from growing point in logarithmic spirals that approximate an Archimedes' spiral, whose numbers invariably match with the terms of the Fibonacci Sequence (see details in Chapter 20). With reference to phyllotaxy, most if not all investigators have included it under the spiral category (Martinez et al., 2016) and words like "genetic spiral," "phyllotactic spiral," and "ontogenetic spiral" are used to emphasize this point. However, there is also the use of the word "foliar helices," implying phyllotaxy as a helical phenomenon. A recent review by Smyth (2016) on helical growth did not include phyllotaxy under it, and another review by Palmer (2005) classifies phyllotaxy under spiral growth. When the leaves are distributed far apart from another on the stem and if their midpoints are connected and a "genetic spiral" is obtained, the resultant curve is a helix rather than a spiral in the real sense, but if the leaves are crowded at the shoot tip and if observed from the top, the curve connecting the midpoints of all these leaves will describe an equiangular figure, characteristic of a typical spiral. It is important to recall here the definitions that distinguish a typical spiral from the atypical helix, the latter having a fixed radius and an unchanged curvature unlike in the former. This point should be kept in mind while discussing phyllotaxy (see Chapter 14). The so-called spiral grain in woods should also discussed only after ascertaining whether it is a helical or spiral phenomenon (see Chapter 4), and it is pertinent to point out here that the spiral grain is included under helical growth in the recent review by Smyth (2016).

The other type of spiral is the equable spiral or the spiral of Archimedes. In this type of spiral the coil in each whorl is of the same breadth. This spiral may be described as follows: "If a straight line revolves uniformly about its extremity, a point which likewise travels uniformly along it will describe the equable spiral" (D'Arcy Thompson, 1969). There is no clear example for the Archimedes spiral in plants (perhaps also in animals).

1.5.2   HELICITY

Although the definitions of spirality given at the beginning of this section are wide enough to include a number of different curves, they exclude, as we already indicated, at least one that "in popular speech we are apt to confuse with a true spiral" (D'Arcy Thompson, 1969). This latter curve, as already stated, is the simple screw, or cylindrical *helix*. Several structures at different levels of biological organization, exhibit helicity. The helical thickening of tracheary elements, the "spiral" thread, is within the tracheal tubes of insects. The "spiral" grain in woods, elaters of some cryptogamous plants, the "spiral" twist, and twining of climbing stems or of tendrils, circumnutation, etc. are not, mathematically speaking, spirals at all but are helices. Less well-known helical phenomena are the specialized twisting leaf and floral stems, coiling valves of dehiscing fruits, resupination (Hill, 1939; Endress, 1999), coiling awns of grasses, etc. (see Smyth, 2016). Helices belong to a distinct, though not very much unrelated, family of curves. Further to the characterization of a helix already made earlier, it may be added that the helix is "a line that turns around a rod-shaped structure or space moving uninterrupted from its one end to the other" (Smyth, 2016). A helix is often defined by its length, radius, or pitch, and it can be simultaneously extended at right angles to the plane.

1.5.3   DIRECTION OF SPIRALITY AND HELICITY

The interpretation of spiral or helical asymmetries is beset with difficulties and is not as simple as one may imagine. These difficulties were reviewed elegantly by Kihara (1972) for plants and by Galloway (1983) and Robertson (1993) for animals. Kihara (1972) had traced the history of confusion that prevailed regarding the direction of spiraling/turning. Based on his analysis, Kihara classified the botanists into *orthodox* and *antiorthodox* categories. Most zoologists, biochemists, physicists, engineers, and anti-orthodox botanists followed one convention, according to which a helix or spiral is termed dextral if the direction traced by a point moving from the near end to the far end is clockwise *as viewed from the near end*, and if not, it is termed sinistral. Dextral and sinistral spirals and helices are very important for morphogenesis in plants. Chapter 4 describes the various issues related to the direction of spiral and helix.

1.6   RELATIONSHIP BETWEEN THE VARIOUS CONCEPTS

Are the concepts of handedness, asymmetry, chirality, spirality, and helicity independent of one another? If so, how are they related to each other? If not independent of one another, how are they different from one another, and are the differences strong enough to differentiate them? These questions are naturally raised because there is a great amount of confusion in the use of these words by earlier investigators. There is often a combination of these five words used: left-right asymmetry, chiral asymmetry, left- and right-hand chirality, chiral handedness, mirror-image asymmetry, asymmetric chirality, chirality as a property of handedness, etc. As already pointed out, *spiral* and *helical* are the most confusedly used words in biological literature. Some researchers have used two or more of these words in one and the same publication. There is also the use of the words *isomerism* and *bioisomerism* (Bahadur et al., 1977a, 1977b).

Realizing the confusion that prevailed, Palmer (2005) first attempted to sort out the terminology. He considered all these terms a spectrum of asymmetric variation and that the terminology used in the past to describe this was cumbersome and confusing. He cited four reasons for this confusion: (i) departures from symmetry are either subtle or very conspicuous. Some technical terms used to refer to the former are also used to refer to the latter, in spite of the fact that their precise meanings are different; (ii) there is a great variety of conspicuous asymmetries, and naming each one of them is rather difficult, for example naming different spirals or helices; terms used for helical asymmetries look strange when used for bilateral asymmetries; (iii) terms relating to the "direction" of a "whole body, spiral or helical symmetry are defined arbitrarily" when compared to those of a "bilaterally paired structure"; and (iv) terms defining symmetries of individuals without any ambiguity lose this property/precision when applied to populations of the same species.

We should appreciate Palmer (2005) for bringing order amongst disorder that prevailed all along in literature relating to existing terminology. However, the authors of this chapter feel that there are subtle differences in the connotations that these five terms really indicate. Already the differences between "spiral" and "helix" have been clearly mentioned earlier in this chapter. Hence, they are different from another. Chirality and handedness are both used to describe any form that is not superimposable on its mirror images. The two mirror images of a chiral molecule are called enantiomers, and pairs of enantiomers are usually designated as right- and left-handed (see Bahadur and Rao, 1981; Bahadur et al., 1983) if there is bias (if there is no bias, the word "achiral" is preferable). We feel that the use of chirality may be restricted to molecules and not to be extended to organic forms bigger than molecules. For the latter, we prefer handedness. In both chirality and handedness, the mirror images have the same overall symmetry except for the directional change. Any deviation or alteration from symmetry is asymmetry; hence, the resultant asymmetric structure need not always be a mirror image of the original structure. For example, an oblique leaf lamina is definitely asymmetric, but one-half of the lamina (on one side of midrib) is not a mirror image of the other half of the lamina.

1.7 EVOLUTIONARY SIGNIFICANCE

Most studies that had been made on handedness, chirality, asymmetry, spirality, and helicity have been largely descriptive, some have highlighted their functional importance, and only very few have dealt with their evolutionary importance (and most of these are on animals—see Palmer, 2005, 2016). While speaking on these aspects,

D'Arcy Thompson (1969) emphasized that the studies should be more analytical, by which terminology he implied not only a mathematical analysis but also a functional and evolutionary analysis. A study of these components would bring forth an explanation of why such phenomena as these five have a functional and a fitness value in terms of standard evolutionary theories such as Darwinism. Hence, it would be better in future studies to begin by describing the phenomena at different levels of biological organization, followed by an analysis of the same based on physico-chemical and mathematical principles, and finally by discussing them in terms of functional importance and of the tenets of evolutionary theories. A mathematical definition/analysis has the quality of precision, which is lacking in mere description. We should supplement with the positive and negative forces that decide the functional adoption of these, which, in turn, forms the basis for evolutionary explanation. For example, most enantiomers in nature exist in the sinistral form (Blackmond, 2010). Finding the origin of the increased sinistral and decreased dextral conditions is vital to the understanding of the origin and evolution of life on this planet. We should also know why most scientists believe that homochirality is a precondition for life.

REFERENCES

Anderson, P.W. 1972. More is different. *Science.* 177: 393–396. doi:10.1126/science. 177.4047.393.

Bahadur, B., Kumar, P.V., Reddy, N.P., and Rao K.L. 1977a. Bio-isomerism in spathes of *Xanthosoma violacium. Curr. Sci.* 46: 869–870.

Bahadur, B. and Rao, M.M. 1981. Seedling handedness in Fabaceae. *Proc. Indian Acad. Sci.* 50: 242–243.

Bahadur, B., Reddy, M.M., Ramaswamy. N., and Narsaiah, G. 1983. Seedling handedness in Gramineae. *Proc. Indian. Acad. Sci.* 92: 279–284.

Bahadur, B., Reddy, N.P., and Kumar, P.V. 1977b. Bio-isomerism in Cycads. *Curr. Sci.* 47: 404–405.

Blackmond, D. 2010. The origin of biological homochirality. *CSH Perspect. Biol.* 2(5): a002147. doi:10.1098/rstb.2011.0130.

Brimness, N. 1999. *Constructing the Colonial Encounter: Right and Left Hand Castes in Early Colonial South India.* Curzon Press, Richmond, VA.

Cook, T.A. 1903. *Spirals in Nature and Art.* London, UK.

Cook, T.A. 1914. *Curves of Life.* London, UK.

Coren, S. and Porae, C. 1977. Fifty centuries of right-handedness: The historical record. *Science.* 194: 631–632. doi:10.1126/science.335510.

D'Arcy Thompson, W. 1969. *On Growth and Form.* Abridges Edition. Ed. J.T. Bonner. Cambridge University Press, Cambridge, UK.

Davis, T.A. 1971. Right-handed, left-handed, and neutral palms. *Principles.* 15: 63–68.

Endress, P.K. 1999. Symmetry in flowers: Diversity and evolution. *Int. J. Plant. Sci.* 160: Suppl. S2–S23. doi:10.1086/314211.

Findly, E.B. 2008. *Plant Lives: Borderline Beings in Indian Traditions.* Motilal Banarsidass Publishers Pvt. Ltd., Delhi, India.

Galloway, J. 1983. Helix through the looking glass. *New Science.* 97: 242–245.

Gardner, M. 1990. *The New Ambidextrous Universe.* W. H. Freeman and Company, New York.

Giemsa, B. 2017. Mirror-image asymmetry, chirality and suttree. *Eur. J. Am. Studies.* 12: 1–29. doi:10.4000/ejas.12336.

Hill, A.W. 1939. Resupination studies in flowers and leaves. *Ann. Bot.* 3: 871–887.

Kazlacheva, Z. 2013. Symmetry in nature and symmetry in fashion design. *EMIT.* 1(4): 266–276.

Kearns, J.F. 1876. The right-hand and left-hand castes. *Indian Antiquity.* 5: 353–354.

Kihara, H. 1972. Right-and left-handedness in plants: A review. *Seiken Ziho.* 23: 1–37.

Krishnamurthy, K.V. 2015. *Growth and Development in Plants.* Scientific Publishers, Jodhpur, India.

Levin, M., Klar, A.J.S., and Ramsdell, A.F. 2016. Introduction to provocative questions in left-right asymmetry. *Phil. Trans Royal Soc. B.* 371: 1–7. doi:10.1098/rstb.2015.0399.

Longo, M.J. 2008. Does the Universe have a handedness? *Astrophysics.* 1–11.

Ludwig, W. 1932. *Das Rechts-Links Problem im Tierreich Und beim Menschen.* Springer, Berlin, Germany.

Martinez, C.C., Chitwood, D.H., Smith, R.S., and Sinha, N.R. 2016. Left-right leaf symmetry in decussate and distichous phyllotactic systems. *Philos. Trans. R. Soc. B.* 371: 20150412.

Mathai, A.M. and Davis, T.A. 1974. Constructing the sunflower head. *Math. Biosci.* 20: 117–133.

Meinhardt, H. 2001. Organizer and axes formation as a self-organizing process. *J. Dev. Biol.* 45: 177–188.

Oster, G. 1969. *The Science of Morie Patterns.* 2nd Edition. Edmund Scientific, Barrington, NJ.

Oster, G. 1974. L' evolution des especes les spirales. *La Reserache.* 44: 316–324. (In French).

Palmer, A.R. 1996. From symmetry to asymmetry: Phylogenetic patterns of asymmetry variation in animals and their evolutionary significance. *Proc. Nat. Acad. Sci. USA.* 93: 14279–14286.

Palmer, A.R. 2004. Symmetry—Breaking and the evolution of development. *Science.* 306: 828–833.

Palmer, A.R. 2005. Antisymmetry. In: *Variation, A Central Concept of Biology.* Benedikt Hallgrimsson, B & Hall, B.K. (Eds.). Elsevier, New York. pp. 359–397.

Palmer, A.R. 2016. What determines direction of asymmetry genes, environment or chance? *Phil. Trans. R. Soc. B.* 371: 20150417.

Piano, F. and Kemphues, K.J. 2000. Cell polarity. In: *Cell Polarity.* Drubin, D.G. (Ed.). Oxford University Press, Oxford, UK.

Pillitteri, L.J., Guo, X., and Dong, J. 2016. Asymmetric cell division in plants: Mechanism of symmetry breaking and cell fate determination. *Cell Mol. Life Sci.* 73: 4213–4229. doi:10.1007/s00018-016-2290-2.

Prusinueurez, P. and Lindenmayer, A. 2004. *The Algorithmic Beauty of Plants.* Springer-Verlag, New York.

Riehl, J.P. 2011. *Mirror-Image Asymmetry: An Introduction to the Origin and Consequences of Chirality.* John Wiley & Sons, Hoboken, NJ.

Robertson, R. 1993. Snail handedness. The coiling directions of gastropods. *Natl. Geogr. Res. Explor.* 9: 109–119.

Sinnott, E.W. 1960. *Plant Morphogenesis.* McGraw-Hill, New York.

Smyth, D.R. 2016. Helical growth in plant organs: Mechanisms and significance. *Development.* 143: 3272–3282.

Srinivasachari, C.S. 1929. The origin of the right and left hand caste divisions. *J. Andhra Hist. Res. Soc.* 1: 77–85.

Venkayya, S.V. 1948. Introduction. *South Indian Inscriptions.* 2: 10.

2 Left-Right Symmetry in Animals and Plants
A Comparison

M. A. Akbar Sha, K. V. Krishnamurthy, and S. John Adams

CONTENTS

2.1 INTRODUCTION

It is a well-known fact that animals and plants co-evolved and that both of them exhibit quite a number of distinctive types of interactions between them. Several characteristic features of both these groups of organisms are being constantly subjected to selection pressures and fitness tests for survival. Symmetry/asymmetry is one such characteristic feature. It is a potential aspect for making a comparative taxonomic and evolutionary study of animals and plants. This feature can be studied fairly easily and in a very meaningful way across many unrelated taxa of both these groups of organisms at all levels of biological organization and integration because of its binary-switch nature (Palmer 2004) and because it allows the character to be scored easily and without much ambiguity. Right from Protista and Plantae to Animalia, all groups possess an array of left-right asymmetries in their body design (Hashimoto 2002; Levin et al. 2016), which have evolved independently in many groups (Ludwig 1932; Kihara 1972; Palmer 2004, 2005, 2009, 2016). In extant organisms, particularly those belonging to higher taxonomic categories, the establishment of body form/shape needs a precise developmental pattern as well as coordination of anterior-posterior (A/P) and dorsal-ventral (D/V) axes, which define the midplane and the left and right sides lying on opposite sides of this midplane. By convention, the right side is the side to the right of the A/P axis (it is defined from the perspective of the animal) (Neville 1976). In flowers, however, the right side is defined from the perspective of the observer facing a mature flower (Endress 2001) (see more discussion in Chapter 1).

This chapter examines comprehensively the phenomenon of left-right asymmetry and its importance in the structure and function of animals and plants. Two specific questions are addressed: (i) what causes/favors asymmetry, and (ii) what causes/favors asymmetry in a specific direction, as raised by Palmer (2016).

2.2 LEFT-RIGHT SYMMETRY

During the course of evolution, developmental systems for all levels of biological organization (from molecules to whole organisms) and integration (cells, organs, organ systems, and whole organisms) in animals and plants have invariably converged to left-right symmetry with two types: helical/spiral and bilateral (Palmer 2016). Both these symmetries have two states: right-handed (i.e., dextral) and left-handed (i.e., sinistral). This convergence was required to respond appropriately to the various forces of nature, such as gravity, wind, tension, stress, strain, solar, and magnetic, that operated on the organism and its parts (Ludwig 1932; McManus 2002; Levin et al. 2016). These forces are generated because the organisms and, therefore, their various parts live in a very mobile medium, such as air or water, or because they are mobile in themselves even if they live in a static medium. Hence, many authors consider bilateral symmetry as the "default developmental mode" for many higher organisms or their various parts (Palmer 2004, 2009, 2016; Levin et al. 2016).

2.2.1 INTERNAL AND EXTERNAL LEFT-RIGHT ASYMMETRY

Organ asymmetries are found throughout the animal kingdom. These asymmetries pertain to positional asymmetries, morphological asymmetries, or both (Blum and Ott 2018). There are also functional asymmetries as in the human brain. Left-right symmetry often affects the internal parts in animals, which are normally protected by virtue of their position. A critical study shows that the internal parts of animals are affected by symmetry/asymmetry to a much greater extent than the internal parts of plants, most probably because most organs of animals are internal and all organs of plants are external. However, there are internal plant tissues that might exhibit asymmetry. The direction of "spiral" grains in wood and helical xylem tracheary elements may be cited as examples. We may mention the positions of heart (on the left side), liver (on the right side), lung with lesser number of lobes on the left than on the right due to space restrictions, stomach and spleen (on the left), and the two brain hemispheres inside the human body as examples of internal organs in animals that exhibit left-right asymmetry (Blum and Ott 2018; Monsoro-Burq and Levin 2018).

Although internal asymmetry, which is invariably of primary origin (i.e., embryo) in animals and meristematic origin in plants (i.e., both are from stem cells), may be common, secondary asymmetry may also be seen in externally located parts of animals. For example, the unequal claws of male fiddler crabs and many other decapods, the anteriorly directed incisors of narwhals, the ear openings of some owls, the position of genitalia in some insects, in the bill of crossbill finches, the mouth of some fishes, and the twisted abdomen of many male insects are solitary medial structures that deflect/radiate to one side causing asymmetry. In some extreme cases, the whole animal body may be asymmetrical, as in some barnacles, flat fishes, and coiled shell snails, although these are secondary to bilaterality during early to fully developed stages (see Palmer 2004 for more details).

Examples of asymmetric external plant parts include zygomorphic and resupinated flowers, some flowers with skewed positions of gynoecium/stamen or twisting/contortions of floral leaves (see Endress 2012), direction of phyllotoctic spirals, anisophylly, heterophylly, unequal lamina on either side of the midrib, seedling handedness, and left-handed and right-handed climbing/twining patterns in whole climbers or their climbing organs like tendrils (Edwards et al. 2007). Circumnutation that gets initiated in the growing embryo itself (see Chapter 9) and appears to manifest throughout the life of the plant (see Chapter 10) is a phenomenon that is exclusive to plants and does not appear to be present in animals. This phenomenon is vital to control the longitudinal growth of plants, phyllotaxy, climbing of plants and plant parts like tendrils, and the left-right direction of phyllotactic (or genetic) spirals. Monsoro-Burq and Levin (2018) discuss the left-right asymmetry from normal and pathological conditions. The cases of heart, lung, brain, liver, spleen, etc., are normal occurrences, *situs solitus*, whereas developmental abnormalities resulting in complete inversion, *inversus totalis*, can occur but do not cause physiological malfunctions. On the other hand, mirror duplication of the left- or right-sided organs, for example, duplication of the right atrium, a condition of isomerism, is pathological. There are cases of right isomerism (duplication of the right heart chamber, of the right lung, and of the liver, loss of spleen, abnormal positioning of the stomach) and cases of left isomerism (multiple non-functional small spleen-like structures, often associated with gastrointestinal abnormal rotation and cardiac anomalies). Many such cases of isomerism end up in pathological conditions to the extent of being reckoned as syndromes.

2.2.2 IMPORTANCE IN REPRODUCTION

Left-right asymmetry is very important to many animals in reproduction and, thus, has great adaptive value. In *Drosophila* (Lang and Orgogozo 2012) and other insects (see Huber et al. 2007; Schilthuizen 2007), as well as in many other animals with internal fertilization, this sexual asymmetry is very important. For example, in the tree snail, *Amphidromus inversus* (Schilthuizen et al. 2012) dextral and sinistral enantiomorphs differ in mating behavior, and these may be very important in maintaining sexual dimorphism. Dextral and sinistral individuals are also important in zoogeographical analysis as has been revealed by Hoso's (2012) work on snails in oceanic islands and in the adjacent mainland. There is a higher percentage of coil-reverse snail species on oceanic islands than on the mainland.

Left-right symmetric features in the floral organs of flowering plants also have great importance in sexual reproduction. Detailed studies have shown that features like herkogamy, heterostyly, enantiostyly, invertostyly, and temporal differences in timing of maturation and/or receptivity promote cross-pollination and out-crossing and avoid self-fertilization (Jesson and Barrett 2002; Iwata et al. 2012). These features, thus, have a great importance in plant population genetics and in controlling allele frequencies.

2.2.3 ORIGIN AND EVOLUTION OF LEFT-RIGHT SYMMETRY

As already stated, bilateral symmetry usually arises as a primary feature of the organism. In animals, it usually originates during embryonic development, for example, as early as the zygote or later with the development of limbs. On the other hand, in some groups of animals the fertilized egg or early embryo itself, such as those undergoing spiral cleavages, is already handed (Henry and Martindale 1999; Davison et al. 2016). Attention may be drawn here to the *Nodal* gene, which is present from radially symmetric animals to vertebrates, including humans (Blum and Ott 2018). This gene is very important in controlling asymmetry through its activity in the early embryo itself (e.g., in chicks) before the beginning of organogenesis. This gene's product, the signaling protein Nodal, is stated to be key for determining left-right asymmetry (Blum et al. 2014). Nodal acts in conjunction with cilia for symmetry breaking during embryonic development. In flowering plants, external bilateral symmetry of the whole plant arises as the embryo matures, either with two cotyledons (in dicots) or with one cotyledon (in monocots). However, there are some instances, both in animals and plants, where bilateral asymmetry may arise as a secondary feature in organisms that normally have symmetry (Palmer 2004).

An analysis of the diversity of left-right asymmetry mentioned above reveals that there are two fundamentally different, but yet easily distinguishable, categories (Palmer 2004): (i) antisymmetry or random asymmetry, and (ii) directional asymmetry or fixed asymmetry. In antisymmetry, left- and right-handed forms are equally common within a population of a species regardless of parental phenotypes. In directional asymmetry, most individuals are asymmetrical in the same specific direction. The difference between the two categories is whether or not it's inherited. In antisymmetry, remarkably, the direction of asymmetry is almost not inherited (see Table S1 of Palmer 2004 for list of plant and animal traits). In contrast, in directional asymmetry it is typically inherited (Harvey 1998) (see Table S1 of Palmer 2004 for list of plant and animal traits). This distinction has been made by a detailed study carried out by Palmer (2004) in which he had reviewed the information available then on both animals and plants.

Hashimoto (2002) speaks of a presumed genetic control for the direction of tendril or stem coiling in climbing plants, which normally shows fixed handedness (dextral or sinistral). As examples, he cites hops and honeysuckle as sinistral and convolvulus as dextral. However, he also cites the exception of *Lygodium*, a climbing fern, in which the rachis of the leaf shows a random handedness while climbing. He also gives the examples of *Datura stramonium* in which the petal contortion is fixed anticlockwise and *Nerium oleander*, which is also fixed but clockwise. However, in the above cases no genetic/mutant studies have been carried out to prove their handedness as genetic.

The emergence of asymmetry (= symmetry breaking) and its direction is explained by two fundamentally different models (Palmer 2016): (i) determination of asymmetry by cell-level chirality (see details in Chapter 3), and (ii) determination of asymmetry by an asymmetric signal on one side of the body during later development. According to Palmer (2016), this signal can arise from at least three different sources: (a) asymmetric gene expression, (b) asymmetric effects of environment (random or biased toward one side), and (c) by change or stochastic factors. If the embryo itself is already handed, there is no need for asymmetry. Both models can co-exist with the first, both in animals and plants.

2.2.3.1   Role of Genetics in the Determination of Direction of Symmetry

While discussing these aspects, Palmer (2016) has dealt with three representative animals: (1) the coiling direction of the gastropod shell, in which three instead of two alternative alleles are involved—left, right, and stochastic. He concludes that genetic determination of the direction of coiling in this group of animals is quite labile; (2) the eye-side of flat fish, which involves a weak or polygenic inheritance; and (3) the bill-crossing direction in crossbill finches (*Loxia* sp.) is not inherited.

With reference to plants, Palmer (2016) stated that plant asymmetries may be either inherited or not inherited depending on the plant species. These random asymmetries include seedling handedness, leaf-rolling direction, spiral/helical orientation, and floral feature (see Kihara 1972). In all these cases, the direction of asymmetry is not inherited. The only known exception is the direction of the style in the flowers of *Heteranthera* (see Chapters 19 and 20). This feature is controlled by a single locus, and right bending is dominant over left bending (Jesson and Barrett 2002).

It is a pity that there are only very a few studies on whether the direction of asymmetry is inherited in plants where asymmetries are not random. Palmer (2016) cites the seed-pod coiling direction in the *Medicago* species as the notable example. In some species of this genus, the pod coils to the right and in others to the left (Lilienfeld 1959); the coiling direction in two dextral species is controlled by two alleles at the single locus with dominant dextrality. The dominant allele corresponds to the predominant phenotype in the natural populations of these species. Some genetic studies have also been carried out in *Arabidopsis* mutants. In these mutants, different alleles for α and β-tubulins or for microtubule-associated proteins result in helical growth in roots, hypocotyls, petioles, petals, etc., (Hashimoto 2002; Buschmann et al. 2009; see also Chapters 3 and 8). Interestingly, these mutants are spiral (dextral or sinistral) while the wild types are symmetric. These mutants are exclusively derived in the laboratory and are not so far known from natural populations. Moreover, this plant is not a climbing taxon.

2.2.3.2   Role of Environment in the Determination
of Direction of Asymmetry

Environment is an important factor in the determination of direction of asymmetry. We often find in the environment some left- and some right-handed individuals. The preferred handedness side varies at random, but the phenotypic effects of this handed behavior—for instance, the greater use of one side (say, right)—works to transform the other side (left side). This is a case where the direction of asymmetry is not simply stochastic but biased toward one side.

Palmer (2016) discusses the examples of lobsters with claw asymmetry where a lobster with two cutter-type claws gets transformed into a lobster with one cutter-type and one crusher-type, after the first five juvenile instars, in the sixth instar. This is a clear case of change due to an asymmetric signal on one side of the body, with the environment playing an important role. The other examples discussed by Palmer (2016) are the katydids (Tettigoniidae)—bush crickets—and their sound-producing structures and earwigs with left- and right-handed penises; in the latter, the use of the right-handed penis had preceded asymmetry.

The environment plays a very important role in the asymmetric structures of plants also. The best examples are plants with heterophylly (i.e., having more than one type of leaf). In partially submerged aquatic plants like *Ceratophyllum* and *Limnophila*, aerial leaves are entire whereas submerged leaves are highly dissected. The aquatic environment plays a very critical role here. However, we do not know if this feature is gene controlled and heritable, and we also do not know if these plants fail to develop heterophylly if grown in a non-aquatic environment. The cases of *Artocarpus* and some *Acacia* species are worth mentioning here. The leaves of juvenile plants of *Artocarpus* are not entire, but leaves of mature plants are entire. Young plants of some *Acacia* species possess compound leaves without a phyllode, but leaves of older plants are only phyllodes. This again has not been proved to possess a genetic base but purely an age-related phenomenon.

Most leaves possess true bilateral symmetry (i.e., laminae on either side of the midrib compared to the other side); we may say that there is a directional asymmetry (Klingenberg 2015; Martinez et al. 2016). Recent sensitive techniques on measuring leaf shape indicate the directional asymmetry along the left or right side of a leaf is likely to be more common than so far thought. For example, Chitwood et al. (2012a) and Martinez et al. (2016) demonstrated it in *Arabidopsis* and tomato, and this asymmetry in leaves is dependent on the handedness of the plants that bear these leaves. Also, leaf morphological features, such as venation, leaf shape, curling, coiling, and resupination, may become asymmetrically skewed in relation to spiral phyllotaxy patterns (i.e., left or right) (Korn 2006; Chitwood et al. 2012b).

2.2.3.3   Role of Stochastic Factors in the Determination of Direction of Asymmetry

Chance factors are also important in the determination of the direction of asymmetry. Unlike the role of the environment and genes, stochastic factors cannot be manipulated at will and, therefore, the direction cannot be biased. Palmer (2016) discusses some instances of stochastic control in the animal system based on the conclusion that stochastic control can clearly occur as part of normal, regulated development.

Based on a number of evidences, Palmer (2016) mentions that genes for the direction of asymmetry "are followers almost as often as they are leaders in the evolution of fixed left-right asymmetry" (see also Palmer 2004). Palmer (2016) further raises the question as to how a genetic variation can "capture" an already existing phenotypic variation related to direction of asymmetry. To explain this, he takes the instance of a plant as an example: the enantiostyly of *Heteranthera multiflora*. In this species, all the styles in all flowers of an individual plant bend only left or right, that is, they bend in the same direction. Some individuals show only left-bend styles, and other individuals show only right-bend styles (see Chapters 19 and 20). It is interesting to note that in most species of this genus, stylar bending within an individual plant is random (Jesson and Barrett 2003), and the direction of this asymmetry is not inherited at all. But in *H. multiflora*, two alleles at a single locus control the style-bending direction, and this must have evolved by a process of genetic assimilation (see Palmer 2004, for detailed discussion on genetic assimilation). Schilthuizen and Gravendee (2012) sum up the status as follows: "Exciting advances in genetic and experimental research on left-right asymmetry of plants have been made over the past 40 years. We now know the twisted growth of corkscrew hazelnut and asymmetry of flowers of *Heteranthera* seems controlled by a single gene." It is likely that within the next decade, the identity of these genes will be discovered and that these discoveries will throw greater light on the genetic basis of symmetric plant traits as well as on their adaptive values.

2.3 CONCLUSIONS

The field of left-right asymmetry is in a vibrant state of advancement. This is propelled by identification of a steady stream of molecular-level work performed in a range of model systems. Models proposed to explain laterality mechanisms remain controversial. There are still a number of unresolved questions about it and its evolutionary history. The microtubule-controlled model appears to be more relevant to plant and ciliary/flagellar left-right asymmetry. The morphogen model and the positional information model appear to be more theoretical models and need to be substantiated by more experimental results. Also, molecular investigations on left-right asymmetry are largely restricted to animals, wherein plants are somewhat neglected.

REFERENCES

Blum, M. and Ott, T. 2018. Animal left-right asymmetry. *Curr. Biol.* 28: R293–R305.

Blum, M., Feistel, K., Thumberger, T., and Schweickert, A. 2014. The evolution and conservation of left-right patterning mechanisms. *Development.* 141: 1603–1613.

Buschmann, H., Hauptmann, M., Niessing, D., Llyod, C.W., and Schaffner, A.R. 2009. Helical growth of the *Arabidopsis* mutant *tortifolia2* does not depend on cell division patterns but involves handed twisting of isolated cells. *Plant Cell.* 21: 2090–2106.

Chitwood, D.H., Headland, L.R., Ranjan, A., Martinex, C.C., Braybrook, S.A., Koenig, D.P., Kuhlmeier, C., Smith, R.S., and Sinha, N.R. 2012a. Leaf symmetry as a developmental constraint imposed by auxin-dependent phyllotactic patterning. *Plant Cell.* 24: 1–10.

Chitwood, D.H., Naylor, D.T., Thamapichai, P., Weeger, A.C.S., Headland, L.R., and Sinha, N.R. 2012b. Conflict between intrinsic leaf asymmetry and phyllotaxis in the resupinate leaves of *Alstroemeria psittacina*. *Front. Plant Sci.* 3: 182.

Davison, A., McDowell, G.S., Holden, J.M., Johnson, H.F., Koutsovoulos, G.D., Liu, M.M. Huliau, P. et al. 2016. Formin is associated with left-right asymmetry in the pond snail and the frog. *Curr. Biol.* 26: 654–660. doi:10.1016/j.cub.2015.12.071.

Edwards, W., Moles, A.T., and Franks, P. 2007. The global trend in plant twining direction. *Glob. Ecol. Biogeogr.* 16: 795–800. doi:10.111/j.1466-8238.2007.00326.x.

Endress, P.K. 2001. Evolution of floral symmetry. *Curr. Opin. Plant Biol.* 4: 86–91. doi:10.1016/S1369-5266(00)00140-0.

Endress, P.K. 2012. The immense diversity of floral monosymmetry and asymmetry across angiosperms. *Bot. Rev.* 78: 345–397. doi:10.1007/s12229-012-9106-3.

Harvey, A.W. 1998. Genes for asymmetry easily overruled. *Nature.* 392: 345–346.

Hashimoto, T. 2002. Molecular genetic analysis of left-right handedness in plant. *Phil. Trans. R. Soc. Lond. B.* 357: 799–808.

Henry, J.J. and Martindale, M.Q. 1999. Conservation and innovation in spiralian development. *Hydrobiologia.* 402: 255–265. doi:10.1023/A:1003756912738.

Hoso, M. 2012. Non-adaptive speciation of snails by left-right reversal is facilitated on oceanic islands. *Contrib. Zool.* 81: 79–85.

Huber, B.A., Sinclair, B.J., and Schmitt, M. 2007. The evolution of asymmetric genitalia in spiders and insects. *Biol. Rev.* 82: 647–698.

Iwata, T., Nagasaki, O., Ishii, H.S., and Ushimaru, A. 2012. Inflorescence architecture affects pollinator behaviour and mating success in *Spiranthes sinensis* (Orchidaceae). *New Phytol.* 193: 196–203.

Jesson, L.K. and Barrett, S.C.H. 2002. The genetics of mirror-image flowers. *Proc. R. Soc. Lond. B.* 269: 1835–1839.

Jesson, L.K. and Barrett, S.C.H. 2003. The comparative biology of mirror-image flowers. *Int. J. Plant Sci.* 164: S237–S249.

Kihara, H. 1972. Right- and left-handedness in plants: A review. *Seiken Zihô.* 23: 1–37.

Klingenberg, C. 2015. Analyzing fluctuating asymmetry with geometric morphometrics: Concepts, methods and applications. *Symmetry.* 7: 843–934.

Korn, R.W. 2006. Anodic asymmetry of leaves and flowers and its relationship of phyllotaxis. *Ann. Bot.* 97: 1011–1015.

Lang, M. and Orgogozo, V. 2012. Distinct copulation positions in *Drosophila pachea* males with symmetric or asymmetric external genitalia. *Contrib. Zool.* 81: 87–94.

Levin, M., Klar, A.J.S., and Ramsdell, A.F. 2016. Introduction to provocative questions in left-right asymmetry. *Phil. Trans. R. Soc. B.* 371: 20150399. doi:10:1098/rstb.2015.0399.

Lilienfeld, F.A. 1959. Dextrality and sinistrality in plants. III. Medicago tuberculata Willd. and M. litoralis Rohde. *Proc. Jap. Acad. Sci.* 35: 475–481.

Ludwig, W. 1932. *Das Rechs-Links-Problem imTierreich und bein Menschen.* Berlin, Germany: Springer.

Martinez, C.C., Chitwood, D.H., Smith, R.S., and Sinha, N.R. 2016. Left-right leaf asymmetry in decussate and distichous phyllotactic systems. *Phil. Trans. R. Soc. B.* 371: 20150412.

McManus, I.C. 2002. *Right Hand Left Hand. The Origins of Asymmetry in Brains, Bodies, Atoms and Cultures.* Cambridge, UK: Harvard University Press.

Monsoro-Burq, A.H. and Levin, M. 2018. Avian models and the study of invariant asymmetry: How the chicken and the egg taught us to tell right from left. *Int. J. Dev. Biol.* 62(1–3): 63–77. doi:10. 1387/ijdb.180047ml.

Neville, A.C. 1976. *Animal Asymmetry.* London, UK: Edward Arnold.

Palmer, A.R. 2004. Symmetry breaking and the evolution of development. *Science.* 306: 828–833.

Palmer, A.R. 2005. Antisymmetry. pp. 359–397. In: *Variation.* Hallgrimsson, B. and Hall, B.K. (Eds.). Elsevier.

Palmer, A.R. 2009. Animal asymmetry. *Curr. Biol.* 19: R473–R477. doi:10.1016/j.cub.2009.04.006.

Palmer, A.R. 2016. What determines direction of asymmetry: Genes, environment or chance? *Phil. Trans. R. Soc. B.* 371: 20150417. doi:10.1098/rstb.2015.0417.

Schilthuizen, M. 2007. The evolution of chirally dimorphic insect genitalia. *Tijdschrift voor Entomologie.* 150: 347–354.

Schilthuizen, M. and Gravendee, B. 2012. Left-right asymmetry in plants and animals: A gold mine for research. *Contrib. Zool.* 81: 75–78.

Schilthuizen, M., Haase, M., Koops, K. Looijestijin, S.M., and Hendrikse, S. 2012. The ecology of shell shape difference in chirally dimorphic snails. *Contrib. Zool.* 81: 95–101.

3 Asymmetry at the Cell Level

K. V. Krishnamurthy and S. John Adams

CONTENTS

3.1   INTRODUCTION

The word "cell" was first introduced by Robert Hooke in the 17th century. It has all along been considered as the smallest structural and functional unit of life (Sitte, 1992). It has also been considered the smallest level of biological integration within the different levels of biological organization (see Krishnamurthy, 2015; Krishnamurthy and Bir Bahadur, 2015). Cells are made of several organelles, which in turn are composed of several molecules. In multicellular organisms, more than one cell makes up the whole organism in simple cases, or of tissues in complex cases. There are two distinct theories that discuss the importance of cells to organisms. The first one is the *cell theory* proposed in 1838–1839, which in its modern form states the following (Evered and Marsh, 1989): (i) all organisms are made of one or more cells; (ii) all chemical reactions, including energy-related events, occur within the cell; and (iii) cells can arise only from pre-existing cells through cell divisions, which are considered as very important in growth, development, and overall organization of the organisms. The second theory is called the *organismal theory*, which was proposed in the later part of the 19th century. According to the modern version of this theory, the whole organism is not merely a group of independent cells but also a "living unit that is subdivided into cells which are connected and coordinated" into a whole entity (Krishnamurthy and Bahadur, 2015). This theory further emphasizes that "cell division is merely a 'marker' of plant growth but does not influence them" (Kaplan and Hagemann, 1991). In the light of the above, it is a paradox that cell division is being viewed as both important and unimportant in the development of organisms (Meyerowitz, 1996; see also Fleming, 2006), giving rise to aphorisms like "it is the plant that forms cells, and not the cells that form plants" (deBary, 1879; Barlow, 1982; Kaplan and Hagemann, 1991; Cooke and Lu, 1992; Kaplan, 1992; Sitte, 1992) and "the plant makes cells and the cells make the plants" (Meijer and Murray, 2001). It is to be emphasized that both theories are not mutually exclusive (Krishnamurthy and Bahadur, 2015). Both these theories place the cell at the forefront of biological organization and integration.

Consistent with the above discussion, exciting recent data have revealed that individual cells (even *in vitro*) have intrinsic chiral properties, that is, directionally oriented asymmetric behaviors that have, at least in some cases, an impact on higher levels of biological organization, such as asymmetry, and that the individual cells contribute to the same (see Levin et al., 2016). In other words, asymmetry "leveraged" in individual cells may be present throughout the tissue and organs of which they are a part. This link between single cell asymmetry and higher organizational level (such as tissues, organs, and organisms) asymmetry is turning out to be one of the very exciting areas of future research in understanding asymmetry. Most of the studies on the above-mentioned linkage have been done on animals (Levin et al., 2016; Inaki et al., 2016). This chapter will analyze whether there is a fundamental cellular basis for asymmetry in plants.

3.2   ASYMMETRIC CELL DIVISIONS

3.2.1   IMPORTANCE OF THE CELL CYCLE IN ASYMMETRIC DIVISION

It is believed that plant meristematic or embryonic cells have with them a greater repertoire of phenotypic programs than their progenies. This is evident from a critical study of cell cycle events in these cells. Increasing evidences are being added to show that there is a very close relationship between the cell cycle and cell/tissue differentiation. There must, therefore, be a "determinative element" in the cell cycle that may be responsible for cell/tissue differentiation (Lawrence, 1975). In a typical cell cycle, the control/regulation is done at two crucial transition points, called **checkpoints**; one at G1-S transition and the other at G2-M transition (Boniotti and Griffith, 2002; see details in Krishnamurthy et al., 2015). The first checkpoint is vital in controlling cell differentiation. Cell-cycle-mediated mechanisms involve core regulators like cyclins, cyclin-dependent kinases (CDKs), CDK inhibitors, retinoblastoma-related (RBR) substances, auxins, and other growth regulators.

Cell divisions in the meristematic or embryonic plant tissues are of various types. In *symmetric* cell division, a meristematic cell **A** divides into two daughter cells, **A1** and **A2**, both of which behave like the parent cell **A**. In *asymmetric* cell division, a meristematic cell **A** gives rise to **A1** and **B** daughter cells, of which cell **A1** again behaves like **A** to renew meristem cell population, while cell **B** proceeds toward differentiation (Horvitz and Herskowitz, 1992; Pillitteri et al., 2016); these two cells are invariably of different sizes also. In some instances, meristematic cell **A** divides into **B** and **C** daughter cells, both of which proceed to have the same type of differentiation fate or two different types of differentiation fates (Gallagher and Smith, 1997, 2000); both daughter cells may be of the same size but differ in size from their parent cell **A**, or the two daughter cells may be unequal but both differ in size from their parent cell **A**. In asymmetric cell divisions, the effect of the above-mentioned "determinative element" is quite conspicuous. What is that determinative element in the cell cycle that, after mitosis, causes the production of two different cells that are unequal as well as with different fates of differentiation? In insects, the period intervening between the end of DNA synthesis (S phase) and metaphase (M phase) is different for symmetric and asymmetric cell divisions (Lawrence, 1975). It should also be emphasized that the differences between the two sister cells derived after asymmetric division need not necessarily be obtained directly from unequal mitosis but can be newly built up through local interactions. Differences may also be set up by the cell destined to differentiate being placed in a different local environment. In other words, studies have revealed that both intrinsic and extrinsic mechanisms are involved in asymmetric division and in determining different fates to the two daughter cells (Pillitteri et al., 2016). These mechanisms are discussed subsequently in this article.

There are cell divisions where the two daughter cells derived from a parent cell are identical to one another and also, often, to the parent cell. Yet, sometimes, these daughter cells subsequently become different from the parent cell. Whether this type of difference, in any way, is dependent on events occurring in the preceding cell cycle, it is in this context that the basic idea of **quantal mitosis** was proposed (Okazaki and Holtzer, 1966). The proponents of quantal mitosis believe that there is an important event in the preceding cell cycle that is essential for the differentiation of the two equal daughter cells and also that "quantal cell cycles make available for transcription in daughter cells regions of the genome that were not available for transcription in the mother cell" (Holtzer et al., 1975). This concept further emphasizes that differentiation cannot occur without such a mitosis; in other words, it is believed that if cell function changes independently of the preceding cell cycle, it is not true differentiation (Holtzer et al., 1975; Lawrence, 1975). Although there is no experimental evidence to prove this beyond a doubt, quantal mitosis may very well happen in plants (and animals).

3.2.2 Changes Accompanying Asymmetric Divisions

3.2.2.1 Polarity Changes

In asymmetric cell division, the division plane is asymmetrically placed so that the two daughter cells receive different amounts of cytoplasm and cytoplasmic organelles. There is always a change in the concentration of cytoplasm and its organelles in the two poles of the about-to-divide cells, and thus polarity determination is done in plant cells through the use of differences that set in (Stebbins and Jain, 1960; Shao and Dong, 2016). This strong relationship between cell polarity and asymmetric division is very elegantly demonstrated by observational and experimental data obtained with centrifugation of differentiating spores of *Onoclea sensibilis* (Bassel and Miller, 1982). The normal division of the spore of this fern is preceded by migration of the nucleus from the center of the spore to one end of it. Asymmetric division follows. The longer cell forms the protonema, and the shorter forms the rhizoid. Under centrifugation, there is no change in the pattern of asymmetric cell division, although there is a displacement of the nucleus and cytoplasmic organelles. Only when centrifugation is carried out just before or during nuclear division or cytokinesis, asymmetric cell division is affected (see also Krishnamurthy, 2015).

In many cases with asymmetric division, the orientation of the division wall is also peculiar (oblique, curved, etc.). This is particularly seen in microspore mitosis to initiate pollen formation (see later for more details on pollen mitosis). Polarity determination is done in plants through the use of unique intrinsic cell polarity proteins along with mobile transcription factors and cell wall components to influence asymmetric divisions and the cell plate. There is a very clear pole for altered auxin distribution and signaling in distinguishing the two daughter cells. There is also an emerging epigenetic modification through chromatic remodelers and DNA methylation (Pillitteri et al., 2016; Shao and Dong, 2016; Zhang et al., 2016).

3.2.2.2 Cytoskeletal Behavior and Cortical Sites

The behavior of the cytoplasm, particularly of its cytoskeleton elements (microtubules, MTs) is shown to be very important in asymmetric cell division. One of the most important arrays of MTs in cells is in the mitotic apparatus or spindle (Satir, 2016). Mitosis becomes asymmetric to result in distinct left and right (as in sieve cell and companion cell production, as seen in longitudinal sections) or distinct anterior and posterior (as in the production of apical and basal cells after zygotic division in flowering plants) daughter cells. The mechanism for setting up spindle asymmetry during such cell divisions seems to have a common basis in plants and animals, that is, polarity proteins that help set up a cortical system of proteins whereby MTs are captured asymmetrically and spindles are oriented by the cortical proteins.

One of the earliest studies on cortical sites was done in marine algae like *Fucus* and *Pelvetia* (Fowler and Quatrano, 1997). The polarity of the cytoplasm in the zygote of these plants is slowly built up when it is subjected to unilateral light. During the first ten hours of development, a localized rhizoidal tip comes out from the shaded side (the first evidence of polarization and preparation for asymmetric division). The light signals establish a cortical site to which the secretory apparatus is redirected to cause a change from symmetry to asymmetry. The cortical site contains dihydroxypyridine (DHP) receptors, membrane domains (which help in inward ion flow), high levels of Ca^{2+}, and a patch of actin microfilaments (MFs). Similar asymmetric cortical sites have been reported in the bud sites of yeasts, although there is a slight difference in their chemical composition during stomatal meristemoid formation during stomatogenesis. It has been suggested that a conserved mechanism involving asymmetric divisions to segregate cell fate determinants is operating in the initial stages of stomatal ontogeny. Studies made in *Tradescantia* by Croxdale et al. (1992), Cleary (1995), and Kennard and Cleary (1997) have thrown some light on this aspect by involving the role of cortical sites as described already.

The first sign of asymmetry in the stoma mother cell (SMC) is cytoplasmic aggregation at a cortical site nearer to the guard mother cell (GMC) followed by the migration of the nucleus toward this site. MFs also accumulate at this cortical site between the nucleus of the SMC and GMC, and simultaneously in the adjacent cortex of the GMC. The nucleus then gets connected to the newly accumulated cortical actin by actin filaments. This polar accumulation of actin persists throughout mitosis. The SMC mitosis takes place with asymmetrically shifted MT and MF preprophase bands appearing at an SMC cortex that predirects the future plane. A similar series of events take place in the asymmetric divisions that produce the GMC. But the signals responsible for these are shown to clearly emanate from the GMC and may be related to ion channels and fluxes. It was further suggested that an extracellular cue originating from the GMC is interpreted by the SMC to orient both its nuclear migration and the local accumulation of cortical actin. Either of these asymmetric events can serve to reinforce the GMC-originating spatial cue, thus establishing a more permanent polarity, which might in turn influence the actin cytoskeleton of the adjacent GMC extracellularly and the SMC cytokinetic apparatus intracellularly (Fowler and Quatrano, 1997).

The molecular aspects involved in cortical site protein formation have been brought to light in *Arabidopsis* (Zhang et al., 2016). According to this study, the breaking of the asymmetry in the stomal lineage (BASL) protein shows a polarized localization pattern in the cell that undergoes asymmetric division that is required for different fates of the derived daughter cells. The polarization of BASL is helped by a possible "feedback loop with a canonical" mitogen-activated protein kinase (MAPK) pathway that "recruits the MAPKK kinases YODA (YDA) and MAPK 6 (MPK6)" to the cortical site. The BASL is associated with the membrane and is slowly replenished at the cortical site; this mobility is closely linked to its phosphorylation status. BASL polarity is exhibited by only one daughter cell after the asymmetric division is over. These authors have further shown that the YDAMAPK cascade "transduces upstream ligand-receptor signaling to the transcription factor SPEECHLESS (SPCH), which controls stomatal initiation." This factor is suppressed by MAPK3/6-mediated phosphorylation. Whether a similar mechanism happens in other examples of asymmetric divisions is not known.

3.2.2.3 Genetical Changes

What happens at the transcription level during asymmetric division? This has not been addressed properly, except perhaps by Zhang et al. (2016) (described later under "formation of stomata") and Mena et al. (2017). According to the latter workers in symmetrically dividing cells, the nascent transcription rate increases along with an increase in cell volume before division. This increase in transcription rate is compensated by maintaining the rate of actual mRNA synthesis constant. In budding yeasts, in the bud produced (i.e., smaller cell), there is a great increase in the mRNA synthesis rate. In contrast to other eukaryotes with asymmetric divisions, budding yeasts keep the nascent transcription rates of their RNA polymerases constant and increase RNA stability.

The mutant for the SCARECROW gene in *Arabidopsis* has only one layer instead of the normal two layers (cortical and endodermal) present in wild-type young roots. The single layer exhibits cell-type markers of both layers (Di Laurenzio et al., 1996). The *Arabidopsis* short-root (shr) mutation also does not show the asymmetric division characteristic of endodermal origin, but it produces a single layer showing only the cortex cell fate (Benfey et al., 1993). These provide genetic evidence for asymmetric division during endodermal development (Scheres et al., 1994).

3.2.3   Most Important Asymmetric Cell Divisions in Plants

The most important asymmetric cell divisions observed in plants are as follows (Krishnamurthy, 2015):

1. Zygotic division in angiosperms results in a larger basal cell and a smaller apical cell, of which the former invariably organizes the suspensor and the latter the embryo proper. Here the zygote is polarized, and the cytoplasm is asymmetric from the beginning. Zygotic asymmetric division is strongly under genetic control. For example, *gnom* and *emb3o* mutants of *Arabidopsis* have near-symmetric divisions (Mayer et al., 1993). Zygotic division in brown algae like *Fucus* and *Pelvetia* is also asymmetric (see Fowler and Quatrano, 1997). In these cases, the zygote is not initially polarized and does so only after being subjected to unilateral illumination. The cytoplasmic changes were described already in an earlier page of this chapter. The division of the zygote and all the subsequent divisions in the rhizoidal cell are asymmetric. Asymmetry induced in the zygote cell wall is preserved in its two daughter cells; the rhizoidal cell wall contains sulphated fucoidans and some binding proteins, while the thallus cell does not contain these. Fowler and Quatrano (1997) insisted that the asymmetries fixed in the zygotic cell wall provide the signal for positional information in the cytoplasm that is crucial for further development.

2. Formation of root hair in certain plant species is the result of asymmetric cell division. A root dermal initial cell undergoes an asymmetric division, resulting in a small trichoblast, which develops into a root hair, and a larger cell, which becomes an ordinary dermal cell (Sinnott, 1960). Laser ablation studies made in *Arabidopsis* have indicated that the trichoblast specification is initiated only before the emergence of the root hair. The expression pattern of the GL2 promoter-β-glucuronidae reporter gene fusion construct in the root meristematic region of *Arabidopsis* seedlings have been studied, and these show that *GL2* promoter directs the expression preferentially in future ordinary epidermal cells, that is, atrichoblasts, located over the cortical cells (Schiefelbein et al., 1997) but not in the prospective trichoblasts, which are located over the junction walls of cortical cells. Histochemical studies made on the trichoblasts of *Phleum pretense* and *Hydrocharis morsus-ranae* show that they have greater activities of one or more enzymes, such as acid phosphatase, succinic dehydrogenase, and cytochrome oxidase, than in atrichoblasts (Avers and Grim, 1959;

Cutter and Feldman, 1970a). Trichoblasts also show a certain degree of endoreduplication of their nuclei (Cutter and Feldman, 1970b).

3. Formation of stomata also involves asymmetric division. A stoma is initiated by the asymmetric division of an initial epidermal cell. The smaller daughter cell, called a *meristemoid*, either may directly give rise (by symmetric division) to two guard cells or may divide a few times before giving rise to the guard cells (Sylvester et al., 1996). In some taxa, there are subsidiary mother cells (SMCs), which divide often asymmetrically to form two to many subsidiary cells, depending on the plant species, around the guard cells. The importance of the cortical site in asymmetry during stomatal ontogeny was already indicated.

4. Pollen formation from the haploid microspore also involves asymmetric division. The first mitotic division of the microspore, called pollen mitosis I (PM1), is asymmetric. Immunofluorescence studies on *Phalaenopsis equentris* show that the bulk of MTs appear in radial arrays surrounding the centrally located microspore nucleus. Later on, their concentration at the distal surface of the microspore between the plasma membrane and the nuclear envelope marks the first indication of polarity. This distal MT system marks the path of nuclear migration to the region before mitosis. During PM1, MTs disappear and their locus marks the pole of the mitotic spindle. Though the actin distribution pattern is unchanged during the premitotic stage, its filaments get co-aligned with spindle MTs during pollen mitosis. These actin filaments are believed to maintain the acentric position of the mitotic spindle at anaphase (Brown and Lemmon, 1992). The resultant smaller daughter cell is the generative cell (GC), and the larger daughter cell is the vegetative cell (VC). In tobacco, the *lat52* promoter is sufficient to direct VC-specific transcription of the nuclear-targeted β-glucuronidase (GUS) fusion, providing a useful marker for VC identity (Twell, 1992).

5. The endodermis forms the innermost cortical layer in roots and separates the stele from the cortex. The origin of this layer through asymmetric division has been studied in *Arabidopsis*. The inner smaller derivatives form the endodermal layer while the outer larger-celled layer forms the penultimate cortical layer outside of endodermis (Scheres et al., 1994). This is supported by histological and clonal evidences (Dolan et al., 1993).

3.3   MODELS ON CELLULAR ASYMMETRY

Four models have been proposed till now to describe cellular asymmetry: morphogen model, positional information model, somatic DNA strand-specific imprinting model, and cytoskeleton model.

3.3.1   MORPHOGEN MODEL

This model is based on Turing's theory (Turing, 1952) or the **diffusion reaction theory**. This theory states that the embryonic/meristematic cells are initially undifferentiated and that a morphogen gradient is generated, from an initially homogenous distribution pattern, to specifically regulate, through a regular and patterned diffusion and distribution, gene networks so that it can undergo asymmetry or initiate asymmetric cell divisions. Thus, the cell can be considered as "a complex, specific, diffusion reaction system." This system functions according to the laws of physical chemistry, physics, and mathematics to help in the patterned distribution of the morphogen. The identity of the morphogen is greatly debated, but many agree now that it may be a signaling molecule (Tello, 2007). Morphogens are believed to be responsible for symmetry-asymmetry changes.

3.3.2 Positional Information or French Flag Model

Wolpert (1970, 1981) and Kerszberg and Wolpert (1998) initially adopted the diffusible morphogen model and stated that the propagation of the morphogen depended on the proximity of cells and that the propagation happened along cell membranes between proximally located cells. Once the morphogen reaches the cell surface it is bound to cell surface receptors (Tello, 2007). According to Wolpert (see Richardson, 2009), positional information and pattern formation are important but to "reliably specify it with a diffusible molecule is out of question." There must be a direct cell-to-cell contact, a sort of polarity, and there is no diffusion involved. According to Wolpert (see Richardson, 2009), "Diffusible gradients are out." It was also emphasized that positional information is very important, and not the morphogen, in all morphogenetic events and particularly in cellular asymmetric behaviors.

3.3.3 Somatic DNA Strand-Specific Model

Levin et al. (2016), while discussing left-right asymmetry, stated that the involved mechanisms remain open, with two, non-mutually exclusive, emerging alternatives to the morphogen model as well as to the positional information model, both of which place the determining element in L/R asymmetry inside the cell, as contrary to the extracellular flow of the concerned determining element. The first of these two alternative models was presented by Klar (2016). This model is based on DNA-strand-specific segregation, which is driven by the "mechanisms of somatic strand-specific imprinting and selective chromatid segregation (SSIS)." This mechanism involves of monochromatid gene expression and is based on the asymmetry of DNA sequences of gene(s) that control development "followed by selective segregation of the thus epigenetically differentiated sister chromatids in the mitosis of a deterministic cell to produce developmentally unequal sister cells" (Levin et al., 2016). By varying the selective strand/chromatid segregation process that functions at the centromere of the relevant chromosome, both regulated asymmetric and symmetric cell divisions can be produced in different cell types by the same SSIS mechanism.

3.3.4 Cytoskeleton Model

Another alternative model for cellular asymmetry is based on the role of cytoskeleton. Cellular polarity and cytoskeleton (made of MTs) behavior are so co-ordinated as to bring about cellular chirality, since intracellular cytoskeletal elements are highly conserved. MTs are formed of heterodimeric α- and β-tubulin sub-units. They are arranged in 13 protofilaments arranged parallelly in each single MT (Nogales, 2015). They exhibit at the plus end polymerization of subunits while shrink at the minus end via depolymerization (Dixit and Cyr, 2004).

The initial investigations on the *spiral* mutants of *Arabidopsis* suggested that MTs are involved in helical growth (Thitamadee et al., 2002). Subsequent research by Ishida et al. (2007) showed 37 root-slanting mutants in *Arabidopsis* roots, which affected tubulin gene family members and caused either left- or right-handed phe-notypes. A number of MT-associated proteins (MAPs) have also been implicated in this helical growth. *SPIRAL1* encodes a protein found at the growing plus end of MTs and the MT lattice (Sedbrook et al., 2004). *SPIRAL2* protein specifically asso-ciates with MTs along their entire length and contains protein–protein interacting domains (Buschmann et al., 2004). A third gene called GCP2, with a right-handed helical growth mutant *spiral3*, codes for a component of the γ-tubulin-containing complex, and this is required for MT nucleation (Nakamura and Hashimoto, 2009). It is difficult to link the orientation of MT arrays lying internal to the plasma membrane to the helical growth of a cell. Early observations of the helical cells of the green alga *Nitella* showed the relatively uniform extension growth of their lateral cell walls (Green, 1954). A close link between cortical MT orientation and that of the overlying cellulose microfibrils was revealed, which led to the align-ment hypothesis (see literature in Baskin, 2001). This hypothesis was also enabled by the discovery that MFs are synthesized by a cellulose synthase complex (CSC) embedded in the plasma membrane as a hexagonal rosette and that the orienta-tion of new MFs depended on the direction of movement of the rosette (Slabaugh et al., 2014). The influence of underlying MTs on this direction of movement was subsequently revealed in transgenic *Arabidopsis* plants carrying MTs and cellulose synthase proteins with fluorescent tags.

Subsequently, cellulose synthase interactive protein 1 (SCI 1), a bridging protein that binds directly to both MTs and cellulose biosynthetic proteins, was discovered, and this revealed a direct mechanical link between these cellular com-ponents (Gu et al., 2010; Li et al., 2012). In mutants of the SCI 1 gene (originally called *POM-POM2*), the orientation of cellulose microfibrils and cortical MTs was uncoupled (Li et al., 2012).

3.3.5 CELLULAR BASIS FOR HELICAL GROWTH IN ORGANS

As discussed above, the MT–MF link is firmly established, but we are not still very clear as to how this link controls helical growth. The elongation of cylindrical plant cells usually occurs by a longitudinal increase of side walls, while the diameter of cells remains almost constant. Cellulose MF bands, particularly those in inner, more recently made cell wall layers, normally lie transverse to the long axis of the cell while intervening cell wall regions apparently loosen evenly during elongation. The MF bands may prevent lateral cell expression, while narrowing of the cell diam-eter is prevented by positive turgor pressure (Frei and Preston, 1961). By contrast, in those cases in which the MFs occur in an oblique, helical pattern in the more inner recently formed cell wall layers, longitudinal growth may be associated with an increase in the pitch of the helices. This would necessarily cause the cell to twist (Buschmann et al., 2009). This would account for the striking observation that the handedness of abnormal organ twisting among MT mutants, and drug-treated cells, is always opposite to the handedness of the distorted underlying cortical MT arrays, and probably also of the overlying cellulosic microfibrils. Right-handed organ twisting is always associated with left-handed MT arrays and *vice versa*.

The physical driver of helical growth of an organ has been attributed to properties of either individual cells or tissues of that organ. Cell interactions are not necessarily involved, as single cultured cells of *Arabidopsis tortifolia 2* mutants (the result of a point mutation in the α-tubulin gene); nevertheless, they exhibit helical twisting (Buschmann et al., 2009). Aggregates of such helically twisted cells, especially in cell files, often cause twisting of the resulting tissues. Helical growth may also be the result of the regulated sequential timing of cell divisions; this mechanism has been reported in root meristems of *Arabidopsis* (Baum and Rost, 1996). Furutani et al. (2000) proposed another tissue-based mechanism. Root spiral twisting in *Arabidopsis spiral* mutants occurs to accommodate the reduced longitudinal extension of internal cells relative to the overlying epidermis.

However, the above explanation does not explain left and right twisting. Mutant studies indicate that this may depend on an inherent structural property of MT arrays, that is, individual molecular tweaks to MTs or MT-associated proteins. These tweaks can result in spiral changes to their intracellular orientation in one or other direction. For example, Ishida et al. (2007) proposed that the MTs of tubulin mutants might be helically twisted by the presence of abnormal amino acids, perhaps as the MTs were now built from abnormal protofilaments members. There is structural evidence that MTs with more or fewer than the normal 13 protofilaments do twist (Sui and Downing, 2010); this however, has not been shown to occur in the MTs of mutant plants.

Mention must now be made about the exceptions to a strict relationship between spirals of MTs or MFs and cell twisting. Spiral MT arrays are often observed with a defined chirality in a non-twisted wild-type tissue, such as in the differentiating root regions, xylem tracheary elements (Barnett and Boham, 2004), and in immature leaf hairs. This may imply that the cellulose MFs can adopt a defined helical structure even in the likely absence of MT function (see literature in Smyth, 2016). For example, left-handed helicity is seen in *Arabidopsis* roots in which MTs are disrupted by chemicals, or by mutations in the microtubule organization 1 (MOR 1) gene, but the cellulose MFs are still arranged in radial bands and are not detectably helical. In many such situations, it is possible that the helical cellulose MF pattern is overridden by residual transverse bands of cortical MT arrays or some other independent factor. As Landrein et al. (2013) aptly state: "The main issue with stem twisting is not what is introducing it because stem twisting can occur by default, but what is maintaining the straight stem in the wild type."

The genetic and cellular studies carried out so far have revealed that there is a strong functional link between MT arrangement and the potential helical twisting of cells and tissues. This appears to be caused by the MT-mediated guidance of cellulose MF patterning in the developing cell wall, although the latter may sometimes exhibit its own independent patterning.

3.4   MOTILITY IN HELICAL BACTERIA

It is important to describe briefly here the motility of some helical bacteria like *Spiroplasma*. It has a diameter of 200 nm, a length of few μm, and lacks a stiff cell wall. Its helical growth motion is generated by the propagation of a pair of perversions or kinks down the length of its body axis. Two inverse rotations of the helices are generated by the kinks, one before and the other after the kink, and these serve as a driving force to propel in the opposite direction of kink propagation at a velocity of a few microns/second (Shaevitz et al., 2005; Roth et al., 2018). The kinks themselves get generated through a transition between left- and right-handed regions of the helical bacterial body. The switch of handedness, that is, perversion, is reported to be caused by a contractile ribbon of many filamentous chains of ellipse-shaped protein fibrils that can switch conformation between two diameters (Trachtenberg et al., 2003). These protein filaments are attached to the cell membrane by another filamentous ribbon, probably the cytoskeleton component called MreB, and involved in overall cell shape. There is a concerted conformational switching of all proteins of a single fibril filament to cause a differential length change relative to the almost static MreB ribbon. This causes the cell to switch handedness and bend. This successive contraction of the ribbon makes up a linear chain motor, which is so far only known from eukaryotic cells.

3.5   HELICAL PLANT CELL WALL THICKENINGS

Helical cell wall thickenings of tracheary elements of xylem tissue is well-known. Helical winding and fibril angle are important in determining the mechanical properties of primary xylem wood and wood pulp. Grey (2014) stated that the helical coils all appeared to be left-handed (see also Silva et al., 2017). In the celery petiole xylem element (Grey, 2014), the helical coils were invariably left-handed, and the individual turns of the coils were in contact with each other. Helical coils in all other plants studied are left-handed on top of the coils, and this changes to an apparent Z helix, as the focus was lowered to the bottom of the coils, but this was an artifact; the helix remained left-handed. Examination of fossil records of primitive vascular plants also show evidence of left-handed coils. It is left-handed in living taxa like lotus, Musa, rhubarb, and squash. The coils may be single, double, or multiple-stranded, are very long, and much longer than the tracheary cells themselves. However, the coils show a periodicity along their length, which may correspond to the original cell length. This behavior contrasts with that of the coils produced by *in vitro* electro-spinning of cellulose derivatives, where both left- and right-handed coils were observed (Goldiho et al., 2009; Silva et al., 2017). Left-handed helical thickenings of xylem elements help the cell to withstand the negative pressure within the plant vascular system and act as tube-containers to carry ligands to the different parts of plants, via capillarity (Silva et al., 2017).

The presence of only left-handed helical coils is recorded by Grey (2014) in celery petioles. On the contrary, the cellulose MFs in the S2 layer of the secondary wall of virtually all wood fibers follow a fairly shallow right-handed helix (Meylan and Butterfield, 1978a). The same investigators found that helical thickenings observed in some dicot vessel elements and conifer tracheids were in the form of S-helices, that is, of the opposite handedness (Meylan and Butterfield, 1978b). This relationship between these macroscopic manifestations of chirality and the intrinsic microscopic chirality of glucose and cellulose at the molecular level is difficult to explain. The molecular and macroscopic difference are at a length scale; experimental and theoretical evidences show that the individual crystalline cellulose microfibrils also form right-handed twisted structures, but suspensions of cellulose nanocrystals only form left-handed helicoidal arrangements. The details of how these molecular-level chiral structures pack together to produce macroscopic-level chiral structures of the same or different handedness remain largely speculative. However, these experimental observations were used by Grey (2014) to rationalize the difference in handedness between the right-handed organization of MFs in the wood cell secondary wall and the left-handed helical structure of the tracheary elements in celery petioles, and also recorded in many other plants. To form the preferred right-handed helical orientation in the S2 layer of the secondary wall, the orientation of the cellulose MFs must depart from a parallel (achiral) orientation and must wrap around the long axis of xylem fiber in a right-handed helix.

Grey (2014) also explained the preferred left-handed helical structure of tracheary elements by suggesting that the orientation of the cellulose in the helical coil must also depart from the achiral organization (in celery petioles shown by individual ring like thickenings in vascular xylem cells) to give a left-handed helical cell, whose orientation is offset from the achiral case. In both cases (i.e., S2 and helical thickenings), the direction of the offset from the straight corresponds to a left-handed helicoidal arrangement of cellulosic material when viewed perpendicular to the axis of the coil. The same left-handed orientation is observed during the organization of chiral nematic cellulose nanocrystal suspensions and films along the helicoidal axis. In other words, according to Grey (2014), the observed alignments of cellulose MFs in the shallow right-handed helices of the S2 layer of the xylem cell secondary wall and in tight left-handed helices in the tracheary thickenings of celery petioles are both in accordance with the *in vitro* observations of a left-handed helicoidal arrangement of cellulose nanocrystals in suspensions viewed base to top.

3.6 CHIRALITY IN EXTRACELLULAR BACTERIAL FLAGELLA AND CILIA

Since helices are intrinsically chiral (see Chapter 1), extracellular structures, such as flagella, are intrinsically chiral (Satir, 2016). The specific helical twist of these flagella is very important in cellular function in organisms possessing these flagella. For instance, the bacterial flagellum consists of an extracellular helical filament made of proteinaceous flagellin attached by a hook to a rotary motor containing a number of intramembrane and peripheral membrane proteins. This motor generates clockwise or anticlockwise at high-speed rotation using ion motive force (Macnab, 1999). When the filament undergoes anticlockwise rotation (viewed looking downward from the filament toward the cell), several neighboring filaments with left-handed helices become coupled together hydrodynamically into a bundle that grows through the medium to cause a run, a straight-line motion. However, the flagella motor can switch to clockwise rotation, causing a transformation from a left-handed to a right-handed helix of the filament. This results in each filament producing directionally uncoordinated motion to result in cell tumbling. As soon as anticlockwise rotation resumes, the filament bundle reforms to propel the cell in a different direction. Thus, various helical forms are produced by the coiling of protofilaments in different conformation and package. Thus, in this case, the chiral helical structure of the flagellum definitely has a function.

With respect to chirality exhibited by cilia, the following may be stated. It is known that cilia grow from basal bodies, which have complex MTs with characteristic chirality; therefore, cilia are expected to preserve the same chirality of the basal bodies, doublet MTs. When cilia assume motility, chirality is reflected in the direction of their effective stroke. The actual positioning of cilia on the cell surface depends on polarity proteins. During motility, differences between non-motile and motile cilia are initiated, particularly two structural differences: dynein arms, which run clockwise when viewed base to tip, and usually a central pair of single MTs, giving rise to the $9 + 2$ pattern. The dynein arm and the central pair of singlet MTs are involved in cilia motility and beat. The model that explains this ciliary motility is called the switch point hypothesis (Satir et al., 2016).

Satir (2016) states that the chirality of the cilia is related to the chirality of the basal body, and the chirality of the latter is related to the polarity (and chirality) of the cell as a whole. The actual positioning of the basal bodies on the cell surface is likely to depend on the proteins that are implicated in planar cell polarity as in vertebrate tissues, such mammalian trachea. But the left-right determination of ciliary movements in different organisms is not dependent on the cilia themselves but rather on the chiralities of MTs and actin filaments of the basal bodies (Satir, 2016).

3.7   BASAL BODIES AND CENTRIOLES

Basal bodies and centrioles are distinctive eukaryotic cell organelles. Modern Protista organisms have only basal bodies, while in advanced eukaryotes basal bodies are associated with the spindle apparatus to form true centrioles. Both basal bodies and centrioles are also chiral (Satir, 2016). The MTs of the basal body normally get polymerized around a cart wheel (a structure assembled from SAS-6-protein dimmers (Van Breugel et al., 2011) in three steps to result respectively in MT subfiber A (a complete 13 protofilament MT) to which a subfiber B is attached and a third partial MT subfiber C is attached to result in a complete triplet at the base of this organelle. When we look at it from base to tip, each triplet appears to spiral outward from the cart wheel in an anticlockwise direction; no organisms show clockwise spiraling triplet MTs when viewed in the same direction (Satir, 2016).

3.8   EVOLUTIONARY ASPECTS

It is suggested that there is a possibility that asymmetry is an ancient property that had appeared even before the evolution of multicellularity, since it has been shown in unicellular ciliates, bacteria, and slime molds (Satir, 2016). Asymmetry has also evolved in higher plant (and animal) cells subsequently. Consistent with this fact,

Asymmetry in Plants

exciting recent data have revealed that individual cells have many directionally oriented asymmetric behaviors that can be explained by chiral properties that have an impact on large-scale structures (such as tissues, organs, and even whole organisms) to which these cells contribute (see Levin et al., 2016). The ability of single cells (even *in vitro*) to determine their L/R axis in the absence of ciliated organs suggested that, in at least some cases, asymmetry is controlled intracellularly and may be present throughout the body of the organism.

REFERENCES

Avers, C.J. and Grimm, R.B. 1959 Comparative enzyme differentiations in grass roots. II. Peroxidase, *J. Exp. Bot.* 10: 341–344. doi:10.1093/jxb/10.3.341.

Barlow, P.W. 1982. The plant forms cells, not cells the plant: The origin of de Bary's aphorism. *Ann. Bot.* 49: 269–271. https://www.jstor.org/stable/42757120.

Barnett, J.R. and Boham, V.A. 2004. Cellulose microfibril angle in the cell wall of wood fibres. *Biol. Rev.* 79: 461–472. doi:10.1017/S1464793103006377.

Baskin, T.I. 2001. On the alignment of cellulose microfibrils by cortical microtubules: A review and a model. *Protoplasma.* 215: 150–171. doi:10.1007/BF01280311.

Bassel, A.R. and Miller, J.H. 1982. The effects of centrifugation on asymmetric cell division and differentiation of fern spores. *Ann. Bot.* 50: 185–198. https://www.jstor.org/stable/42756840.

Baum, S.F. and Rost, T.L. 1996. Root apical organization in *Arabidopsis thalianacl.* Root cap and protoderm. *Protoplasma.* 192: 178–188. doi:10.1007/BF01273890.

Benfey, P.N., Linstead, P.J., Roberts, K., Schiefelbein J.W., Haüser, M.T., and Aeschbacher, R.A. 1993. Root development in *Arabidopsis* four mutants with dramatically altered root morphogenesis. *Development.* 119: 57–70.

Boniotti, M.B. and Griffith, M.E. 2002. "Cross-talk" between cell division cycle and development in plants. *Plant Cell.* 14: 11–16. doi:10.1105/tpc.000000.

Brown, R.C. and Lemmon, B.E. 1992. Pollen development in Orchids. 4. Cytoskeleton and ultrastructure of the unequal pollen mitoses in *Phalaenopsis. Protoplasma.* 167: 183–192. doi:10.1007/BF01403382.

Buschmann, H., Fabri, C.O., Hauptmann, M., Hutzler, P., Laux, T., Llyod, C.W., and Shäffner, A.R. 2004. Helical growth of the Arabidopsis mutant *tortifolial* reveals a plant-specific microtubule-associated protein. *Curr. Biol.* 14: 1515–1521. doi:10.1016/j.cub.2004.08.033.

Buschmann, H., Hauptmann, M., Niessing, D., Llyod, C.W., and Schaffner, A.R. 2009. Helical growth of the Arabidopsis mutant *tortifolia2* does not depend on cell division patterns but involves handed twisting of isolated cells. *Plant Cell* 21: 2090–2106. doi:10.1105/tpc.108.061242.

Cleary, A.L. 1995. F–action redistributions at the division site in living *Tradescantia* complexes as revealed by microinjection of rhodamine-phalloidin. *Protoplasma* 185: 152–165. doi:10.1007/BF01272855.

Cooke, T.J. and Lu, B. 1992. The independence of cell shape and overall form in multicellular algae and land plants: Cells do not act as building blocks for constructing plant organs. *Int. J. Plant Sci.* 153: S7–S27.

Croxdale, J., Smith, J., Yandell, B., and Johnson, J.B. 1992. Stomatal patterning in *Tradescantia*: An evaluation of the cell lineage theory. *Dev. Biol.* 149: 158–167. doi:10.1016/0012-1606(92)90272-I.

Cutter, E.G. and Feldman, L.J. 1970a. Trichoblasts in *Hydrocharis*, I. Origin, differentiation, dimensions and growth. *Am. J. Bot.* 57: 196–201.

Cutter, E.G. and Feldman, L.J. 1970b. Trichoblasts in *Hydrocharis*, II. Nucleic acids, proteins and a consideration of cell growth in relation to endopolyploidy. *Am. J. Bot.* 57: 202– 211. doi:10.1002/j.1537-2197.1970.tb09808.x.

deBary, A. 1879. Lehrbuch der Botanik für mittlere und höhere Lehranstalten. *Bot Ztg.* 37: 221–223.

Di Laurenzio L., Wysocka–Diller, J., Malamy, J.E., Pysh, L., Helariutta, Y., Freshour, G., Hahn, M.G., Feldmann, K.A., and Benfey, P.N. 1996. The SCARECROW gene regulates an asymmetric cell division that is essential for generating the radial organization of the Arabidopsis root. *Cell* 86(3): 423–33. doi:10.1016/S0092-8674(00)80115-4.

Dixit, R. and Cyr, R. 2004. The cortical microtubules array: From dynamics to organization. *Plant Cell.* 16: 2546–2552. doi:10.1105/tpc.104.161030.

Dolan, L., Janmaat, K., Willemsen, V., Linstead, P., Poethig, S., Roberts, K., and Scheres, B. 1993. Cellular organization of the *Arabidopsis thaliana* root. *Development* 119: 71–84.

Evered, D., and Marsh, J. 1989. The cellular basis of morphogenesis. John Wiley & Sons, Chichester, UK.

Fleming, A.J. 2006. The coordination of cell division, differentiation and morphogenesis in shoot apical meristem: A perspective. *J. Exp. Bot.* 57: 25–32. doi:10.1093/jxb/eri268.

Fowler, J.E. and Quatrano, R.S. 1997. Plant cell morphogenesis: Plasma membrane interactions with the cytoskeleton and cell wall. *Annu.Rev. Cell. Dev. Biol.* 13: 697–743. doi:10.1146/annurev.cellbio.13.1.697.

Frei, E. and Preston, R.D. 1961. Cell wall organization and wall growth in the filamentous green algae cladophora and chaetomorpha. II. Spiral structure and spiral growth. *Proc. R. Soc. B. Biol. Sci.* 155: 55–77. doi:10.1098/rspb.1961.0057.

Furutani, I., Watanabe, Y., Prieto, R., Masukawa, M., Suzuki, K., Naoi, K., Thitamadee, S., Shikanai, T., and Hashimoto, T. (2000). The SPIRAL genes are required for directional control of cell elongation in *Arabidopsis thaliana*. *Development* 127: 4443–4453.

Gallagher, K. and Smith, L.G. 1997. Asymmetric cell division and cell fate in plants. *Curr. Opin. Cell Biol.* 19: 842–848. doi:10.1016/S0955-0674(97)80086-5.

Gallagher, K. and Smith, L.G. 2000. Roles of polarity and nuclear determinants in specifying daughter cell fates after an asymmetric cell division in the maize leaf. *Curr. Biol.* 10: 1229–1232.

Goldiho, M.H., Canejo, J.P., Pinto, L.F.V., Borgesa, J.P., and Teixeira, P.I.C. 2009. How to mimic the shapes of plant tendrils on the nano and microscale: Spirals and helices of electrospun liquid crystalline cellulose derivatives. *Soft Matter.* 5: 2772–2776. doi:10.1016/S0960-9822(00)00730-2.

Green, P.B. 1954. The spiral growth pattern of the cell wall in *Nitella axillaris*. *Ann. J. Bot.* 41: 403–409. doi:10.2307/2438730.

Grey, D.G. 2014. Isolation and handedness of helical-coiled cellulose thickenings from plant petiole tracheary elements. *Cellulose.* 21: 3181–3191. doi:10.1007/s10570-014-0382-4.

Gu, Y., Kaplinsky, N., Brigmann, M., Cobb, A., Carroll, A., Sampathkumar, A., baskin, T.I., Persoon, S., and Somerville, C.R. 2010. Identification of a cellulose synthase-associated protein required for cellulose biosynthesis. *Proc. Natl. Acad. Sci. USA.* 107: 12866–12871. doi:10.1073/pnas.1007092107.

Holtzer, H., Rubinstein, N., Fellini, S., Yeoh, G., Chi, J., Birnbaum, J., and Okayama, M. 1975. Lineages, quantal cell cycles, and the generation of diversity. *Q. Rev. Biophys.* 8: 1–34. doi:10.1017/S0033583500001980.

Horvitz, H.R. and Herskowitz, I. 1992. Mechanism of asymmetric cell division: Two Bs or not Bs, that is the question. *Cell.* 68: 237–255. doi:10.1016/0092-8674(92)90468-R.

Inaki, M., Liu, J., and Matsuno, K. 2016. Cell chirality: Its origin and roles in left-right asymmetric development. *Phil. Trans. R. Soc. B.* 371: 20150403. doi:10.1098/rstb.2015.0403.

Ishida, T., Thitamandae, S., and Hashimoto, T. 2007. Twisted growth and organization of cortical microtubules. *J. Plant Res.* 120: 61–70. doi:10.1007/s10265-006-0039-y.

Kaplan, D.R. 1992. The relationship of cells to organisms in plants: Problem and implications of an organismal perspective. *Int. J. Plant Sci.* 153: S28–S37.

Kaplan, D.R. and Hagemann, W. 1991. The relationship of cell and organism in vascular plants: Are cells the building blocks of plant form? *Bioscience.* 41: 693–703. doi:10.2307/1311764.

Kennard, J.L. and Cleary, A.L. 1997. Pre-mitotic nuclear migration in subsidiary mother cells of *Tradescantia* occurs in G1 of the cell cycle and requires F-actin. *Cell Motil. Cytoskeleton.* 36: 55–67.

Kerszberg, M. and Wolpert, L. 1998. Mechanisms for positional signaling by morphogen transport: A theoretical study. *J. Theor. Biol.* 191: 103–114. doi:10.1006/jtbi. 1997.0575.

Klar, A.J.S. 2016. Split hand/foot malformation genetics supports the chromosome 7 copy segregation mechanism for human limb development. *Phil. Trans. R. Soc. B.* 371: 20150415. doi:10.1098/rstb.2015.0415.

Krishnamurthy, K.V. 2015. *Growth and Development in Plants.* Scientific Publishers, Jodhpur, India.

Krishnamurthy, K.V. and Bahadur, B. 2015. Plant organization at the cellular level. In: *Plant Biology and Biotechnology* (Vol. 1) Bahadur, B., Rajam, M.V., Sahijram, L., and Krishnamurthy, K.V. (Eds.). Springer, New Delhi, India. doi:10.1007/978-81-322-2286-6_2.

Krishnamurthy, K.V., Bahadur, B., Adams, S.J., and Venkatasubramanian, P. 2015. Development and organization of cell types and tissues. In: *Plant Biology and Biotechnology* (Vol. 1) Bahadur, B., Rajam, M.V., Sahijram, L., and Krishnamurthy, K.V. (Eds.). Springer, New Delhi, India. doi:10.1007/978-81-322-2286-6_3.

Landrein, B., Lathe, R., Bringmann, M., Vouillot, C., Ivakov, A., Boudaoud, A., Persson, S., and Hamant, O. 2013. Impaired cellulose synthase guidance leads to stem torsion and twists phyllotactic patterns in Arabidopsis. *Curr. Biol.* 23: 895–900. doi:10.1016/j. cub.2013.04.013.

Lawrence, P.A. 1975. The cell cycle and cellular differentiation in insects. In: *Cell Cycle and Cell Differentiation.* Reinert, J. and Holtzer, H. (Eds.). Springer, Berlin, Germany. pp. 111–121.

Levin, M., Klar, A.J.S., and Ramsdell, F. 2016. Introduction to provocative questions in left-right asymmetry. *Phil. Tran. R. Soc. B.* 371: 20150399. doi:10.1098/rstb.2015.0399.

Li, S., Lei, L., Somerville, C.S. and Gu, Y. 2012. Cellulose synthase interactive protein 1 (CSI1) links microtubules and cellulose synthase complexes. *Proc. Natl. Acad. Sci. USA.* 109: 185–190. doi:10.1073/pnas.1118560109.

Macnab, R. M. 1999. The bacterial flagellum: Reversible rotary propellor and type III export apparatus. *J. Bacteriol.* 181: 7149–7153.

Mayer, U., Büttner, G., and Jürgens, G. 1993. Apical-basal pattern formation in the *Arabidopsis* embryo: Studies on the role of *gnom* gene. *Development.* 117: 149–162.

Meijer, M. and Murray, J.A.H. 2001. Cell cycle controls and the development of plant form. *Curr. Opin. Plant Biol.* 4: 44–49. doi:10.1016/S1369-5266(00)00134-5.

Mena, A., Medina, D.A., Pérez-Ortín, J.E. García-Martínez, J., Begley, V., Singh, A., Chávez, S., Muñoz-Centeno, M.C., and Pérez-Ortín, J.E. 2017. Asymmetric cell division requires specific mechanisms for adjusting global transcription. *Nucleic Acids Res.* 45(21): 12401–12412. doi:10.1093/nar/gkx974.

Meyerowitz, E.M. 1996. Plant development: Local control and global patterning. *Curr. Opin. Genet. Dev.* 6: 475–479. doi:10.1016/S0959-437X(96)80070-0.

Meylan, B.A. and Butterfield, B.G. 1978a. Helical orientation of the microfibrils in tracheids, fibres and vessels. *Wood Sci. Technol.* 12: 219–222. doi:10.1007/BF00372867.

Meylan, B.A. and Butterfield, B.G. 1978b. Occurrence of helical thickenings in the vessels of New Zealand woods. *New Phytol.* 81: 139–144.

Nakamura, M. and Hashimoto, T. 2009. A mutation in the Arabidopsis γ-tubulin-containing complex causes helical growth and abnormal microtubule branching. *J. Cell Sci.* 122: 2208–2217. doi:10.1242/jcs.044131.

Nogales, E. 2015. An electron microscopy journey in the study of microtubules structure and dynamics. *Protein Sci.* 24: 1912–1919. doi:10.1002/pro.2808.

Okazaki, K. and Holtzer, H. 1966. Myogenesis: Fusion, myosin synthesis and the mitotic cycle. *Proc. Nat. Acad. Sci. U.S. A.* 56: 1484–1490.

Pillitteri, L.J., Guo, X., and Dong, J. 2016. Asymmetric cell division in plants, mechanisms of symmetry breaking and cell fate determination. *Cell Mol. Life Sci.* 73: 4213–4229. doi:10.1007/s00018-016-2290-2.

Richardson, M.K. 2009. Diffusible gradients are out—An interview with Lewis Wolpert. *Int. J. Dev. Biol.* 53: 659–662. doi:10.1387/ijdb.072559mr.

Roth, J., Koch, M.D., and Rohrbach, A. 2018. Dynamics of a protein chain motor driving helical bacteria under stress. *Biophys. J.* 114: 1955–1969. doi:10.1016/j.bpj.2018.02.043.

Satir, P. 2016. Chirality of the cytoskeleton in the origins of cellular asymmetry. *Phil. Trans. R. Soc. B.* 371: 20150408. doi:10.1098/rstb.2015.0408.

Scheres, B., Wolkenfelt, H., Willemsen, V., Tenlouw, M., Lawson, E., Dean, C., and Weisbeek, P. 1994. Embryonic origin of the *Arabidopsis* primary root and root meristem initials. *Development* 120: 2457–2487.

Schiefelbein, J.W., Masucci, J.D., and Wang, H. 1997. Building a root: The control of patterning and morphogenesis during root development. *Plant Cell.* 9: 1089–1098. doi:10.1105/tpc.9.7.1089.

Sedbrook, J.C., Ehrhardt, D.W., Fisher, S.E., Scheible, W.-R. and Somerville, C.R. 2004. The Arabidopsis *SKU6/SPIRAL 1* gene encodes a plus end-localized microtubule-interacting protein involved in directional cell expansion. *Plant Cell* 16: 1506–1520. doi:10.1105/tpc.020644.

Shaevitz, J.W., Lee, J.Y., and Fletcher, D.A. 2005. Spiroplasma swim by a progressive change in cell helicity. *Cell* 122: 941–945. doi:10.1016/j.cell.2005.07.004.

Shao, W. and Dong, J. 2016. Polarity in plant asymmetric cell division: Division orientation and cell plate differentiation. *Dev. Biol.* 19(1): 121–131. doi:10.1016/j.ydbio.2016.07.020.

Silva, P.E.S., Vistulo de Abreu, F., and Godinho, M.H. 2017. Shaping helical electrospun filaments: A review. *Soft Matter.* 13: 6678–6688. doi:10.1039/C7SM01280B.

Sinnott, E.W.1960. *Plant Morphogenesis*. McGraw-Hill, New York.

Sitte, P. 1992. A modern concept of the "cell theory." A perspective on competing hypotheses of structure. *Int. J. Plant Sci.* 153: S1–S6. doi:10.1086/297059.

Slabaugh, E., Davis, J.K., Haigler, C.H., Yingling, Y.G., and Zimmer, J. 2014. Cellulose Synthases: New insights from crystallography and modeling. *Trends Plant Sci.* 19: 99–106. doi:10.1016/j.tplants.2013.09.009.

Smyth, D.R. 2016. Helical growth in plant organs: Mechanisms and significance. *Development* 143: 3272–3282. doi:10.1242/dev.134064.

Stebbins, G.L. and Jain, S.K. 1960. Developmental studies of cell differentiation in the epidermis of monocotyledons. I. *Allium, Rhoeo*, and *Commelina*. *Dev. Biol.* 2: 409–426. doi:10.1016/0012-1606(60)90025-7.

Sui, H. and Downing, K.H. 2010. Structural basis of inter-protofilament interaction and lateral deformation of microtubules. *Nucleic Acids Res.* 18: 1022–1031. doi:10.1016/j.str.2010.05.010.

Sylvester, A.W., Smith, L., and Freeling, M. 1996. Acquisition of identity in the developing leaf. *Annu. Rev. Cell Dev. Biol.* 12: 257–304.

Tello, J.I. 2007. Mathematical analysis of a model of morphogenesis: Steady states. pp. 1–8. *XX Congreso de Ecuaciones Diferenciales Y Aplicaciones X Congreso de Matemática Aplicada*, pp. 115–152, Sevilla, 24–28 Septiembre 2007.

Thitamadee, S., Tuchihara, K., and Hashimoto, T. 2002. Microtubule basis for left-handed helical growth in *Arabidopsis*. *Nature* 417: 193–196. doi:10.1038/417193a.

Trachtenberg, S., Andrews, S.B., and Leapman, R.D. 2003. Mass distribution and spatial organization of the linear bacterial motor of spiroplasma citri R 8A2. *J. Bacteriol.* 185: 1987–1994. doi:10.1128/JB.185.6.1987-1994.2003.

Turing, A.M. 1952. The chemical basis of morphogenesis. *Phil. Trans. Roy. Soc. B.* 237: 37–72. doi:10.1098/rstb.1952.0012.

Twell, D. 1992. Use of a nuclear-targeted β-glucuronidase fusion protein to demonstrate vegetative cell-specific gene expression in developing pollen. *Plant J.* 2: 887–892. doi:10.1111/j.1365-313X.1992.00887.x.

van Breugel, M. et al. 2011. Structures of SAS-6 suggest its organization in centrioles. *Science* 331: 1196–1199. doi:10.1126/science.1199325.

Wolpert, L. 1970. Positional information and pattern formation, In. *Towards a Theoretical Biology* (ed.) C.H. Waddington, Edinburgh University Press, Edinburgh, UK, 198–230.

Wolpert, L. 1981. Positional information and pattern formation. *Philos. Trans. R. Soc., London Ser. B.* 295: 441–450.

Zhang, Y., Guo, X., and Dong, J. 2016. Phosphorylation of the polarity protein BASL differentiates asymmetric cell fate through MAPks and SPCh. *Curr. Biol.* 26(1): 2957–2965.

4 Handedness Events in Vascular Cambium and Their Relation to the So-Called Spiral Grains in Wood

K. V. Krishnamurthy, Bir Bahadur, and T. N. Manohara

CONTENTS

4.1   INTRODUCTION

The term "spiral grain" indicates the alignment of the longitudinal elements of the secondary xylem (= wood) at an angle to the main axis of the tree. The direction of "spiral" alignment may be to the left (S) or right (Z) of the upper extremity of the longitudinal tree-axis as observed by anyone standing on the ground (Harris 1981). Since spiral grain has a great effect on timber quality and utilization, its study has attracted the attention of many people for the last two hundred years. Wood is a product of vascular cambial activity; hence, spiral grains have been reported from both dicot and coniferous woods. Any attempt to study these grains must start in the vascular cambium, as it is in this meristem that the origin and development of the grains should be traced. Since spiral grains are handed structures, it is ideal to examine whether vascular cambium also shows handedness, and if so how this handedness is manifested to result in grains.

4.2   STRUCTURE OF VASCULAR CAMBIUM

Growths in vascular plants are of two kinds: (i) longitudinal, vertical, or growth in length; and (ii) transverse, radial, or growth in girth or thickness. Both happen in the axial organs (stem and root) of the plant. In most dicotyledons and gymnosperms, growth in girth is due to a specialized lateral meristem called the vascular cambium. (It is popularly known as cambium, and this term is commonly used in this chapter). The cambium is a thin strip of meristematic tissue, often seen as three to four layers of cells, sandwiched between the secondary xylem and secondary phloem (= bast), both of which are the products of its activity. Cambium is made up of a vertical system of fusiform initials that produce the vertically aligned system of vascular tissues and a horizontal system of ray initials that produce the horizontally aligned ray cell elements of the vascular system (Figure 4.1) (Krishnamurthy 2015; Krishnamurthy et al. 2015).

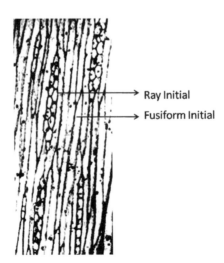

Ray Initial

Fusiform Initial

FIGURE 4.1   Tangential longitudinal section of vascular cambium of *Tectona grandis.*

4.3 HANDEDNESS IN VASCULAR CAMBIUM

Several developmental/morphogenetic events in the cambium show handedness (Hejnowicz 1990; he uses the word "chiral" instead of "handed"), that is, they occur in two alternative orientations: left (S) and right (Z). These events are classified into **cyclic** and **non-cyclic**. The former events conform to the domain type (see Section 4.1) and include an intrusive growth of fusiform initials, pseudotransverse type of anticlinal divisions, fusion of ray initials, and fission of ray initials. Intrusive growth of fusiform initials results in overlapping of these initials (Figure 4.2). The pseudotransverse divisions show highly inclined separation walls between the two resultant fusiform initials (Figure 4.3). Fusion and fission of ray initials, respectively, result in the formation of a single and two ray initials (Figure 4.4). All these events may happen to the left or right side (Figures 4.2 through 4.4). The direction of these events causes left-handedness or right-handedness that have an important effect on grains directions (see Section 4.3.1.1). To the non-cyclic events belongs the periclinal division of the fusiform initials. The twisting of the periclinal division partitions occurs when there is an incongruity between them and the domain types (Hejnowicz 1990).

Although these events happening in the cambium can be detected in the cambium itself (as shown in Figures 4.2 through 4.4), their effects are very well recorded in its products, the wood, and bast. When these events are cumulative and frequent, they affect the orientation of the longitudinal axes of wood cells and

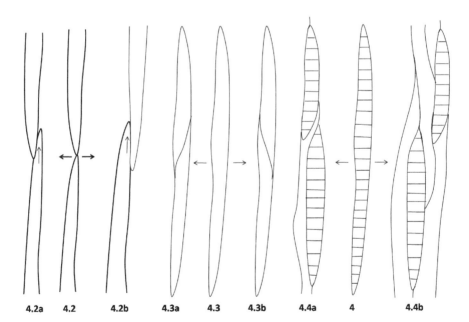

4.2a   4.2   4.2b   4.3a   4.3   4.3b   4.4a   4   4.4b

FIGURES 4.2–4.4 Diagrammatic representations of events in vascular cambium that cause domain patterns. 4.2. Right (2a) and left (2b) overlapping growth of fusiform initials result in right and left handedness respectively; 4.3. Right and left pseudotransverse divisions of fusiform initials; 4.4. Right and left side spitting of existing ray through right and left side inclusive growth of fusiform initials respectively.

results in wood grains. Therefore, it may be stressed here that inferences about cambial events can be made on studying xylem cells and the grains they make (see Section 4.3.1.2).

4.3.1 CYCLIC HANDEDNESS EVENTS AND DOMAINS IN CAMBIUM THAT PRODUCE WAVY AND INTERLOCKED GRAINS

In the cambium, there are extensive regions or zones, each characterized by one type of handedness, either S or Z, of events (Hejnowicz and Romberger 1973). These regions are called **domains**, and the form of handedness realized in a domain determines its type (S or Z type). Neighboring fusiform cells in a domain invariably tend to produce pseudotransverse types of anticlinal divisions in the same direction (S or Z) but not at random. The same direction of cell division is not maintained for a very long time, and as a result there are changes in domain pattern with time. Thus, if a sheet of cambium is examined, one may be able to see both types of domains (S and Z) occupying regions that can be demarcated from one another (Figure 4.5).

Typically, the borders between these two domains run transversely so that the domain patterns appear in the form of transverse stripes. It should also be stressed here that the fascinating property of the domains is that their borders migrate, so as to belong to cyclically changing domains (Hejnowicz 1990). The transverse stripe pattern migrates along the stem, usually upward, but there are trees where a reverse direction of migration is noticed. In general, the longer the axial dimension of the domains, the faster their migration, either upward or downward. This results

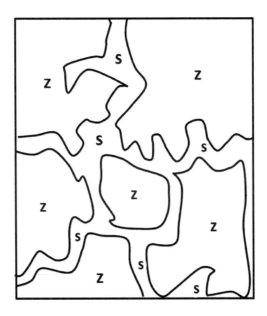

FIGURE 4.5 Diagrammatic sketch of Z and S domains caused by respective anticlinal divisions of fusiform initials. The thick black lines separate the areas demarcated by high occurrence of handedness types.

in causing the span of time a domain covers a particular site relatively constant (in most taxa about 12 years). In conifer and dicot taxa that produce interlocked or wavy grains (see later for definitions of these grain types), this migrating domain pattern occurs. In these cases, the frequency of S or Z events is fairly high so as to affect the inclination of fusiform initials. If there are no periodic reversals in the direction of pseudotransverse divisions grain angle, changes would be kept low while the reverse situation occurs if reversals take a longer time. Since the domain pattern migrates, the grain undulations seen on the radial facet of wood are inclined with reference to ray initials. The length of undulations (i.e., the distance between two crests) is two times the domain length; the angle of undulations (α), however, is related to the rate of domain migration. The ratio domain length: velocity or undulation length: tan α varies only slightly; λ/tan a = constant (Hejnowicz 1990).

As already indicated, because of the migrating cambial pattern, the handedness of cambial events in the same group of cambial initials changes cyclically with time as well as with distance. Thus, handedness is a "cyclic function of time as well as of spatial distance" (Hejnowicz 1990). The first function can be converted into the second function. It is already known that a cyclic function of time and distance is a very characteristic feature of any traveling wave. The detailed researches of Hejnowicz and his group of investigators have clearly shown that there are traveling waves of handedness in the cambium. The domain pattern, described above, represents the spatial aspect of the traveling wave while the periodic reversal of handedness at a particular location represents the temporal aspect. As is well-known, a traveling wave is characterized by length λ, period (time) **T**, and velocity **V**; these three parameters are interrelated by the equation $\lambda = \mathbf{VT}$ (Hejnowicz 1990).

The handed wave in the cambium can be considered as a morphogenetic wave since it affects the arrangement cambial initials and consequently the cambial form. This wave is unusual because the period is extremely long (as already stated, about 12 years). Also, the period is the least variable parameter of the wave. If one takes the lowest value of the parameter as a unit, the variation, is as follows: the wave length λ varies from a few millimeters to several hundred millimeters or even a few meters (e.g., *Nyssa*). The velocity varies from a fraction of a millimeter per year to a few hundred millimeters. Since, as already stated, the longer the wave the faster it travels, the period given by λ/v is the least-variable parameter of the wave and is almost invariant (Hejnowicz 1975).

One interesting property of the cambial waves is that it may have, at the same time, two or more waves that have the same period but different λ and v; these waves may even travel in opposite directions (Hejnowicz and Romberger 1979). Another important property of these waves is that they are **additive**, that is, they undergo **interference** (as is common in light and sound waves). This is evident from the presence of complex grain figures, such as moiré, quilted or chequered, or from the pulsation of domain patterns (Hejnowicz 1974; Hejnowicz and Romberger 1979).

Another important property of grain that is decided by the cambium is the **amplitude** of undulations of the grain. It is defined as "the maximum angle by which the grain deviates from the longitudinal axis of the grain" (Hejnowicz and Romberger 1973). The amplitude of undulation is related to the frequency of handedness events in the cambium producing wavy grains in wood. If it is assumed that the wave underlies the grain undulation, then the amplitude of the wave can be considered as manifested by the frequency of handedness event in the cambium.

The derivatives of ray initials of cambium, that is, the xylem rays, seem to play only a passive role in changing grain angles; they lag behind that of the derivatives of fusiform initials (Harris 1969, 1973). As the cambium grows outward (due to more and more production of xylem cells on its inside), each ray initial grouping appears to rotate about its center in adjustment to the changing orientation of fusiform initials rather than behaving as a separate, dynamic component of changing grain angles (Harris 1973; Hejnowicz 1968; Jones 1963). The relative positions of wood rays remain remarkably constant across successive growth rings, despite the fact that a ray may be split by intrusive growth of fusiform initials, or that two rays may become united either by loss, retraction, or parenchymatization of the part or the whole of a fusiform initial.

4.3.1.1   Cyclic Handedness Events and Domains in Cambium That Produce Straight or Spiral Grains

The cambium that produces straight or spiral-grained woods (see later for definitions) often shows only a low frequency of handedness events (Hejnowicz 1990). Such cambia are also very common in trees, but less in number than the trees with inter-locked grains. In these cambia, handedness events are often non-random; domains are often difficult to delineate on the map of events. Such maps of handedness events on the cambium of *Aesculus hippocastanum* have been produced by Hejnowicz and Krawczyszyn (1969). The number of **S** and **Z** partitions in pseudotransverse divisions in unit area of cambium was counted; it was found by the authors that the probability of **Z** events was smaller than that **S** events, and their distribution was also found to be non-random. There was also a grouping of both **S** and **Z** events, thus indicating a tendency toward domains in the cambium. Similar groupings are also observed in other cambia, including stratified cambia. These groups do not share domain migration; their handed-ness changes were described as "sparkling" (Wloch and Bieczewska 1987).

However, in those cambia that produce straight-grained wood, there may occur regions that can be called "migrating domains." In the cambium of a 60-year-old spruce tree with normal wood, most of its cambium was characterized by only a **Z** event but for a very small area (stripe) with an **S**-type event. This stripe migrated from left to right at a rate of about 7 mm per year or about 2 mm during the produc-tion of 1 mm of xylem. In 170-year old *Abies* trees, in an area of 380 × 160 mm cambium, the total number of **S** and **Z** cell division walls was nearly the same. Yet most of the unit areas within this total area showed predominance of only one type of handedness in the cell partition wall.

4.3.1.2   Working Hypothesis

In the light of the above discussion, a working hypothesis has been developed by Hejnowicz (1990). The details of this hypothesis are as follows: it may be stated that in cambia that produce wavy-grained wood there is a cyclic change of handed-ness both temporally and spatially. This results in the migration of a striped domain pattern along the stem, either upward or downward depending upon the species. In cambia that produce straight-grained wood, S and Z regions of different sizes appear in the cambium; there is also a change of handedness with time in these groups, although not synchronously with all cells within a group. In addition, the period taken for change is not as long as in the cambium that produces wavy grain.

The cambium is a system of regions in which handedness changes with time. At one extreme are the traveling waves of changes in handedness with a marked regular cyclic change and relatively long cycle duration. At the other extreme is the occurrence of sparkling handedness groups exhibiting shorter duration of handedness type at a site than in the previous case. In each case, the cambium can be considered as a system of handed oscillators, in fact a system of coupled handed oscillators. An elementary oscillator appears to be a small group of cambial cells or cell ends, but not as a single cell (or cell end). The oscillation pertains to an unknown process. The oscillation relates to some unknown process within the cell group that is manifested in the two handedness types of cyclic events. It is probable that something in this process changes continuously with time, but only signs of this change are reflected in the handedness. If so, there is a continuous flow of phase in the process. The coupling between different oscillators affects the phase in the process. The coupling between different oscillators affects the phase and period of the oscillators, and the degree of this coupling depends on the amplitude (i.e., frequency of events). High amplitude favors coupling. Strong coupling results either in synchrony (the cambium producing wood with interlocked grains) or in a constant gradient of phase, that is, a linear change in gradient with distance (the cambium producing wavy-grained wood). The steepness of the gradient determines the wavelength without affecting the period. Hence, it determines also the velocity: the less steep the gradient the longer and faster the wave. Strong coupling is said to increase the duration of the oscillation cycle, which explains the long period taken by the traveling waves. Weak coupling, on the other hand, makes the neighboring oscillators differ in phase. If the phase difference is different from 180°, the differently phased oscillators may not be discernible as an alternative handedness at a particular instant. Thus, the neighboring oscillators showing phase difference may still exhibit the same handedness instantaneously, but their handedness does not change synchronously. As a consequence, any cambial area, distinguished by the same handedness at a certain instant, changes its shape and size with time. The cyclic change in handedness with a short period inherent in individual oscillators and the other changes in the area showing handedness are probably the basis of the observed sparkling of handedness in the cambia producing straight-grained wood. The sparking may get additionally complicated if there is instability of the borders between oscillators due to phase shifting (Hejnowicz 1990).

Thus, according to this working hypothesis proposed by Hejnowicz (1990), the normal cambium producing straight-grained wood is in a state of weak coupling between the handed oscillators, while the one producing wavy-grained wood and characterized by a high frequency of events is in a state of strong coupling.

4.3.2 NON-CYCLIC HANDEDNESS EVENTS

The second group of handedness events is represented by a twist of periclinal cell wall partitions; the partition is not strictly flat but is distorted (in the radial rows of derivatives) at its ends in the manner of a screw propeller (Wloch 1981). The twist may be of a S or Z type. However, it does not cycle as in anticlinal pseudotransverse divisions, and its handedness may not conform to the type of domain in which peri-clinal division occurs. These twists happen to be of the S type in straight-grained wood of spruce and larch, and probably this type of twist might be a feature of conifers where it may be a basis for developing an S-spiral grain in young trees. This S-type inclination of grains may be counteracted by Z events.

4.4 GRAINS IN WOOD

Casual references to wood grains and their types were made in reference to cam-bium in earlier sections of this chapter. Here, a detailed account on wood grains is provided.

4.4.1 Definition and Types of Wood Grains

Wood grains are common features of any wood and may well be part of the normal process of wood production in trees (Northcott 1957). They refer to patterns of xylem elements (particularly fibers, tracheids, and vessels) seen in a cut surface of wood, both transverse and longitudinal. Wood grains are classified into straight, spiral, interlocked (ribbon figure), wavy, and irregular categories (Panshin and De Zeeuw 1970). Straight grains refer to distribution of secondary xylem elements parallel to the long axis of the stem. Straight-grained wood is of relatively rare occurrence. This type of wood is gen-erally the easiest to work and machine with the least complications. Spiral grains refer to the disposition of wood elements at a slight incline from the straight vertical line with spiral pattern circling the trunk (Figure 4.6). Spiral grains, however, are a common characteristic of almost all trees. Interlocked grains refer to grains that "spiral back and forth" throughout the trunk and thus alternate between left- and right-hand "spirals." These are formed when successive growth rings run in a different direction. There are weakly and strongly interlocked grains, depending on the tree. These grains are very clearly seen in quarter-sawn surfaces and create a ribbon-stripe figure. Woods with this type of grain also create problems when machining. A wavy grain, as the name implies, has a wavy outline and is best seen in flat-sawn sections of wood. It results when the direction of wood elements alternates. Irregular grain is an ambiguous grain type, and this is often twisted or swirled in irregular ways. This is caused by knots, burls, etc.

4.4.2 Spiral Grains

With particular reference to spiral grains, deviations from straight alignment nor-mally happen at small angles. Braun (1854) and Hartig (1895) were perhaps the first to record that trees markedly show changes both in spiral and directions throughout the life of a tree. This angle may be calculated as the angle between the central vertical longitudinal line of the stem and the spirals that are visible when the stem is debarked (Figure 4.7). This angle ranges from 30° to 50° and in extreme cases up to 90°. The other way is to represent the spiral angle in radians and for this transverse disc and used of stems and the angle is measured in the growth ring border region between adjacent growth rings. The spiral angle in radians usually ranges from −0.2 to +10.5. The spiral grain generally decreases the strength of the wood. Champion (1929) and

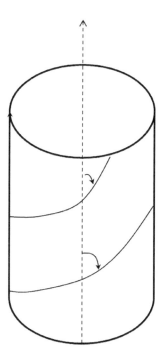

FIGURE 4.6 Diagrammatic sketch of wood cylinder showing the spiral grains and the spiral angles (arrows) that these grains make with the vertical axis (dotted arrow line).

Dadswell and Nicholls (1959) used the Figure 4.7 (radian) spirality as being the upper limit of an acceptability for quality solid wood products. In sawn timber, these grains can result in twisting on drying. Again, spirally grained wood poles often tend to twist and untwist due to changing water content. Spiral grains also cause problems when plywoods are used due to dimensional instability (Harris 1981). In view of the above, "spiral" grains have been studied by foresters, botanists, and wood technologists for more than 150 years (Harris 1981, 1988; Noskowiak 1963).

4.4.3 SPIRAL GRAIN OF CONIFERS

A critical review of work done on spiral grains reveals that there is clear agreement as to which of the different categories of grain deviation should come under spiral grains, because the different ways of spiral grain development are quite unlike in different species; some suggest that each of these different ways should be treated separately. These suggestions have come, for example, after injecting dye along a radius at the bottom of a living conifer tree and observing at least five different distribution patterns of the dye (Vité 1967; Vité and Rudinsky 1959). There are also fundamental differences between hardwoods and softwoods in grain inclination developing over the first year (Priestley 1945). However, a number of recent researches on hardwoods suggest that almost similar structural mechanisms can be used to explain simple "spiral" grain and other grains (Hejnowicz and Romberger 1979).

FIGURE 4.7   Photograph of a debarked trunk of *Pinus albicaulis* showing left-handed spiral grains. (Courtesy of Earle, C.J., The Gymnosperm Database, https://www.conifers.org/topics/spiral_grain.php.)

4.4.3.1   Typical Spiral Grain Patterns

According to Noskowiak (1963), there is a general pattern of spiral grain development in conifers, which is agreed upon by several researchers. In this pattern, the tracheids of the first-formed primary xylem adjacent to the pith are laid down roughly parallel with the stem's long axis, but subsequent growth layers develop increasing left spiral grain, which invariably reaches its maximum within the first 10 years of growth (Harris 1981). Once spiral grain gets initiated, the time taken to reach the maximum left-handed spirality is extremely variable, both within the same species and often between species. Northcott (1957) showed in *Pseudotsuga menzies* (from 10 different localities that the average time taken to reach maximum left-handed spirality varied from 20 to more than 200 years. After 10 years, the left grain angle decreases until it becomes straight-grained once again. In some trees, this subsequently becomes inclined to the right "spiral" and continues to increase in magnitude. Northcott (1957) showed in different populations of *Pseudotsuga menziesii* that the average time taken to change over to right-handed spirality varied from 100 to 300 years. If this general pattern of grain direction is observed down a single growth ring from stem apex to the base of a medium-aged tree, it will be noticed that the straight grain at the stem apex becomes increasingly left spiral in the upper stem, the grain angle diminishes in the mid-stem, passes on to a straight grain, and finally a right-hand spiral at the lower stem. Thus, within a single growth ring a more or less complete "spiral" pattern develops.

A detailed study of Tian et al. (1995) in a 25-year-old plantation of *Pinus radiate* showed that the spiral grain exhibits a more complex pattern than many other properties, although there was still an overall tendency forward radial and vertical symmetry.

4.4.3.2 Deviations from Typical "Spiral" Patterns

Significant derivations from the typical spiral pattern have been described by many researchers. Of the *Pseudotsuga menziesii* tree populations examined by Northcott (1957), 64% of the trees followed a typical pattern in which left-handed spirality changed to a right-handed one with age, 19% were continuously increasing in left-handed spirality, and 17% exhibited most other possible combinations. In *Larix kaempferi*, left-handed to right-handed spirality was found in 53% of population, 41% remained left-handed till up to 40 years, and 6% were more or less straight grained (Nakagawa 1972; Mikami 1973). In some softwoods, the pattern of spiral grains was highly unpredictable. Northcott (1957) noticed that in 19% of Douglas fir stems there was a small left "spiral" in the early years to an increasing left spiral in later life without any right spiral at all. In some trees like *Pinus balfowriana*, there was a spiral angle as high as 40° (Noskowiak 1963), while in *Pinus roxbrughii* tracheids are sometimes oriented almost horizontally around the stem (Venkataramanan 1967). Grain angle may vary markedly within an annual growth ring or across different radii in the same stem section (Misra 1939; Paul 1955). In some trees, a pattern of interlocked grain may be superimposed upon the more generalized "spiral" pattern (Harris 1988; Misra 1939).

4.4.4 SPIRAL GRAINS IN HARDWOODS

With reference to spiral grains of hardwoods (dicots), the pattern is less predictable and more complex than in softwoods. Noskowiak (1963) concluded that it is frequently the reverse of that found in conifers (softwoods). As a result, grain angles that develop to the right in early years change to the left in a mature tree. However, there are too many exceptions to this. The change of spiral grain direction is less frequent in hardwoods than in softwoods (Ohkura 1958). In *Sambucus nigra*, *Syringa vulgaris*, *Castanea sativa*, etc., Priestley (1945) reported a destructive development of "spiral" grain, which the pitch of the spiral is determined in the first year's wood, and this further increased purely geometrically in girth. This apparently "passive" spirality development appears to arise through total inability to restore axial polarity once some small initial twist has developed in the apical shoot. In all of the forms of spiral grain, there seems to be at least some response by way of cambial reorientation.

4.5 CAUSES OF SPIRAL GRAIN

4.5.1 EXTERNAL CAUSES

Amongst the external environmental causes that may lead to spiral grain formation, altitude, aspects, soil, injuries, solar movement, earth's rotation, wind action, etc. have been suggested in the past. There are reports that it is caused by Coriolis force or that trees always spiral in one direction in the northern hemisphere and in the other direction in the southern hemisphere of the earth. These aspects have been reviewed by Noskowiak (1963), Nicholls (1965), Kubler (1991), and Skatter and Kucera (1998). Nicholls (1965), however, concluded that none of these factors can adequately account for the changes in spiral grain direction observed, in nature. However, these external factors may modify "spiral" grain patterns established by other means and also determine the level of their development. Unfortunately, experimental work is lacking to support this categorically. One factor that has been experimentally tested is the effect of torsion on the spiral grains of *Aesculus hippocastanum* (Pyszynski 1977). This plant has a spiral grain direction always to the right, although wavy grain may sometimes to be superimposed on this. Pyszynski (1977) subjected the stems to alternate left- and right-handed torsion, and found the residual torsions to the right to be larger than those on the left. Since the fibrillar helix in cell walls is also predominantly oriented to the right, he conjectured that asymmetry of the stem's mechanical properties may be the reason for the formation of right spiral grain in the stem subjected to various torsions under wind action. Skatter and Kucera (1998) have established that the spiral grain is an optimized growth character in trees exposed to combined bending and torsion.

4.5.2 EFFECT OF GROWTH RATE AND GROWTH PATTERN

For any initial grain deviation, rapid growth in girth would tend to produce wider grain angles than slower growth per unit time (Priestley 1945). This is designated by him as the "geometrical effect." If this effect is a very important feature of developing spirality, then it would be expected that rapid growth in girth would be closely associated with severe spirality. But this is not so (Krempl 1965; Sachsse 1965; Mikami et al. 1972). Hence, spiral growth and growth rate are totally independent of each other.

Trees often have asymmetric crowns with the side facing the sun having a larger crown than the side facing away from the sun. Skatter and Kucera (1998) asserted that crown asymmetry combined with prevailing westerly winds produces spiral grains with a predominant right-handed spiral in the northern hemisphere and a predominant left-handed spiral in the southern hemisphere. But his has also been proved to be incorrect. There is no hemispheric effect.

Spiral growth is stated to allow water from each individual root to reach every branch on the tree. In addition, trees become less stiff and bend more easily because of the spiral grain (Kubler 1991). Excessive spiraling not only reduces the stiffness of the tree but also weakens the strength of the tree. Hence, there must be a limiting point on how much the stiffness can be reduced in order for a tree to stand up straight. The grain angle should be below 37° and should not increase beyond it. The grain angle is often greater at the bottom of the trunk and becomes smaller and smaller toward the tip. Spirals in the grain are developed for the transport of water to all branches.

4.5.3 GENETIC CONTROL

Early researches indicated that spiral grains are a highly heritable feature of at least some species (Cown et al. 1991). This conclusion was made on the basis of breeding work undertaken in *P. roxbughii* in India. This species often produces severely spiralled grains (Champion 1924, 1927, 1930). This type of work was not followed up by further investigations for a very long time until the 1960s when it was renewed in fast-growing *P. radiata* and *Pinus taeda*. Some of these studies revealed an essentially additive nature of inheritance and a broad-sense heritability for spirality in *P. radiata* clones of 0.66 for annual rings 2 to 8 (counted from pith) but was gradually reduced to only 0.28 when rings 7 to 8 were considered by themselves (Dadswell et al. 1961). Similarly Nicholls (1965) noticed that the broad-sense heritability decreased gradually from 0.6 for the first ring (counted from pith) to 0.1 for the 21st ring. In *P. radiata*, spirality usually reached its maximum value within the first three growth rings but became very small outside the tenth ring. Pederick (1971) found heritability of spiral grain in the second annual ring as 0.40, and the non-additive component of the genetic variance was extremely small (almost zero). Zobel et al (1968) studied, 1,043 four-year-old trees belonging to 52 controlled-pollinated *P. taeda* "families" and found definite "familial" difference in grain spirality at 20 cm height with the smallest average grain angle for a "family" of 21 trees 1.7° and the largest average angle 4.8°. They also found that the genetic variation for spiral grain was non-additive. Mikami et al. (1972) analyzed 400 trees (belonging to 80 clones) of eight- and nine-year-old *Larix leptalepis* and got gross heritability values from 0.35 to 0.42 for grain angles in individual growth layers and 0.49 for a maximum grain angle (see also Zobel 1965).

4.6 ANATOMY OF SPIRAL GRAIN DEVELOPMENT

As already discussed, wood spirality arises in most, if not all, instances from the reorientation of cambial elements. Although it is often convenient to follow cambial reorientation by reference to changes in successive xylem layers (= growth rings), it should be kept in mind that it is in the cambium that most of these changes first take place. As already discussed in detail earlier, the fairly simple anatomy of conifer wood often ensures a continuous record of cambial reorientation that is often left behind in the secondary xylem. However, in direct woods, the presence of vessel elements and the often intermittent production of axial parenchyma makes the interpretation very difficult and to rebuild the record problematic. In the latter, the terminal zone of xylem in the annual ring is often used to interpret changes (Hejnowicz and Krawczyszyn 1969).

4.7 ARE THE SPIRAL GRAINS REALLY SPIRAL?

In the first chapter of this volume, it was pointed out that there are conceptual differences between spirality and helicality and that a number of instances described by some workers as spiral are in fact helical. Spiral grains are one such phenomena, and they should in fact be appropriately described as helical grains. The so-called spiral grains neither show the characters nor of equiangular spiral, but really show features of helical structures. Smyth (2016) in his review clearly designates these as showing a helical pattern.

Like many other typical helical structures of plants, helical grains (other than straight) of both conifer and dicot woods often show handedness, which frequently changes in relation to time, position, and velocity of cambial waves.

REFERENCES

Braun, A. 1854. Uberden schiefenVerlauf der Holzfasern und der dadurch bewirkten Drehung der Baume. *Ber Berlin Akad* 432–484.

Champion, H.G. 1924. Contributions towards knowledge of twisted fibre in trees. *Indian For. Rec* 11:11–80.

Champion, H.G. 1927. An interim report on the progress of investigations into the origin of twisted fibre in *Pinus longifolia* Roxb. *Indian For.* 53:18–22.

Champion, H.G. 1929. More about spiral grain in conifers. *Indian For.* 55:57–58.

Champion, H.G. 1930. Second interim report on the progress of investigation into the origins of twisted fibre in *Pinus longifolia* Roxb. *Indian For.* 56:511–520.

Cown D.J., McConchie, D.I. and Young, G.D. 1991. Radiata pine wood properties survey: New Zealand Ministry of Forestry, FRI Bulletin No. 50 (Revised edition).

Dadswell, H.E. Fielding, J.M., Nicholls, J.W.P. and Brown, A.G. 1961. Tree-to-tree variations and the gross heritability of wood characteristics of Pinus radiate. *TAPPI* 44:174–179.

Dadswell, H.F. and Nicholls, J.W.P. 1959. Assessment of wood qualities for tree breeding 1. In *Pinus elliottii* var. *elliottii* from Queensland. CSIRO *Aust. Div. For Prod. Tech. Pap.* 4, 16 pp.

Earle, C.J. The Gymnosperm Database, https://www.conifers.org/topics/spiral_grain.php.

Harris, J.M. 1969. On the causes of spiral grains in corewood of radiate pine. *N Z J. Bot.* 7:189–213.

Harris, J.M. 1973. Spiral grain and xylem polarity in radiata pine: Microscopy of radial reorientation. *N Z For. Service* 3:363–378.

Harris, J.M. 1981. Spiral grain formation. In: Barnett, J.R., (Ed.) *Xylem Cell Development.* Castle House Publications, Tunbridge Wells, UK. pp. 256–274.

Harris, J.M. 1988. *Spiral Grain and Wave Phenomena in Wood Formation.* Springer-Verlag, Berlin, Germany, 215 pp.

Hartig, R. 1895. Uber den Drehwuchs der Kiefer. *Forst. Naturwiss Z* 4:313–326.

Hejnowicz, Z. 1968. The structural mechanism involved in the changes of grain in timber. *Acta Soc. Bot. Polon.* 37:347–365.

Hejnowicz, Z. 1974. Pulsations of domain length as support for the hypothesis of morphogenetic waves in the cambium. *Acta Soc. Bot. Pol.* 43:261–271.

Hejnowicz, Z. 1975. A model for morphogenetic map and clock. *J. Theor. Biol.* 54:345–362.

Hejnowicz, Z. 1990. Phenomena of orientation in the cambium. In: *The Vascular Cambium*, Iqbal., M. (Ed.). Research Studies Press, Taunton, UK.

Hejnowicz, Z. and Krawczyszyn, J. 1969. Oriented morphogenetic phenomena in cambium of broadleaved trees. *Acta Soc. Bot. Pol.* 38:547–560.

Hejnowicz, Z. and Romberger, J.A. 1973. Migrating cambial domain and the origin of wavy grain in xylem of broadleaved trees. *Am. J. Bot.* 60:209–222.

Hejnowicz, Z. and Romberger, J.A. 1979. The common basis of wood grain figures is systematically changing orientation of cambial fusiform cells. *Wood Sci. Technol.* 13:89–96.

Jones, B.E. 1963. Cell adjustments accompanying the development of spiral grain in a specimen of *Pseudotsuga taxifolia* Britt. *Commonw. For. Rev.* 43:151–158.

Krempl, H. 1965. Preliminary research results on spiral grain in Norway spruce. *Proc. IUFRO Melbourne, Sect 41*, Vol. 1, 17 pp.

Krishnamurthy, K.V. 2015. *Growth and Development in Plants.* Scientific Publishers, Jodhpur, India.

Krishnamurthy, K.V., Bahadur, B., Adams, S.J. and Venkatasubramanian, P. 2015. Development and organization of cell types and tissues. pp. 73–112. In: *Plant Biology and Biotechnology Volume I: Plant Diversity, Organization, Function and Improvement*, Bahadur, B., Rajam, M.V., Sahijram, L., Krishnamurthy, K.V. (Eds.). Springer, New Delhi, India. doi:10.1007/978-81-322-2286-6_4.

Kubler, H. 1991. Function of spiral grains in trees. *Trees* 5:125–135.

Mikami, S. 1973. Preliminary studies on the genetic improvement in spiral grain in Japanese larch. *Proceedings of IUFRO Div* 5 S Afr, 11 pp.

Mikami, S., Watanabe, M., and Ohta, N. 1972. Clonal variation in spiral grain of *Larix leptolepis* Gord. *J. Japan For. Soc.* 54:213–217.

Misra, P. 1939. Observations on spiral grain in the wood of pinus Longifolia Roxb. *Forestry* 13:118–133.

Nakagawa, S. 1972. Distribution of spiral grain within stem and the spirality pattern on *Larix leptolepis* Gordon. *Bull. Govt. For. Expt. Stn. Meguro.* 248:97–120.

Nicholls, J.W.P. 1965. The possible causes of spiral grain. *Proceedings of IUFRO Melbourne Sect 41*, Vol. I, 7 pp.

Northcott, P.L. 1957. Is spiral grain the normal growth pattern? *For Chorn.* 33:335–352.

Noskowiak, A.F. 1963. Spiral grain in trees: A review. *Forest Prod J.* 13:266–275.

Ohkura, S. 1958. On the macroscopic features of twisted fibre in trees. *J. Fac. Agri. Shinshu Univ.* 8:59–100.

Panshin, A.J. and De Zeeuw, C.1970. *Textbook of Wood Technology*, 3rd edition. Vol. 1. McGraw-Hill, New York.

Paul, B.H. 1955. Importance of wood quality in tree breeding. *J. Forest.* 53:659–661.

Pederick, L.A. 1971. Inheritance of spiral grain in young radiate pine. Presented to XVth congress, IUFRO, Gainesville, FL.

Priestley, J.H. 1945. Observations on spiral grain in timber. *Am. J. Bot.* 32:277–284.

Pyszynski, W. 1977. Echanism of formation of spiral grain in Aesculus stems: Dissymmetry of deformation of stem caused by cyclic torsion. *Acta. Soc. Bot. Polon.* 46:501–522.

Sachsse, H. 1965. The effect of the rate of growth on the occurrence of spiral grain. *Proceedings of Meeting of Section.* 41:1.

Skatter, S. and Kucera, B. 1998. The causes of the prevalent directions of the spiral grain patterns in conifers. *Trees* 12:265–273.

Smyth, D.R. 2016. Helical growth in plant organs: Mechanisms and significance. *Development* 143:3272–3282. doi:10.1242/dev.134064.

Tian, X., Cown, D.J. and Lausberg, M.J.F. 1955. Modeling of *Pinus radiate* wood properties. Part 1: Spiral grain. *New Zeal. J. For Sci.* 25:200–213.

Venkataramanan, S.V. 1967. Spiral grain in chir (*Pinus roxburghii* Sargent). *Proceedings of XVI IUFRO Congress*, Munich, IX. 484–497.

Vité, J.P. 1967. Water conduction and spiral grain: Causes and effects. *Proceedings of XIV IUFRO Congress*, Munich, IX. 338–351.

Vité, J.P. and Rudinsky, J.A. 1959. The water conducting systems in conifers and their importance to the distribution of trunk-injected chemicals. *Contrib. Boyce Thompson Inst.* 20:27–38.

Wloch, W. 1981. Nonparallelism of Cambium Cells in neighbouring rows. *Acta. Soc. Bot. Pol.* 50:625–636.

Wloch, W. and Bieczewska, E.1987. Fibrillation of events in the cambial domains of Tilia cordata. Mill. *Acta Soc. Bot. Pol.* 56:19–35.

Zobel, B.J. 1965. Inheritance of spiral grain. *Proceedings of Meeting of Section 41: IUFRO.* Melbourne, Australia, 1.

Zobel, B.J, Stonecypher, R.W., and Browne, C. 1968. Inheritance of spiral grains in young Loblolly pine. *Forest Sci.* 14:376–379. doi:10.1093/forestscience/14.4.376.

5 Broken Symmetry and Handedness in Cryptogams

Robert W. Korn

CONTENTS

5.1 INTRODUCTION

The concept of broken symmetry was recently introduced into developmental biology (Piano and Kemphues, 2000; Pillitteri et al., 2016), although it has a long history in physics (Anderson, 1972). Consider a ball balanced between two valleys; a slight influence is introduced at the resting point causing the ball to fall into one of the two valleys. Since the valleys were originally balanced or were symmetrical, the influence breaks symmetry so the two valleys become different, one with and one without the ball. Symmetries are of relatively low degree of order and asymmetries are of greater degree of order and so are of greater interest.

Biology has experienced a two-fold history of breaking symmetry. In one, a morphological description (form or shape) has mirror imagery or symmetry, such as dichotomous branching of a liverwort gametophyte. Symmetry is broken by greater growth in one over the other. The second approach is anatomical with a difference between parts, whereby mother-cell division is equal and daughter cells are identical; therefore, cells are symmetrical. Introducing a signal into the mother cell in order to break symmetry leads to an unequal division, so the two daughter cells are different or asymmetric. At the cell level, breaking symmetry is a four-fold process involving tissue polarity, cytoplasmic localization of a determinant, orientation of the mitotic spindle, and asymmetric cell division (Horvitz and Herskowitz, 1992).

Examples of symmetry breaking in plants have been taken from model systems, such as zygote development, stomate formation, and root cell kinetics, all three in *Arabidopsis*, namely, from phanerogam material (Petricka et al., 2009). While phanerogams have multicellular meristems for most of cell proliferation, and, as will be shown, most cryptogams grow by involving apical cell division and resultant merophyte (Douin, 1923; Barlow, 1994) formation, this quantal difference has differences on how symmetry breakdown occurs.

5.2 TYPES OF APICAL CELLS

Plant material in this section is from the fern *Dryopteris thelypteris* (L.) Gray unless specified otherwise. The first apical cell to appear in development is in the spore, which grows in one dimension with one cutting face (ac-1cf), especially in darkness and far-red light (Cooke et al., 1987) (Figure 5.1a). This one cutting face is disc shaped. In red or white light, this apical cell undergoes several unequal divisions to form a triangular apical cell with two cutting faces (ac-2cf) (Figure 5.1b). Each of these special divisions produces a new apical cell and a vegetative cell that proliferates into a group of cells called a merophyte (Douin, 1923), that is, a clone of cells

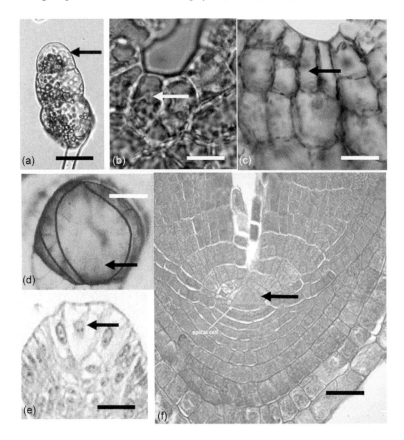

FIGURE 5.1 Fern apical cells. Arrows point to apical cells (a) filamentous stage of thallus. Bar is 30 μm; (b) thallus with two cutting faces. Bar is 34 μm; (c) thallus with three cutting faces and cushion. Bar is 30 μm; (d) leaf sporophyte with two cutting faces (Bierhorst, 1997) with added lines to emphasize cutting plates. Bar is 40 um; (e) stem sporophyte with three cutting faces. Bar is 60 μm; (f) root sporophyte with four cutting faces. Bar in 110 um.

derived from the sister cell to a daughter apical cell. Next to appear is the apical cell with two cutting faces that divides in a highly regulated left-right manner, thereby producing merophytes with the same total areas on both sides of the gametophyte; namely, there is an almost morphological mirror-image symmetry.

This two-faced apical cell lineage is replaced by a lineage of apical cells with three cutting faces changing from a triangular to a rectangular shape (Figure 5.1c). Two of the three facets produce rectangular merophytes to either side of the thallus, and the third division produces a central merophyte that later includes some periclinal divisions to form a several-celled thick cushion on which archegonia form.

In *Dryopteris*, the two lateral merophyte initials are not contiguous but are separated by the central merophyte initial, and so directionality or handedness is recognized as either clockwise or counterclockwise.

The three types of thallose apical cells, recognized here by the number of cutting faces, are appropriate for their function. Spores are sown on the ground and settle between soil particles in the dark. Here filamentous growth provides minimal resistance when growing between soil particles. Apical cells with two cutting faces, to the left then right then left, etc., generate a mirror-image symmetry of merophytes, so there is uniform distribution of photosynthetic cells for supporting later sporophyte development. Finally, an apical cell with a third cutting face leads to a several-celled thick central cushion that upon fertilization of an egg fills with stored photosynthate tissue around the new sporophyte.

Sporophytic apical cells relate to appropriate organs. Following fertilization, the zygote develops into four cells, those of a foot, which anchors the new sporophyte into the gametophyte, a stem, a leaf, and a root. Apical cells of each organ are seen at the 40-celled stage (Shaw, 1898; Campbell, 1918), and they probably form much earlier.

The apical cell of leaves has two cutting faces (Campbell, 1918, p. 409) (Figure 5.1d), that of the stem three cutting faces (Bierhorst, 1977) (Figure 5.1e), and the apical cells of roots with four cutting faces (Korn, 1993) (Figure 5.1f). Of the four cutting faces of roots, three lead to merophytes of proper root tissue and that of the fourth produces the root cap.

Leaves of the fern *D. thelypteris* have anisocytic stomata with two to five adjacent cells. In cases of three adjacent cells, cycles are recognized by relative cell sizes, so the handedness of coiling can be determined. In a small sample of 20 stomata, 11 adjacent cells ran clockwise and 9 ran counterclockwise, values sufficient to assume an expected 1:1 ratio for a random pattern of orientation.

The sporophyte forms sporangia that begin as an apical cell that divides in a peculiar manner. It divides into a daughter apical cell and a merophyte initial cell, with the latter not dividing until much later to form a stalk (Figure 5.2a). The sporocarp, an outgrowth of the leaf petiole of some ferns such as *Marsilia* (Johnson, 1898), has an apical cell with three cutting faces, which most likely came from broken symmetry during the first division of the apical cell lineage. The initial apical cells in these cases involve broken symmetry and the beginning of handedness.

At least three types of apical cells in the gametophyte, the number depends on the fern, and of those four or more in the young sporophyte are characterized by repetition of the same division, a new distinct apical cell and a merophyte initial, that is, an apical cell lineage that produces the bulk of cells of an organ.

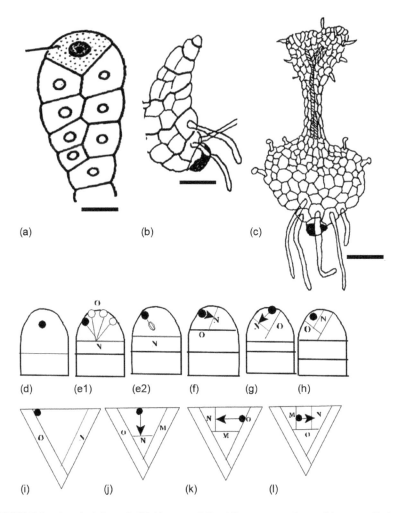

FIGURE 5.2 Atypical fern thalli (a) normal *Pteridium* sporangium with one-celled mero-phytes. After Vishnupriya. Bar is 100 µm; (b) half a thallus with papillae on one side instead of two. Bar is 60 µm; (c) apogamous sporophytic leaf. Error in 2D apical cell in which leaf apical cell is turned on leaf instead of an apical cell that makes 2D with cushion. Bar is 120 µm.

Additional types of apical cells can be found in mosses and liverworts. The erect gametophyte of mosses has "leaves" spiraling around the "stem" ("leaves" and "stems" superficially look like leaves and stems but have no vascular tissue). In *Fissidens* and *Bryoxiphium*, "leaves" are arranged in two rows (1/2 of 360°) or consecutive "leaves" are 180° apart. In *Fontinalis* and *Tetraphis* consecutive "leaves" are 1/3, or are 120° apart. Interestingly, most mosses are either 2/5 (144°) or 3/8 (135°) (Crandall-Stotler, 1984). In a *Bryum* sp., 100 angles were measured with an average of 138.5° (Korn, 1993), close to the Fibonacci angle of 137.5°. This feature is most likely generated by precise orientation of the mitotic spindle.

In the liverwort, *Marchantia*, the thallus occasionally branches dichotomously from an apical cell, dividing equally, giving rise to an apical notch with a lineage of apical cells (Figure 5.2a–d). Apical cells then come from two equally likely types of division; one is unequal producing a new apical cell and an initial cell of a merophyte while the other produces two daughter apical cells, the basis for dichotomous branching.

Meristemoids and apical cells appear to be similar as both are proliferating cells, but their differences are important. Meristemoids are single cells that occur in clusters like a polka dot pattern and are superficial (Bunning, 1953) whereas apical cells occur as marginal, solitary structures (Korn, 2017).

5.3 SYMMETRY BREAKING

The arrangement of leaves in ferns has long been associated with three-faced apical cells based on interpreted drawings (Bower, 1923), but with experimental results from surgery, Wardlaw (1949) implicated a substance originating from stem apical cells that induces leaf siting. Later, Kuehnert (1967), from studies on cultured leaf apices, implicated the previous leaf primordium as the source of the signal. Bierhorst (1977) agreed with Wardlaw in explaining phyllotaxy of sporophytic leaves. In moss gametophytes, the phyllotaxy ratios are 1/2, 1/3, 2/5, and 3/8 (Crandall-Stotler, 1984), apparently by orientation of spindle during apical cell division but how these specific orientations are achieved is unclear. To explain the left-right alteration of merophytes in the fern gametophyte, Korn (1984) proposed an aging effect on facets where apical cell cutting faces are in states new (N), middle aged (M), or old (O), but these, as with other earlier explanations, have shortcomings such as how an apical cell arises in the first place in a lineage. Recently, Solly et al. (2017) studied the dichotomous branching pattern in the liverwort *Marchantia* using the apical notch as the unit of growth analysis. The apical notch is composed of apical cells, merophytes initial cells, and merophytic cells, collectively a continuous hodgepodge that is difficult to evaluate compared to a discrete analysis where individual cells are analyzed by their unique set of features.

This idea of facet age cited above can be reinspected to more precisely dictate types of cell division of apical cells. As noted, the fern gametophyte passes through a series of stages—filamentous, 2D, and 2D with cushion. The first option open to development is the shift from 1D to 2D growth leading to the left-right alternation of cell division (Figure 5.2d–h). It is the first 2D division as to whether it is a left or right direction that is of interest. Of 100 thalli scored, 55 were left-handed (initial merophytic cell to the left when observing apical cell to the front of the thallus and spore wall backwards and closest to the observer), and 45 had the merophyte initial cell to the right, indicating orientation of this cell is random. The influence that dictates which of the two options occurs creates broken symmetry. What this influence is on positioning the dividing nucleus has been unclear but that there is an influence on the location of a proposed cytoplasmic particle (CP) is clear. Given whether the direction of the first step is left or right, all subsequent directions of division is then determined (Figure 5.2i–l).

The next case of broken symmetry is in the transition from 2D to 2D with cushion. Cell activity has to satisfy at least five requirements. First, in creating new spatial features the model utilizes only known structures, such as cell facets (lines in 2D or surfaces in 3D), vertices (points in 2D or lines in (3D), the mitotic spindle, microtubule band, and a cytoplasmic factor. Second, changes during successive iterations is in the location of the CP with respect to facets. Third, there can be different or modified models for each type of apical cell such that the algorithms for structures with two cutting facets is not necessarily the same as those with three or four cutting facets while conserving some other features, such as determining facet age. Fourth, a cytoplasmic particle is invoked to address the problem of complex geometry. The mutant thallus in Figure 5.2b found four times in a batch of spores from the same leaf is asymmetric in cutting to the left because it cannot form a second, old cutting face and only forms new facets in the same direction leading to a curved half thallus with marginal papillae only on the convex side. A second mutant encountered seven times is an apogamous leaf coming from what would have been an apical cell in the thallus with three cutting sides (Figure 5.2c). This anomaly suggests development is a genetic hopping from one type of apical cell algorithm to another.

A good starting point for explaining handedness is that for the fern thallus with two cutting facets (Figure 5.2i–l). Presence of a CP determines the host cell status as an apical cell, and the absence of it is any other type of cell. This CP passes from one daughter apical cell to the next apical cell generation. The 2D thallus has two cutting faces, in states O and N. The new facet is then the cell plate formed during mitosis with the facet formed in the previous cell division now being an old facet. The kinetics of the system then is the CP associates with then center of an old facet with the cell plate, the new facet, and the new apical cell is that daughter cell with a CP. The CP is conserved over the apical cell lineage and moves from one old facet to another during mitosis.

The initial filamentous stage of the fern gametophyte has parallel cross-walls and a CP located in the domed apical cell (Figure 5.2d). The CP locates somewhere over the domed surface, that is, an old wall. Here, broken symmetry takes place, namely, at any of the possible locations on the domed surface (Figure 5.2e1). Next, the spindle attaches to the CP, and a cell plate forms orthogonally to the spindle (Figure 5.2e2–f). Where the CP attaches to the domed surface, left or right, determines handedness and therefore broken symmetry. The influence that determines broken symmetry is the irregularity of the movement of the CP.

Shifting development of the fern thallus from 2D to 2D with a cushion and archegonia involves a replacement of one algorithm by another and the introduction of a cyclic pattern of orientation of apical cell division. This phase of development starts with a cell plate that is an N plate followed by a cell division where the N facet becomes type M facet and at the next division M becomes O; N becomes M with the cell plate as type N. This sequence is completely deterministic, so there is no broken symmetry or change in handedness.

The fern leaf apical cell is of interest because it occasionally bifurcates producing dichotomous branching. The apical cell is lenticular (doubly convex from surface view (Figure 5.3a–d). Dichotomy begins when a new cell plate with an N status is on both sides instead of only on one side, that facing the apical cell. Another interesting apical cell is that of the fern root with four cutting facets, three for making root cells and one for producing a root cap. As with other cases of apical cells, handedness

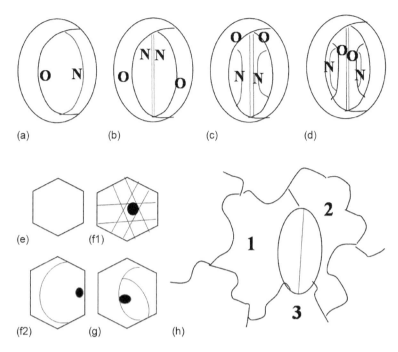

FIGURE 5.3   (a–d) Twin apical cells that form dichotomous thallus. (e–h) Formation of subsidiary cells in stomatal development of *Dryopteris* (e–h). Size of subsidiary cells decreases, which can be used to determine sequence of formation (1st, 2nd, and 3rd in Figure 5.3h) and so handedness.

is determined when the orientation of the first N facet of a lineage is determined. Stomata with three subsidiary cells (Figure 5.3e–h)—the largest is placed first, the next largest is second, or the smallest is third—give handedness of either left, 11 of 20 cases examined (Figure 5.3h) or right, 9 of 20 cases, in a ratio close to 1:1 or are orientated randomly.

5.4   CONCLUSION

It is of interest that most cases of apical cells in cryptogams are replaced by meristems in phanerogams. This difference is important in how the bulk of organs is produced. Persistent proliferation in cryptogams is achieved by apical cells dividing into two different cells, one daughter is an initial merophytic cell, and the other is another apical cell. Meristems in phanerogams are constant in size while proliferating by no obvious strategy as found for apical cells. It has been suggested that perhaps phanerogams have apical cells without the characteristic morphology and size (Korn, 1993), although several genes have been identified as associated with regulating meristem size (Leibfried et al., 2005), but no explanation has been proposed as to how meristem size is regulated during proliferation.

Breaking symmetry and its consequence of handedness occurs when a developmental event has more than one way to proceed, and each way is associated with a probability.

Developmental steps are usually deterministic and only rarely are stochastic, such as left- or right-handed spirals in phyllotaxy in multicellular plants. Which cells in a field become meristemoids, for example, stomatal initials, is also probabilistic.

Apical cell behaviorism is suggestive of coming from loci in genetic networks (Middleton et al., 2012). The apogamous formation of a leaf from an apical cell of a 2D gametophyte (Figure 5.2b) can be interpreted as the result of an error in switching genetic regulatory elements for apical cell formation. If so, at least this switching explains apogamy and apospory. Probability is normally involved in handedness is explained.

ACKNOWLEDGMENT

This work is dedicated to W. N. Steil, mentor and pioneer in research on apogamy.

REFERENCES

Anderson, P.W., 1972. More is different. *Science* 177(4047), pp. 393–396. doi:10.1126/science.177.4047.393.

Barlow, P.W. 1994. From cell to system: Repetitive units of growth in the development of roots and shoots. In M. Iqbal (Ed.), *Growth Patterns in Vascular Plants*, Dioscorides Press, Portland, OR, pp. 19–58.

Bierhorst, D.W. 1977. On the stem apex, leaf initiation and early leaf ontogeny in filicalean ferns. *Am. J. Bot.* 64:125–132.

Bower, F.O. 1923. *The Ferns (Filicales)*. Vol. 3. Cambridge University Press, Cambridge, UK.

Bunning, E. 1953. *Entwickleungs-und Bewegungsphysiologie der Pflanze. 3. Aufl.* Springer-Verlag, Berlin, Germany. doi:10.1007/978-3-662-02138-5.

Campbell, D.H. 1918. *The Structure and Development of Mosses and Ferns*, 3rd ed. The MacMillan & Company, New York.

Cooke, T.L., Racusen, R.H., Hickok, L.G. and Wayne, T.R. 1987. The photocontrol of spore germination of the fern *Ceratopteris richardii*. *Plant Cell Physiol.* 28:753–759.

Crandall-Stotler, B. 1984. Musci, hepatics and anthocerotes: An essay on analogues. In R.M. Schuster (Ed.), *New Manual of Bryology*, Vol. 2. Hattori Botanical Laboratory, Ann Arbor, MI. pp. 1093–1129.

Douin, C. 1923. Researches sur la gamétophyte des Marchantiees. III. Le thalle stèrile des.

Horvitz, H.R. and Herskowitz, I. 1992. Mechanisms of asymmetric cell division: Two Bs or not two Bs, that is the question. *Cell* 68:237–255.

Johnson, D.H. 1898. On the development of the leaf and sporocarp in *Marsilia quadrifolia* L. *Ann. Bot.* 12:119–145.

Korn, R. 1984. Cell shapes and tissue geometries. In P.W. Barlow and D.J. Carr (Eds.), *Positional Controls in Plant Development*. Cambridge University Press, Cambridge, UK. doi:10.1086/414600.

Korn, R. 1993. Apical cells as meristems. *Acta Biotheoretica* 42:175–189.

Korn, R. 2017. Anatomy of prepatterns in plants: A survey. *J. Plant Sci.* 6:89–102. doi:10.5539/jps.v6n1p89.

Kuehnert, C.C. 1967. Developmental potentialities of leaf primordia of *Osmunda cinnamomea* The influence of determined leaf primordia on undetermined leaf primordia. *Can. J. Bot.* 45:2109–2113.

Leibfried, A., To, J.P.C., Busch, W. and Stehl, S. 2005. WUSCHEL controls meristem function by direct regulation of cytokinin-inducible regulators. *Nature* 438:1172–1175. doi:10.1038/nature04270.

Middleton, A.M., Farcot, E. and Owen, M.O. 2012. Modeling regulatory networks to understand plant development: Small is beautiful. *Plant Cell* 24(10):3876–3891.

Petricka, J.J., Van Norman, J.M. and Benfey, P.N. 2009. Symmetry breaking in plants. Molecular mechanisms regulating asymmetric cell divisions in Arabidopsis. *Cold Spring Harbor Perspect Biol.* 1(5):a000497. doi:10.11101/cshperspect.a000497.

Piano, F. and Kemphues, K.J. 2000. In D.G. Drubin (Ed.), *Cell Polarity* (*Frontiers in Molecular Biology*). Oxford, Oxford University, pp. 28.

Pillitteri, L.J., Guo, X. and Dong, J. 2016. Asymmetric cell division in plants: Mechanism of symmetry breaking and cell fate determination. *Cell Mol. Life Sci.* 73:4213–4229. doi:10.1007/s00018-016-2290-2.

Shaw, W.R. 1898. The fertilization of *Onoclea*. *Ann. Bot.* 12:261–2865.

Solly, J.E., Cunniffe, N.J. and Harrison, C.J. 2017. Regional growth rate differences specified by apical notch activities regulate liverwort thallus shape. *Curr. Biol.* 27:16–26. doi:10.1016/j.cub.2016.10.056.

Vishnupriya, R. Pteridium: Habitat, external features and reproduction. Biology Discussion.

Wardlaw, G.W. 1949. Experiments on organogenesis. *Growth* (suppl.) 9:93–131.

6 Biology of Handedness in Fungi

C. Manoharachary and D. Nagaraju

CONTENTS

6.1   INTRODUCTION

Left- and right-handedness is a common phenomenon during plant growth. It has been observed in tendrils or twining of plants; circumnutation during growth of axial organs is the commonest phenomenon because they alternate between clockwise and anti-clockwise directions. Molecular genetic analysis of handedness has been worked out by Hashimoto (2002) in *Arabidopsis thaliana* (L.) Heynh.

Bacterial flagella are helical thread-like filaments that turn clockwise or anti-clockwise. The beating pattern of eukaryotic flagella exhibit left-right handedness in some cases. Venation pattern in some leaves are of an alternate type, including the branching of veins. In scorpoid and helicoid cymes, there is also left- and right-handed arrangement. Left- and right-handedness behavior has also been reported in many animals. This article deals with handedness in fungi.

6.2   SPIRAL GROWTH

Asymmetric movement was observed in myxomycetes like *Physarum polycephalum* (Figure 6.4d). Under experimental condition using a T-shape, the above slime mold has turned right in more than 74% trials (Dimonte et al. 2016).

Spiral growth in fungi has been coined to describe the rotation of the cytoplasmic wall in sporangiophore of *Phycomyces blakesleeanus* (Figure 6.4a). It is also found in some cases that the hyphae of some fungi grow with a bend instead of growing radially from the point of inoculation. The hyphae may bend in a left-handed or right-handed direction. This is dependent upon the fungal species and most of the fungal colonies as a whole showing a spiral growth pattern (clockwise or anti-clockwise). The extension zone wall at tip in the hyphae of young mycelia of *Mucor hiemalis*, *Aspergillus giganteus*, *Emericella nidulans*, and *Neurospora crassa* have displayed clockwise (right-handed) curvatures while that of *P. blakesleeanus* exhibited left-handed (anti-clockwise) curvatures.

Spiral growth of filamentous fungi has been encountered in the colonies that are grown on solid media in a Petri dish. The spiraling of the colony in a clockwise direction was noticed in S*ordaria fimicola* grown on malt extract agar, though angle versus spiral growth was very small. Out of 157 colonies tested, 21% showed marked spiraling when viewed from upper surface of the agar plate (Madelin 1978). Further spiraling depends on the incubation period, temperature, pH, and composition of agar media.

Chitin is the major cell wall component in fungi, and this character along with others allowed the fungi to obtain the status of a kingdom. Cell walls of stipe cells in *Coprinus cinareus Pers.* contain chitin microfibrils that occur as shallow helices, which are either left-handed (2/3) or right-handed. Shallow helicity has been the same in young and mature stipe cells of the mushroom (Kamada et al. 1991).

Sporanogiophores of *P. blakesleeanus* during the stage IVa growth phase rotate in a counterclockwise manner when viewed from above and spontaneously reverse to a clockwise direction in stage IVb. This rotation change lasts for 24-48 h (Goriely and Tabor 2011).

In *Stylopage anomala* and *Drechshlera*, single-celled conidia, are borne either singly or successively in a zig-zag manner on erect conidiophores. However, in *Zoopage* and *Bipolaris* (Figure 6.2b), the conidia are pale brown, elongated, or catenulate and are borne on conidiophores on either side or on left- and right-hand sides of sparingly branched conidiophores on mycelia. Conidia are produced in a spiral manner (left- and right-hand sides) on upright conidiophores forming a hand held in slime of *Helicocephalum* (Barnett and Hunter 1972).

In *Coemansia*, conidiophores are upright, slender, septate, and sparingly branched at intervals bearing sporocladia, which produce conidia on the lower (outer-oneside) surface only (either left- or right-handed side) (Barnett and Hunter 1972).

Conidia are formed on sympodially branched or geniculate conidiophores in *Bipolaris* (Figure 6.2b), *Drechslera* (Figure 6.2c), *Helminthosporium*, and *Curvularia*. Helicoid conidia are produced in *Helicosporium* (Figure 6.3e), *Helicoma, Helicomina, Helicoon* Morgan (Figure 6.3c and d), *Xenosporium, Helicodendron, Helicomyces* (Figure 6.3b) etc. Conidiophores are either small or tall, hyaline or brown, bearing conidia apically or laterally. The conidiophores may be zig-zag and coiling in conidia showing left- or right-handedness. Branched conidia are found in *Varicosporium, Dendrospora, Tricladium*, etc. Branching of conidia is of left- and right-handedness or of one type only (Barnett and Hunter 1972).

In *Hobsonia* (Figure 6.3f), the asexual fruit body is a sporodochium. Conidiophores are hyaline and slender, conidia are hyaline, many celled, and coiled in a loose spiral, which show both left- and right-handedness (Barnett and Hunter 1972).

Tharoopama synnemata with its well-defined stalk and head, bear left- and right-handedly branched sub hyaline to brown conidiophores with apical sporogenous cells bearing hyaline, one-celled conidia produced in a left- and right-handed manner (zig-zag) on small teeth-like structures. The fungus looks like a small tree under the microscope (Barnett and Hunter 1972).

In some fungi like *Pestalotia* (Figure 6.2f) and *Pestalotiopsis*, the conidia bear appendages that also show left- and right-handedness (Barnett and Hunter 1972).

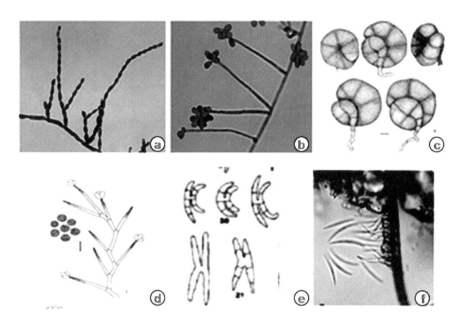

FIGURE 6.1 *Cladophialophora* sp. (a) conidial chain on right side; *Trichothecium* sp. (b) mostly conidiophores are produced on left side of main branch of Hyphae; *Cirrenalia nigrospora* (c) left-handedness of conidial body forms in circle; *Stachybotrys globosa* (d) left- and right-handedness in conidiophore branching; *Cornutispora* sp. (e) repeated conidia produced on the right side; and (f) conidiogenous cells and conidia produced on left side of fertile setae in *Chaetospina unilateralis*.

In *Cirrenalia nigrospora* (Figure 6.1c) and *Cirrenalia longipes* (Figure 6.2e), the conidia are helicoid, three to seven septate, and rounded at the apex, and the curving is either on the left- or right-hand side. Similarly, terminal coiling is either on the left- or right-hand side in *Dwyabeeja ethiopica* (Barnett and Hunter 1972).

In *Chaetospina unilateralis* (Figure 6.1f), the conidiophores are mononematous, sometimes in small groups, setiserous, erect, straight but slightly curved at the apical part, fertile in the above half, unbranched, with conidiogenous cells monophialidic produced in a palisade manner mostly at left side, rarely on right side, bearing falcate and aseptate conidia. It is almost the same case in *Cryptophiale apicalis* (Figure 6.2a), which produces conidia on the right-hand side (Barnett and Hunter 1972).

In *Allomyces* Butler, the zoosporangia are oval, terminate, and sympodially (left- or right-handed) arranged (Aronson and Preston 1960). In *Chaetomium* (Figure 6.2d), the hairs are terminal and lateral, stiff or coiled, branched or unbranched, smooth or ornamented, and the coiling of hairs may show left- or right-handedness in branching, which may be in zig-zag fashion. Sometimes, the hairs are monopodial and dichotomously branched in a left- and right-handed fashion (Manoharachary et al. 2015).

The left- and right-handedness production of conidia or conidiophores, and sometimes, branching of conidia-conidiophores, have been shown in *Cladophialophora* (Figure 6.1a), *Trichothecium* sp. (Figure 6.1b), *Stachybotrys globosa* (Figure 6.1d),

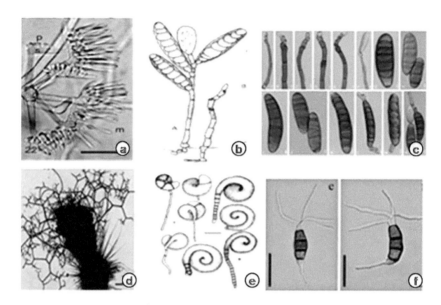

FIGURE 6.2 *Cryptophiale apicalis* (a) produces conidia on right-handed side; *Bipolaris* sp. (b) left- and right-handed conidia; *Drechslera* sp. (c) geniculate conidiophore in branched left and right; *Chaetomium* sp. (d) repeated branches of appendages leads to left- and right-handedness; *Cirrenalia longipes* (e) right-handed conidial formation; and *Pestalotia* sp. (f) conidiophore appendages.

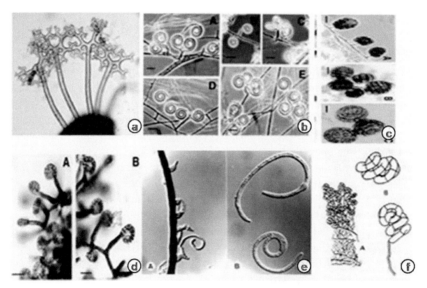

FIGURE 6.3 *Erysiphe* sp. (a) left- and right-handedly branched appendages; helicoid spores showing left-handedness in *Helicomyces* sp. (b) *Helicoon* sp. (c) right-spiralness helicoid fungi; *Helicoon* sp. (d) left- and right-handedness of conidiophore and conidia, left-handed spiraling of conidia in *Helicosporium* sp. (e); and left-handed and right-handed conidia in *Hobsonia* sp. (f).

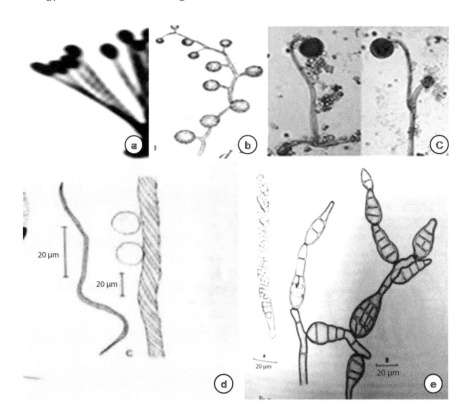

FIGURE 6.4 *Phycomyces blakesleeanus* (a) right-handed and left-handed sporangiophore twisting formation, left- and right-handed sporangia in *Mucor* sp. (b and c), Myxomycetes (d), *Alternaria* (e) conidial state–left- and right-handed branching.

Cornutispora (Figure 6.1e), *Erysiphe* sp. (Figure 6.3a), *Mucor* sp. (Figure 6.4b and c), *Alternaria* sp. (Figure 6.4e), etc. Sometimes the appendages of fungal fruit bodies have also shown left- and right-handed branching, for example, *Chaetomium* sp. (Figure 6.2), *Erysiphe* sp. (Figure 6.3a), and *Chaetomella* sp.

6.3 CONCLUSION

Biology of handedness in living organisms is an important aspect. Asymmetry is not only the morphological variation but may bear a genetic base in many organisms. However, this character of asymmetry is also a kind of adaptation in some to suit the environmental variables and needs of the living biota. Little or no information is available on asymmetry/handedness of fungi. Spiral growth of fungi on solid agar media is the frequently observed character. The helical pattern in chitin fibrils, conidiophores/conidial production, arrangement, and branching in some fungi show left/right handedness. Sympodial or monopodial arrangement in reproductive units of fungi has been observed. Asymmetry/handedness in fungi needs in-depth studies along with their molecular base.

ACKNOWLEDGMENT

Authors are thankful to Dr. Bir Bahadur, Retired Professor of Botany, Kakatiya University, Warangal, for suggesting the topic to write this chapter.

REFERENCES

Aronson, J.M. and Preston, R.D. 1960. The microfibrillar structure of the cell walls of the filamentous fungus-*Allomyces*. *J. Biophys. Biochem. Cytol.* 247–256.

Barnett, H.L. and Hunter, B.B. 1972. *Illustrated Genera of Imperfect Fungi*, 3rd ed., Burgess Publishing Company, Minneapolis, MI. 241pp.

Dimonte, A., Adamatzky, A., Erokhin, V. and Levin, M. 2016. Corrigendum to "On chirality of slime mould". *BioSystems* 140:23–27. doi:10.1016/j.biosystems.2016.04.003.

Goriely, A. and Tabor, M. 2011. Spontaneous rotational inversion in phycomyces. *Phys. Rev. Lett.* 106. doi:10.1103/PhysRevLett.106.138103.

Hashimoto, T. 2002. Molecular genetic analysis of left and right handedness in plants. *Philos. Trans R. Soc, London. B. Biol. Sci.* 357(1422):799–808. doi:10.1098/rstb.2002.1088.

Kamada, T., Takemaru, T., Prosser, J.I. and Gooday, G.W. 1991. Right and left-handed helicity of chitin microfibrils in stipe cells in *Coprinus cinereus*. *Protoplasmsa* 165(1–3):64–70.

Madelin, M.F. 1978. Spiral growth of fungus colonies. *Microbiology* 106:73–80.

Manoharachary, C., Tilak, K.V.B.R., Mallaiah, K.V., and Kunwar, I.K. 2015. *Mycology and Microbiology*. Scientific Publishers, Jodhpur, India. 607pp.

7 The Nature of Right-Left Asymmetries in Plants

K. Tennakone

CONTENTS

7.1 RIGHT-LEFT ASYMMETRY AND CHIRALITY

Right-left asymmetry is best understood by considering the image of an object in a plane mirror. In Figure 7.1, I is the image of an object O formed by reflection at a mirror in the YZ plane. On reflection, the point $P(x, y, z)$ in the object corresponds to the point $P(-x, y, z)$ in I with respect to the same reference coordinate system. If $P_1(x_1, y_1, z_1)$ and $P_2(x_2, y_2, z_2)$ are any two points in the object, the distance $P_1 P_2 = [(x_1-x_2)^2 + (y_1-y_2)^2 + (z_1-z_2)^2]^{1/2}$ will also be the distance between the corresponding points in the image. The invariance of distances is an important property of images formed by reflection at plane surfaces. This accounts for the fact that the image resembles the object, and, in many cases, the object can be brought to automorphism (coincidence) with the image. Another obvious but important property of mirror reflection is that the image of an object resembling I would be identical to the original object. If R denotes the mirror reflection operation, $RO = I$ and $RI = O$. There are also objects where automorphism of the object and image is not

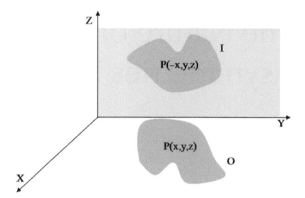

FIGURE 7.1 An object O and its image I formed by reflection at a mirror in the YZ plane. The point P(x, y, z) is transformed to the point P(−x, y, z) during reflection.

possible by translation and rotation, and such objects are said to be chiral. A planar object can never be chiral, unless its dorsal and ventral surfaces differ. They can be brought to coincidence by rotation around the third dimension. Chiral objects exist in two distinct forms; one is the mirror image of the other, termed right-handed and left-handed. Here, right and left are mere conventions. Experience tells us that if some chiral design is crafted, the opposite form could also be made. As humans are generally right-hand biased, right-left symmetry is broken in everyday affairs. Tools, implements, and constructions are frequently of one chirality. There exists no *a priori* reason to consider that right-left symmetry must always remain preserved.

Chirality also appears in elementary constituents of matter, molecules, crystals, and living systems—both plants and animals. Elementary particles could be in chiral states because they possess both linear momentum (**P**) and spin angular momentum (**S**). If the spin vector **S** is parallel to the linear momentum vector **P**, the state is termed right-handed, and the mirror image situation (Figure 7.2) where these vectors set antiparallel corresponds to the left-handed state. If the particles are massive, the above handedness has no Lorentz invariant meaning, because the direction of the momentum vector can be altered by a transformation. If the rest mass happens to be zero, there will be distinct right-handed and left-handed states. In the case of photons, these two states correspond to right and left circular polarized light. Right and left circularly polarized light have identical properties when they interact with an achiral system—a consequence of the right-left symmetry of the electromagnetic interaction. However, substances containing chiral entities feel the two forms of light differently. As plane-polarized light is an equal mixture of right and left circular polarized light, the optical activity results from thenon-identical interaction between the two forms causing a rotation of the plane of polarization (Konnen, 1985).

Neutrinos moving at speeds approaching that of light are left-handed particles, whereas the antineutrinos are right-handed particles. Unlike photons, the left and right-handed neutrinos observed in nature (particle and antiparticle) behave differently—a consequence of the fact that weak interaction does not respect the symmetry of right and left.

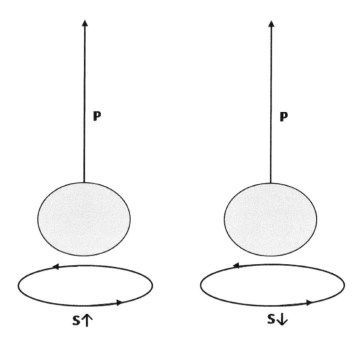

FIGURE 7.2 Right-handed and left-handed particles indicating the directions of linear momentum (P) and angular momentum vector (S).

Since discovery of optical activity by Louis Pasteur (Flack, 2009), it is well-known that some molecules exist as chiral L (levorotatory, left) and D (dextrorotatory, right) enantiomers, and achiral synthesis procedures yield equal proportions of L and D molecules. Except for insignificant and extremely minute contributions from weak interaction, the chemical forces being electromagnetic conserve parity or the symmetry of right and left. Consequently, the achievement of asymmetric synthesis requires the presence of chiral catalysts, agreeing with the saying that the "asymmertry" could be begotten only from asymmetry, implying that a small right-left asymmetry can be amplified.

7.2 BIOCHEMICAL RIGHT-LEFT ASYMMETRY

A long-standing puzzle has been, why biochemistry on Earth, prefers only one stereotype of the organic molecule, generally, L amino acids and D sugars (Bailey, 2000; Blackmond, 2010; Hein and Blackmond, 2012)? A related question that naturally arises would be, is such asymmetry essential for life?

Attempts have been made to attribute minuscule differences in bond energies of two isomers due to weak interaction as the cause for selection of one-handedness (Tennakone, 1984; Kondepudi and Nelson, 1985). However, there is no experimental evidence whatsoever to support this hypothesis. Another plausible cause, again involving weak interaction, could be the biased helicity of electrons produced by radioactive decays in the environment. Although some experiments support this hypothesis (Dreiling and Gay, 2014), the question remains whether the achievable stereo-selectivity would be sufficient to maintain a hemochorial molecular environment. A similar effect could occur during neutrino bombardment to prebiotic oceans by neutrinos originating from supernova bursts (Boyd et al., 2011). Energetic neutrinos interacting with matter produce polarized electrons.

The problem with models that invoke weak interaction to explain biochemical right-left stereo selection points to the difficulty of achieving an appreciable selection defined by the quantity $\{[L] - [D]\} / \{[L] + [D]\}$, where quantities under the square brackets are the respective concentrations. Racemization (conversion of L form into D form and vice versa until the thermodynamic equilibrium $[L] = [D]$ is reached), prevents buildup of one isomer into a sufficient concentration. An alternative possibility would be that a self-replicating molecule of one type (say L) accidently created propagates consuming the L form originally present and derived from the racemization of the D form (Tennakone, 1991). Removal of L molecules by self-replicating species enhances racemic conversion of D molecules to L.

It is interesting to note how living systems had continued to maintain biochemical right-left asymmetry. Plants in their asymmetric (with respect to L and D) chemical environment synthesize proteins based on L-amino acids, which is the source of amino acids for animals as well. When plant and animal products decay, living plants assimilate racemic compounds produced and convert them into chiral forms via asymmetric synthesis. The sequence of these events, projected far into early evolution, could imply that autocatalytic chiral synthesis at the molecular level by self-replicating molecules had existed at the beginning of biochemical evolution.

The chiral properties of liquid crystals could also yield clues towards understanding of the biochemical right-left asymmetry (Stewart, 1966; Lubensky et al., 1998). Liquid crystals made of chiral molecules set into macroscopic chiral assemblies of definite handedness. A recent work demonstrates that achiral molecules form chiral liquid crystal structures of either handedness as a result of spontaneous symmetry breaking (Nayani et al., 2015).

7.3 MACROSCOPIC RIGHT-LEFT ASYMMETRIES IN ANIMALS

Some animals and plants possess readily observable macroscopic right-left asymmetries in arrangement of internal and/or external organs. Such asymmetries in the animals are more thoroughly studied compared to the plants (Palmer, 1996; Levin, 2005; Sherry and Levin, 2008; Schilthuizen and Gravendeel, 2012; Vandenberg and Levin, 2013). Many species of invertebrates and vertebrates possess bilateral symmetry defining an axis of symmetry. As their dorsal and ventral morphologies differ, breaking bilateral symmetry makes them chiral. Subtle right-left differences have been noted in insects. Several workers have studied developmental right-left asymmetries in Drosophila (Klingenberg et al., 1998; van Eeden and St. Johnston, 1999; Ligoxygakis et al., 2001). Although statistically significant deviations in right and left symmetry of the wings and the gut have been noted, possibly implicating genetic basis, there is no evidence to the effect that all individuals have same handedness. Such asymmetry may have an evolutionary advantage in helping the insect to orient itself.

Molluscs have no bilateral symmetry, and the sense of spirality of the shell determines their handedness. Generally, all members of a given species of Mollusca manifest the same handedness. A rare exception is the pond snail *Lymnaea*, where both dextral and sinistral individuals occur, but the dextral being predominant (Wandelt and Nagy, 2004). The handiness of the molluscs shell is maternally inherited, and embryogenesis leading asymmetric development is fairly well understood (Kurita and Wada, 2011), excluding the very primary signal that distinguishes right from left.

Vertebrate right-left asymmetries have been investigated more extensively (Levin et al., 1995; Capdevila et al., 2000; Komatsu and Mishina, 2013; Vandenberg and Levin, 2013; Blum et al., 2014a, 2014b; Grimes and Burdine, 2017). The involvement of nodal cilia in embryogenesis established an important step in the process of differentiation into asymmetry. Although the mechanism is genetically determined, the primary steps remains to be elucidated (Aw and Levin, 2008). Cells could also possess chirality (Inaki et al., 2016). A question that naturally arises is how cells get the asymmetry. A plausible answer would be, chiral molecular species induce cell asymmetry. According to Brown and Wolpert (1990), a chiral molecule, termed an "F molecule," primarily decides the handedness in morphogenesis. Although this entity (entities) has not been identified, the hypothesis defines a primary cause for the developmental right-left asymmetry of a species.

In animals, right-left asymmetry generation in the developmental process is more optimized, and genetic mechanisms have evolved to implement the program because of the advantages of the asymmetry. Food assimilation require organs of large surface area and therefore long intestines. The long span of an animal obviously doesn't fit the environment, necessitating coiling defining sense of right or left. Placement of other organs need to commensurate with this asymmetry.

Handedness helps an organism in movement and locating positions. In everyday life, when one drives, he is instructed to turn right or left at different to points to reach the destination. The majority of individuals in a species possessing the same handedness (right or left) would also facilitate their collective social activities promoting advancement. Needless to mention the chaos that would be created if right-left symmetry is respected with 50% of the human population right-handed and the other 50% left-handed. Here the convention of right and left in matters of everyday life will have to be abandoned.

A discussion on right-left asymmetry of animals would be incomplete without a reference to this aspect in relation to the brain. Difference in morphological features and neurological functioning of the right and left lobes of the brain is well-known (Zaidel, 2001). Computational cognitive functions in two sectors enhances the information processing capability. The original morphogenetic asymmetry had been enabled, acquiring this differentiation. The brain asymmetry seems to be essential for computational purposes as well as interpretation of spatial perception.

7.4 RIGHT-LEFT ASYMMETRIES IN PLANTS

Right-left asymmetries seen in plants differ distinctively from those of animals (see however Chapter 2). In the discussion that follows, plants will be classified into four groups (*Types I–IV*) based on clear differences with respect to the way in which the right-left attributes manifest. In some families, genera, or species, each individual can be characterized as either right-handed or left-handed, judged from the chirality of an external organ or organs. This category will be referred to as *Type I*. In the second category termed *Type II*, individuals possess external chiral organs or organ arrangements distributed with equal probability, but an individual cannot be clearly distinguished as right-handed or left handed. A large majority of plants, named *Type III*, have no discernable macroscopic right-left attributes so as to assign a handedness to any of the external organs. In the category defined as *Type IV*, the entire species, genera, or family has a unique handedness (right or left) determined by the chirality of an organ or organs.

A trend that is clearly evident in plant right-left asymmetries is that those more primitive or older on the phylogenetic ladder are more symmetric than recent advanced species, as indicated here:

$$\frac{Type\ I\ \rightarrow Type\ II\ \rightarrow Types\ III\ and\ IV}{Decreasing\ Symmetry -----\rightarrow}$$

Even in physical systems, as the complexity and size increase, the symmetries are repeatedly broken until they are almost undiscernible (Anderson, 1972). An organized assembly of entities need not possess symmetries that conform to the laws governing those entities. Although the electromagnetic interaction governing the interplay between atoms is parity conserving, there exist chiral forms of bulk assemblies of atoms. The existence of one chiral form somewhere would not be a violation electromagnetic parity. The theorem only tells that the possibility of the opposite form is not ruled out.

Animals generally depend more on right-left asymmetry for optimized morphological development and survival in environment, with the extreme case being man. Although the former seems to apply to plants (as in animals), to some degree but it is not an essential requirement. Thus, plants have not exploited this asymmetry in evolving, except for incidental use of this trait for some advantage.

7.5   SPIRAL PHYLLOTAXIS

Most right-left asymmetries in plants relate to spiral phyllotaxis (Snow and Snow, 1934; Dixon, 1989; Douady and Couder, 1992; Jean, 1994; Rutishauser and Peisi, 2001; Korn, 2008). The spiral phyllotaxis dominates other patterns of leaf arrangement in plants (Mirabet et al., 2012). Widely distinct families, primitive as well as advanced, adopt this scheme in varying geometries. Botanists classify the spirality by the fraction m/n ($n > m$), where n and m are numbers in the Fibonacci sequence 1, 1, 2, 3, 5, 8, 13, 21, 34…, when every n the leaf emerges at the same angular position after going through m turns. The spiraling can be right-handed or left-handed. Phyllotaxy of any Fibonacci classification occurs in both chiralities; rarely are there spirals not definable within this scheme. Despite the extensiveness of the literature on phyllotaxy, comparatively few workers have looked into its right-left asymmetry aspect (Petch, 1911; Schouts, 1921; Imai, 1927; Skutch, 1927; Allard, 1946; Davis, 1962; Bible, 1976; Tennakone et al., 1979; Tennakone et al., 1982; Kanahama et al., 1989; Jean, 1994;

Fredeen et al., 2002; Martinez et al., 2016). Identification of the right-left spiraling and determination of chirality of leaf arrangement present confusion. Instead of right and left, words like clockwise and counterclockwise are also used. The usage of the terms clockwise or counterclockwise depends on how you look at the leaf pattern, whether it is from top or bottom. Scarcity of literature on right-left asymmetry of phyllotaxis compared to the voluminous literature on the other aspects of the subject in journals of botany, physics, and mathematics could be partly due to the confusion in assigning an unambiguous handedness to phyllotactic spirals.

An easy way to determine the sense (right or left) of a generic phyllotactic spiral is as follows (Tennakone et al., 1982): Observe the stem or branch under examination so that the growing leaf apex is positioned facing upwards (Figure 7.3). Focus the eye on any of the leaves directly in the front, and if the next leaf immediately above this is on right-hand side, then the spiral is defined as right-handed; otherwise (i.e., left-hand side), it is left-handed. The right-handedness according to above definition corresponds to anticlockwise sense if one observes the whorl from above.

In many instances, especially the taller trees with the leaf pinnacle position atop (e.g., palms, papaya), a sense of phyllotaxy could be more easily judged looking at the leaf scars, instead of the leaves.

In cylindrical coordinates $(r, \varnothing, z)$, the equations of right- and left-handed spirals can be respectively written as Equations 7.1 and 7.2.

$$r = ar, \varnothing = \eta t, z = ht \tag{7.1}$$

$$r = ar, \varnothing = -\eta t, z = ht \tag{7.2}$$

where a, η, h are constants, and t is a variable parameter, conveniently considered as time of tracing points along the curve. In spiral phyllotaxis, the leaf primordia

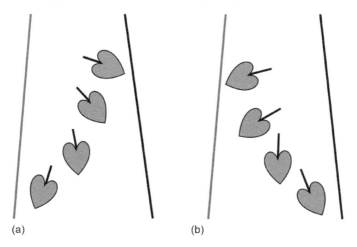

(a)   (b)

FIGURE 7.3 Identification handedness in spiral phyllotaxis. The stem pointed vertically upwards and the eye directed to a leaf (or leaf scar) directly in front (second leaf from bottom). If the leaf (leaf scar) immediately above is on right as in (a), the spiral is defined right-handed, and if it is an in (b) it is left-handed.

are generated one at a time consecutively. Thus, by replacing continuous variable t by a discrete variable n ($n = 1, 2, 3...$), the position of primordia could be represented by:

$$r = ar, \varnothing_n = \eta n, z_n = hn \tag{7.3}$$

$$r = ar, \varnothing_n = -\eta n, z_n = hn \tag{7.4}$$

From (7.3) or (7.4)

$$\varnothing_{n+1} - \varnothing_n = \eta \tag{7.5}$$

The quantity, η, is the angle between consecutive primordia, referred to as the divergence angle close to golden angle (~137.5°) proportioning 2π into the golden ratio $\phi = 1 + (\sqrt{5} - 1)/2$, so that $\eta = 2\pi(2 - \phi)$. The golden angle is more closely approached, when the Fibonacci fraction corresponds to larger consecutive integers in the series. Extensive observations indicate that large number of plants belonging to widely different families adopt this pattern (Okabe, 2015). Despite more than two centuries of discussion (King et al., 2004 and references therein) on this subject, convincing generally accepted explanations has not emerged yet (Pennybacker et al., 2015; Watson, 2017).

In many instances, phyllotactic right-left asymmetry intimately correlates with chirality of other organs, as if it is the primary cause of asymmetry transcending to branching, flowers, and fruits. Rare examples also exist where some organs have chirality, although a handedness could not be assigned to phyllotaxis. Notable examples are the flowers of Apocynaceae, which are always imbricated in, but the phyllotaxis is non-spiral.

7.5.1 RIGHT-LEFT ASYMMETRY TYPE I

Plants classified as *Type I* have right-handed and left-handed individuals defined by the sense of spiral phyllotaxis. The members of both chiralities occur in equal abundance independent of progeny. Crossings right × right, right × left, left × right, and left × left yield offspring of either handedness with equal probabilities (Louis and Chidambaram, 1976; Venkateswarlu, 1982). Coconut and almost all other palms belong to this category, where the handedness can be readily ascertained by examining the leaf scars. Petch (1911) was first to observe the existence of right-handed and left-handed coconut trees, and this could also be one of the earliest observations noticing the separate existence of two chiral forms in spiral phyllotaxis. Davis (1962, 1963, 1974, 1987) extensively studied the handedness of phyllotaxy in coconut and noted that it was not determined genetically. He also reported a slightly excess number of right-handed (according to the definition of the previous section) plants in cultivated populations and the difference in yield of nuts. In a seminal work, Louis and Chidambaram (1976) conducted coconut breeding experiments, crossing all possible combinations. They concluded that there was no bias, and irrespective of the crossing combination, right and left offspring occur with equal probability.

The papaya plants are also either right-handed or left-handed as determined from the phyllotaxy (here again, it is very easy to fix sense of handedness looking at leaf scars). The author found that in the population of papaya trees, both types occur in 50% abundance, and deviations are not statistically significant. Furthermore, when seeds are planted, right- and left-handed trees were produced in equal proportion irrespective of handedness of the parent. A difference in fertility of seeds or yield of fruits was not observed. Occasionally, papaya gives rise to lateral branches (notably after an injury). The handedness of such lateral branches was found to be of either kind with a preponderance of the handedness of the main stem. In papaya, a reversal of the sense of phyllotaxy of the main stem or branch was never observed.

An interesting variation of chiral phyllotaxis is seen in some cultivars of citrus family. Here, phyllotaxy of the leaves appears hard to discern judging from the leaves as the leaf attachments bend and twist. However, thrones, which grow and remain intact, show a clear handed spiral pattern. Trees planted from seeds are either right- or left-handed, whereas those rooted from cuttings continue the original pattern of the mother plant during branching. Unlike in papaya, lateral branches stick to the chirality of the main stem.

The study of phyllotaxy and of the imbrication of petals in flowers of Malvaceae provide much insight into the nature of handedness in plants (Tennakone et al., 1978, 1979, 1982). The seeds of many species of plants belonging to this family, when germinated, produce stems with either right- or left-handed spiral phyllotaxis with equal probability. The lateral branches that develop subsequently are found to be either right-handed or left-handed with a bias towards the type of handedness of the main stem. The degree of bias was found to depend on the species. The effect is conspicuously seen in the fast-growing herb *Hibiscus furcartus* L (Tennakone et al., 1982). In *H. furcartus* and many other species of Malvaceae, the flowers are right- or left-imbricated, which define a spiral (Figure 7.4). When the convention used for

(a)   **(b)**

FIGURE 7.4 Diagrammatic representation of (a) left imbricated flowers and (b) right imbricated flowers.

determining the sense (right or left) of the flower spiral is taken to be the same as that of phyllotaxy, the sense of flower imbrication is found to be opposite to that of phyllotaxy. In other words, stems and lateral branches of right-handed phyllotaxy bear left-handed flowers at the leaf nodes and vice versa, and this rule is never violated. The same behavior is found in all the other examples of spiral phyllotaxy. Flowers of some species also exhibit another type of right-left asymmetry, referred to as enantiostyly, that determines the direction of the deflection (right or left) of the floral axis (Jesson and Barret, 2002). It is not clear how this asymmetry correlates to flower imbrication or leaf phyllotaxy.

When the seeds of *H. furcartus* were germinated, handedness of the leaf arrangement in the main stem did not depend on progeny but varied randomly with 50% right and 50% left. The main stems of one kind (right or left) produced lateral shoots of both kinds, but those of the same kind occurred more frequently by a factor of about 4 to 5. Right-left symmetry was preserved, as both kinds of main stems produced branches of the same kind with equal probability. Suppose L and R are the number of right-handed and left-handed lateral branches observed at time *t* in a parent with a left-handed main stem. The time development of the shoots of the two kinds are represented by the equations:

$$\frac{dL}{dt} = aL + bR \tag{7.6}$$

$$\frac{dR}{dt} = aR + bL \tag{7.7}$$

where $a > b$ and Equations (7.6) and (7.7) are symmetric with respect to the interchange of R and L.

The solution of Equations (7.6) and (7.8) subject to the initial condition $R = 0$, $t = 0$, can be expressed as:

$$L = A[\exp(a+b)t + \exp(a-b)t] \tag{7.8}$$

$$R = A[\exp(a+b)t - \exp(a-b)t] \tag{7.9}$$

Giving $$\frac{R}{L} = \tan h(bt), \; t > 0. \tag{7.10}$$

Thus, $R/L \rightarrow 1$ as $t \rightarrow \infty$, and theoretically the original asymmetry continues indefinitely (Figure 7.5). In reality, during latter periods of growth, branches of both types randomly sports, leveling-off the original excess of L. Nevertheless, if the history of branching is carefully examined, chiral identity of the plant could be established.

Type I plants have all the characteristics of a spontaneous breaking of a symmetry. In this phenomenon, an initial symmetry of a system breaks, owing to a suddenly developed instability, when some parameter governing the dynamics takes a critical value. After symmetry breaks, the system settles into one of the many possible

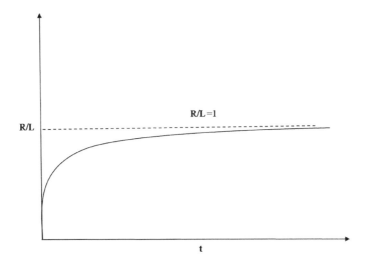

FIGURE 7.5 The variation of ratio R/L (No. of right-handed branches/No. of left-handed branches) in a *Hibiscus furcartus* L grown from a right-handed seedling as predicted by the model.

equilibrium states governed by the symmetry of the system. The original symmetry is preserved in the sense of the equal likelihood of the occupation of all the possible states. In the present case, there are two symmetric states (right and left), and as expected two kinds occur with equal probability. The cause of instability leading to bifurcation remains unresolved.

Some success has been achieved explaining phyllotaxis on basis of cellular stress fields and auxin transport (Smith, 2008; Pennybacker et al., 2015). Nonlinear equations governing these processes lead to instabilities enabling modeling of morphogenesis as pattern formations. To further the progress in this direction, one needs to explore the dynamics that lead to breaking of right-left symmetry as a primary instability.

At the point of instability just before spontaneous symmetry breaking, any system would sensitively react to internal or external biases, preferring settling down of the system into one specific state. Chiral molecules present in plant tissue could serve as such an internal bias. Magnetic fields and Coriolis force are external right-left distinguishing factors. Davis et al. (1987) reported that there is a statistically significant difference in right-handed and left-handed coconuts in the two hemispheres of the earth. It is most unlikely that at dimensions of plant cellular structure, Coriolis force has an effect. Although, a bias caused by the geomagnetic field is not completely ruled out (Minorsky, 1998; Minorsky and Bronstein, 2006), controlled experiments conducted by the author failed to establish the effect of a magnetic field on biasing of the handedness in spiral phyllotaxy, at least in this particular case. In this experiment, *Urena lobota* (Malvaceae) seeds were germinated in a uniform vertical magnetic field between two Helmholtz coils. Although the magnetic field here had been orders of magnitude larger than the geomagnetic field, handedness in phyllotaxy and sense of imbrication of petals in flowers were seen to preserve right-left symmetry.

Davis (1963) also reported that yield of coconuts varies in the right- and left-handed trees, the yield being significantly more in right-handed ones. Just like propellers, which are chiral objects, right- and left-handed leaf whorls of coconut may react differently to clockwise and anticlockwise cyclones. The cyclonic wind movements in the northern hemisphere rotate anticlockwise, and one kind of coconut tree would be more vulnerable to such atmospheric disturbances (Denis and Jean, 1998). Plausibly the difference in detachment of young nuts due to wind could explain the observation of Davis (1987).

7.5.2 RIGHT-LEFT ASYMMETRY OF TYPE II

Plants classified *Type II* in this work possess external chiral organs or organ arrangements distributed with equal probability, but an individual cannot be distinguished as right-handed or left-handed. Some species of Malvaceae and Dipterocarpaceae are prototype examples of this group. In Malvaceae, the *Type II* character appears notably in species that grow as sizable trees. Here, when seeds are planted, the main stem starts with a definite handedness, but reversals develop subsequently. Lateral branches also display the same character to more or less the same degree. Flowers have right or left imbrication, always opposite to the handedness of the stem on which they are born. Conifers also belong to this category where twigs of right-handed and left-handed phyllotaxy occur with equal probability. Cones also possess a handedness, corresponding to the twig on which they are borne. No statistically significant difference in yields of the two types of cones has been observed.

7.5.3 RIGHT-LEFT ASYMMETRY OF TYPE III

In plants classified as *Type III*, there are no macroscopicchiral organs. The phyllotaxy with erratic spiral behavior or other patterns forbids a unique assignment of handedness. Similarly, the pattern of petal and sepal arrangement in flowers happens to be non-chiral. A vast number of perennial trees fall into this group. The *Type II* is a transition from *Type I*; similarly, *Type III* seems to be transition from *Type II*, as complexity and functions of the organism diversifies. The *Type III* character appears more frequently in species placed later in the phylogenetic ladder.

7.5.4 RIGHT-LEFT ASYMMETRY OF TYPE IV

In the class of plants defined as *Type IV*, one or more organs possess a distinct chirality (right or left) common to the entire species. A notable chiral signature of this type of right-left asymmetry is the sense of imbrication of the flowers. In Apocynaceae, most species have left-handed flowers (e.g., *Plumeria*). Turneraceae have left-imbricated flowers, whereas those of Melastomaceae are always right-handed. After counting more than 10^6 flowers of *Plumeria rubra* (Apocynaceae) and *Turnera ulmifolia*, even a single case of deviation from left-handedness was not observed (Tennakone et al., 1978). The *Type IV* handedness seems to be genetically determined and not always correlated to phyllotaxy. In some xerophytic Apocynaceae, the spirality seen in the arrangement of thrones displays a unique handedness. Again, some vines of the same family twist into helices of the same handedness in climbing. These examples suggest that in *Type IV*, some intrinsic genetic factor causes macroscopic right-left asymmetry.

An example of unique chirality in spiral phyllotaxis are banana cultivars, all having right-handed spiral phyllotaxis. Wild banana grown from seeds develop into both varieties. Thus, handedness in cultivars that are vegetatively propagated should be a cytoplasmic inheritance.

7.6   THE COILING OF CLIMBERS AND VINES

The following insightful lyrics by Flanders and Swan (1950) bring attention to the sense of coiling of some climbers and whether it is determined by external or internal factors:

> *The fragrant honeysuckle spirals clockwise to the sun, and many other creepers do the same, But some climb anticlockwise; the bindweed does, for one, Or Convolvulus, to give its proper name. Said the right-hand-thread honeysuckle to the left-hand-thread bindweed, 'Oh! Let us get married if my parents don't mind. Be loving and inseparable, inexplicably entwined. We'd live happily ever after,' said the honeysuckle to the bindweed. A bee who was passing remarked to them then: 'I've said it before and I'll say it again: Consider your offshoot, if offshoots there be. They'll never receive any blessings from me. Poor little sucker! How will it learn, when it's climbing which way to turn? Right? Left? What a disgrace! Or it may go straight upwards and fall flat on its face.'*

The spiral phyllotaxis seems to generate a torsion in the stem; such torsion necessarily decides a sense of coiling. *Dioscorea* species belong to this category. Phyllotactic handedness that is not inherited can be right- or left-handed, and the two types coil in opposite senses during climbing. In *Dioscorea*, the helical structure and torsion in the stem is clearly visible. Climbers where the leaf arrangement displays no helicity also coil in a preferable sense; here, the presence of an intrinsic torsion in the stem is not ruled out. Many thickly grown climbers manifest visible torsion; however, the statistics available to the author was inadequate to determine whether the sense of torsion is intrinsic to the species.

7.7   ADAPTATION OF RIGHT-LEFT ASYMMETRY BY PLANTS

For animals, right-left asymmetry serves as an advantageous trait. Asymmetry has been utilized to position organs optimizing the available space and develop a body structure to suit the environment. It facilitates locomotion and a sense of direction in higher animals. One could easily argue that man would not have advanced so much if not for his right-left asymmetry. Plants being sessile and devoid of organized sensory organs, this asymmetry may mean less profitable for their adaptation to the environment. However, some plants have utilized the trait of right-left asymmetry incidentally. The asymmetry-driven imbrication of leaves and petals in flowers buds ease the developmental process. A recent study has revealed that bees possess handedness—about 25% left, 25% right, and rest with no preference for right and left (Ong et al., 2017). The study also suggests that the above-mentioned difference in handedness distinction helps the insect to choose the travel route and to avoid obstacles. It is very likely that the handedness also helps bees to circumnavigate the hollow of imbricated petals in search of nectar. Navigating against the sense of folding offers more resistance to motion. Flowers are right- or left-imbricated or non-imbricated, and bees seem to follow the same asymmetry pattern. For bees there exists situations in the environment where: (1) right-handedness is preferable; (2) left-handedness is preferable; and (3) having no handedness is preferable. In the case of humans, there are no factors in the natural environment to give a preference for either right or left. However, for a large majority, having one-handedness is a tremendous advantage for coordination of social activities. The obstacle-traversal choice would be much easier for the entire swarm of bees, if all of them had one handedness. The above argument suggests that the observed distribution of handedness among bees has more to do with supplying food to the entire colony than maneuvering through obstacles. Again, torsion resulting from right- or left-handed spiral phyllotaxy helps some vines to climb up around the host.

7.8   WIND DISPERSAL OF SEEDS IN PLANTS OF TYPE II RIGHT-LEFT ASYMMETRY

The clearest example of utilization of right-left asymmetry by plants is in wind dispersal of seeds. Before evolution of birds, wind dispersal would have been one of the most effective modes of long-distance seed dispersal. The main requirement for success of wind dispersal is the reduction of terminal velocity of free-falling seeds via autogyration and/or drag. Seed appendages endowed with right-left asymmetry generate autogyration by the aerodynamic propeller effect. Many species of tall trees and vines climbing high and belonging to widely different families have evolved this strategy.

The author has not found any evidence to the effect that *Type I* plants adopt right-left asymmetry to disperse their seeds. However, *Type II* plants frequently utilize right-left asymmetry for seed dispersal by evolving wing-like appendages on the seed. A prototype example of this phenomenon is seen in Dipterocapaceae. Plants in this family have differentiated organs with these peculiar characteristics. The beauty of right-left symmetry seen in dipterocarps is fascinating and perhaps rarely seen in other families (Tennakone, 2017). They have spiral phyllotaxy and shoots with right and left leaf spirals distributed in equal probabilities and induce this symmetry to other organs developed in the twig. In most dipterocarps, flowers are borne on axial racemes popping out at the leaf brackets (Figure 7.6). The sense of spiral arrangement in axils of racemes is the same as that of the parent twig on which they originate. Again, spirality is generated in the setting of flower buds on the raceme axis (Figure 7.6). The handedness of this spiral changes alternatively; if the pattern of flower buds in a raceme is to the right, the ones above and below them are to the left. Progression of spirals continues to other organs. The sepals and petals contort and imbricate in the same sense, defining a handedness.

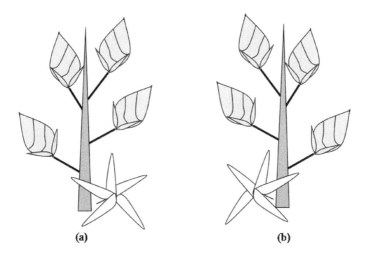

FIGURE 7.6   The pattern of racemes in the axis of a typical dipterocarp: (a) right-handed and (b) left-handed.

Here again, the sense of the spiral alternates from one flower bud or flower to the next. All sepals do not develop equally in most dipterocarps, and a few of them elongate as wings attached to the nut. In *Dipterocarpus zeylanicus* and related species, only two sepals grow and elongate to about 10 cm, while maintaining the original twist giving the handedness (Figure 7.7). Because of the handedness, the dipterocarp nut with wings acts as propellers and rotates (about the axis shown as dotted lines in Figure 7.7) while falling. The right-handed fruit will rotate counterclockwise, when viewed from above as it falls, and the left-handed fruit will rotate clockwise. If V and Ω are the translational and angular velocities of the nut, after falling through a distance H, a part of the potential energy MgH is converted

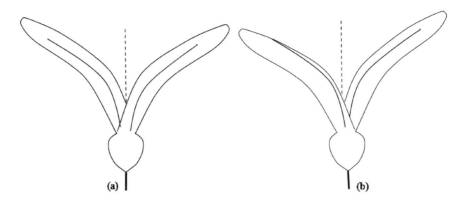

FIGURE 7.7   Fruits of *Dipterocarpus zeylanicus*: (a) right-handed and (b) left-handed.

into kinetic energy 1/2 MV² + 1/2 IΩ² and the rest into heat generated due to the drag. Thus, the condition MgH > 1/2 MV² + 1/2 IΩ², leads to the result

$$V < \sqrt{\left[2gH - \left(\frac{I}{M}\right)\Omega^2\right]}$$ (7.11)

where I = moment of inertia about the axis of rotation. Thus, rotation reduces the velocity of fall, allowing more time for the seed to be carried away by the wind and deposited afar from parent tree. Dipterocarps have clearly evolved to optimize and utilize the handedness of the sepals to disperse the seeds. In order to reduce V, the angular velocity Ω needs to be increased. This is an aerodynamic issue. Here wing span, shape, and torsion are important. If there are N wings, each with a mass m and length L and if the mass of the nut is small compared to the seed, then to a good approximation $M \sim mN$, $I \sim mNL^2$ and (7.11) takes the form,

$$V < \sqrt{[2gH - (L^2)\Omega^2]}$$ (7.12)

The condition (Equation 7.12) is independent of N, implying that having few long wings (<5) would be more advantageous than elongating all the five sepals, which requires production of more biomass.

   Hiptage benghalens (Malpighiacea) has evolved an even more ingenious seed-wind dispersal strategy. Here the pericarp of the fruit develops into three wings. A large, (4–5 cm) long more or less bilaterally symmetric wing and two asymmetric small wings (2–3 cm) placed opposite to each other (Figure 7.8). Small wings are twisted in

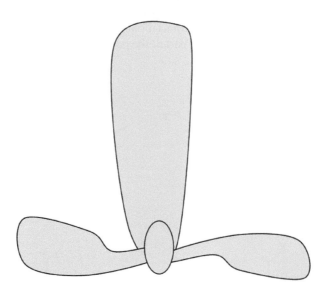

FIGURE 7.8   Schematic sketch of the winged seed of *Hiptage benghalensis*.

opposite senses. The seed is embedded at the point of intersection of the wings. As the large wing is slightly curved transversely (dorsal surface convex) and the ventral and dorsal sides being different, this object acquires chirality. If one samara is assigned right-handedness, the other with interchanged sense of twisting of small wings is left-handed.

Just like the arrangement of leaves, flowers and the fruits are placed alternate on the inflorescence axis. Fruits on either side have opposite handedness, and the pattern reverses from one internode to the next. When the samara detaches and falls, it rotates about the center of gravity located very close to the position of the seed. The axis of rotation is vertical, with the plane of wings inclined to the horizontal at a small angle and the larger wing pointing upwards. Wing arrangement generates high aerodynamic lift and rapid rotation.

Many other plants of *Type II* belonging to families distantly separated adopt similar seed-wind dispersal mechanisms based on autogyration during falling caused by chiral arrangement of wing-like appendages. Notable examples are: (1) *Gyrocarpus jatrophifolius* and *Gyrocarpus americanus* (Hernandicaceae) with right-handed and left-handed fruits where wing structure is very similar to *Dipterocarpus zeylanicus*; (2) *Melanorrhoea usita* (Anacardeaece) five-winged fruit with right as well as left chirality that autogyrates during descent (Figure 7.9); (3) Conifers also possess right-left symmetries of *Type I*. Examination of pines revealed that twigs have or *Type II* check handedness. The leaf fascicles in pine are arranged in two spirals winding in opposite directions. However, the number of winding turns that the spiral takes to reach the same angular position of the fascicle immediately above or below are different in the two spirals, giving a net handedness to the twig, and the twigs of both kinds are equally distributed. Cones have the same feature. There again, the winged seeds are arranged following the same pattern as the fascicles, so that there are two types of cones, right-handed or left-handed, borne in equal abundance. The bilateral symmetry of pine winged seeds breaks to form chiral objects; all of them are similar in one kind of cone (right or left) except the mirror image in the other type (left or right). The ventral view of the winged seeds of right- and left-handed cones of Eastern White Pine (*Pinnus strobus*) is shown in Figure 7.10. As both lobes are folded upwards, the objects are geometrically different and cannot be superposed on each other. Because of chirality, they autogyrate when dropped from the

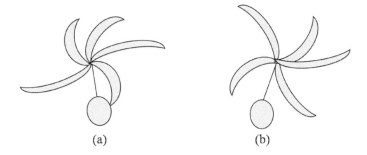

(a)   (b)

FIGURE 7.9   *Melanorrhoea usita*: (a) left-handed and (b) right-handed.

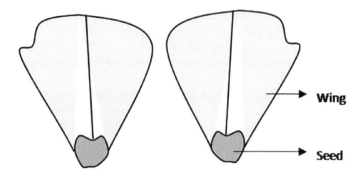

FIGURE 7.10 The ventral view of the winged seeds of right- and left-handed cones of Eastern White Pine.

tree. Wind dispersing ancient species of pine and other conifers generally have more asymmetric seeds. In prehistoric pine species now extinct, the sole mode deposition of seeds afar may have been wind dispersal (Looy and Stevenson, 2014; Stevenson et al., 2015).

(4) Another interesting example illustrating different ways in which plants have utilized right-left asymmetry for the purpose wind dispersing the seeds is seen in some plants of the family Magnoliaceae. Here the right-left symmetry resembles that of dipterocarps and pine. In the species examined *Michelia nilagirica* (found in Sri Lanka), tulip tree (*Liriodendron tulupifera* L). and ornamental magnolias (in United States), it was observed that all these taxa have twigs of spiral phyllotaxy, both handedness occurring in equal proportions. Sepals and petals are almost indistinguishable, but the handedness of the twig passes on to the flower and the fruit. In flower buds, two opposite unequal spirals are seen (normally 5/8 Fibonacci) correlated to the sense of leaf phyllotaxy. Just as in dipterocarp and pine, there are right- and left-handed fruits. The right and left fruits of the tulip tree have respectively right- and left-winged fruits, readily distinguishable from the direction of twist in the ridge placed midway and in the lengthwise direction (Figure 7.11). Winged fruits of right and left types are mirror images of each other, and no automorphism exits because of ventral-dorsal difference. During descent, the two fruits gyrate in opposite senses with the middle ridge as the axis. R-L asymmetry provides torque for the rotational motion, but the lift created is minimal. The moment of inertia is minimum about the mid axis, and motion remains stable. It has been suggested that this stability greatly helps dispersal, although the terminal velocities attained are smaller when compared to other wind-dispersed seeds (McCutchen, 1977).

FIGURE 7.11 The dorsal view of the winged seeds of right- and left-handed fruits of the tulip tree.

7.9 WIND DISPERSAL OF SEEDS IN PLANTS OF TYPE IV RIGHT-LEFT ASYMMETRY

In *Type IV* plants, all individuals of the family or species have one-handedness determined by an organ or organs. Frequently, the chiral character is the sense of imbrication of petals in the flower. It is entertaining to see how some flowers spin when dropped from a height. The left-handed flowers of many species of Apocyanaeceae rotate clockwise (when viewed from above) if released, corolla pointing upwards. Sometimes in *Type IV* plants, handedness is expressed in seed appendages so that the seed can autogyrate. However, unlike in the case of *Type II* plants previously discussed, all fruits gyrate in the same sense when they fall. A classic example of a plant of this category is found in Sri Lanka, sometimes grown as an ornamental plant is shown in Figure 7.12. The star-like purple flower of this vine, probably belonging to Nyctaginaceae, measures about 0.5 cm, whereas the sepals of almost the same color are 3–4 cm. After the flower fades and detaches, sepals, turning green, elongate and a symmetrically positioned oval fruit grows below the sepals. When the fruit ripens, the dried sepals turn brown and remain attached to the fruit. Falling fruits always rotate clockwise (when viewed from above) and accurately maintain stability.

Apocynaceae possess an intrinsic handedness in the flowers, which sometimes transcend to the fruit and the seeds. Because of this asymmetry, plants of this family possess the potential of evolving wind dispersal via autogyration. However, only few tall trees of the family seemed to have utilized the above strategy. The seeds of *Aspidosperma spruceanum*, a tree of the Apocynaceae family rising up to about 20 m, are dispersed by wind (Mönchengladbach, 2010). Here the wing-like seeds are asymmetric and are capable of rotation in one sense during falling. Most of the shorter trees and climbers of Apocynaceae have plumo seeds dispersed by wind. As a result of the chirality, the parachuting seeds sometimes slowly rotate. Here the advantage of rotation seems to be not the generation of a lift but an enhancement of the drag. The centripetal force of rotation spread the plumes in the horizontal plane, thus increasing the drag. The appearance of the long hairs in a parachuting *Wrightia* (*Wrightia antidysenterica)* is shown in Figure 7.13.

FIGURE 7.12   Flower, sepals, and fruit of a species of Nyctaginaceae.

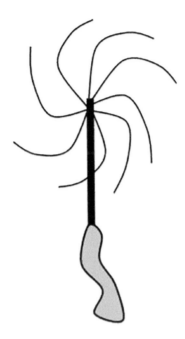

FIGURE 7.13 Slow rotation of a parachuting seed of *Wrightia antidysenterica*.

7.10 AUTOGYRATING SEEDS OF PLANTS OF THE RIGHT-LEFT ASYMMETRY TYPE III

For any falling object to rotate, it must decide which way to turn (i.e., clockwise or anticlockwise as viewed from above). An inherent geometrical asymmetry present in the object enables this decision. If a propeller generates a forward thrust when rotated in one sense, turning in the opposite sense will reverse the direction of the thrust. Thus, a propeller without chirality would not generate either a forward or backward thrust, if forward and backward directions are symmetrical. With respect to the vertical, up and down directions are not symmetrical because of the earth's gravitational field. Consequently, the possibility exists that an achiral object rotates about the vertical and generates a thrust (lift). However, symmetry demands that both senses of rotation are equally likely. It is amazing that some *Type III* plants have resorted to exotic mode of seed wind dispersal. A notable example is maple and related species.

The maple tree has no macroscopic right-left distinguishing characters. Phyllotaxy being opposite, no handedness is defined. Similarly, flowers and fruits are constituted achirally. The maple fruit is a bilaterally symmetric structure with two conjoined winged seeds (samaras), readily separating from each other and from the stalk as the fruit ripens and dries (Figure 7.14). As ventral and dorsal sides of samaras are identical, they possess no chirality. When a samara is dropped from height of several meters, it moves erratically for a while and starts spinning, with the length inclined to upward vertical at an acute angle and seed pointing down. If the same seed is experimented number of times, both senses of rotation (i.e., clockwise or anticlockwise when viewed from above) occur

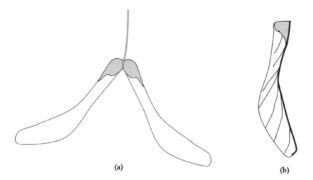

FIGURE 7.14   (a) Schematic diagrams of a fruit of Norway maple showing two conjoined samaras, seeds, and the stalk; and (b) wing shape and veins in the fruits of Norway maple.

with equal probability. The mechanism is clearly a spontaneous breaking of the right-left asymmetry owing to dynamical instability of the non-rotating mode of motion.

A schematic sketch showing the motion of a samara in the two senses is shown in Figure 7.15. When it reaches a critical velocity V_c, rotational motion initiates with the thicker ridge (the dark line marked in Figure 7.14b) as the leading edge (the edge that hits air), taking either of the helical paths.

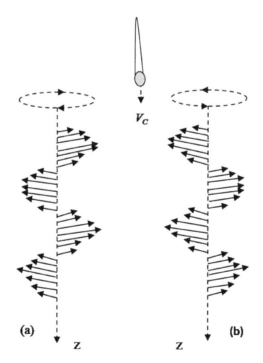

FIGURE 7.15   Schematic sketch illustrating: (a) clockwise and (b) counterclockwise spiraling of a falling maple samara once it reaches the critical velocity (*VC*). The patterns of arrows drawn to the vertical (z-axis) are the projections of the leading edge of the samara.

7.11 TOY PAPER HELICOPTERS SIMULATING WIND DISPERSING SEEDS

Although theoretical models, computer simulations, and wind tunnel experiments have been attempted (Azuma and Yasuda, 1989; Lentink et al., 2009; Camposano et al., 2015; Fang et al., 2017), the aerodynamics of wind-dispersing seeds is complicated and not readily amenable for mathematical analysis. Furthermore, most discussions on the subject pay little attention to the right-left asymmetry aspect of the problem (Tennakone, 2017). Simple toy models constructed out of paper illustrate the salient features of seed-wind dispersal, especially the role played by the asymmetry of right and left. All the models described below are constructed from ordinary printing sheets thickness ~0.2 mm. Models are scalable, and most experiments were performed with samples cut to dimensions of the order 10–15 cm.

> *Model I*: As indicated in Figure 7.16, the Model I is constructed by cutting out a T with a broader transverse arm, removing the rectangular portion from the middle, and folding the flaps forward and backward. One could construct two structures in this way, and they are shown in Figure 7.16a and b. They are non-superimposable mirror images of each other, and hence are chiral. If one is termed right-handed, the other is left-handed. The darkened small rectangle marked at the longitudinal arm is a paper clip serving as a load. If the objects are dropped from a height of few meters, they rotate in opposite senses. The free ends of the flips will act as the trailing edge (rare edge of an airfoil where the air stream moves away) and therefore, the object (a) will rotate clockwise when viewed from above and (c) anticlockwise. The model simulates right-left symmetry aspect dipterocarps (paper clip representing nut). Obviously, natural systems are optimized to minimize the terminal velocity by enhancing the lift and the torque causing the rotational motion.
> *Model II*: In the alternative models shown in Figure 7.17, here two half-glued rectangular paper strips are pasted together and the unpasted flaps folded

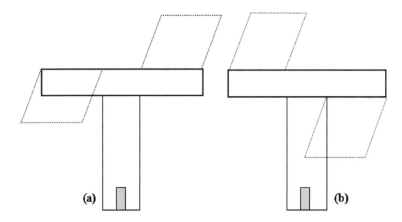

FIGURE 7.16 Model I where paper is folded as in (a) and (b) to form two copters of which are mirror images of each other. The dark rectangle indicates a paper clip load.

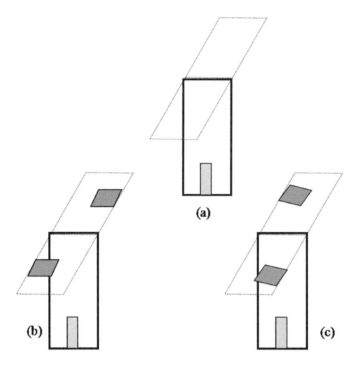

FIGURE 7.17   The copters (a) and (b) are mirror images of each other, and (c) is symmetric and has no handedne.

as in (a) and the base of the longitudinal arm loaded with a paper clip (darkened small rectangle). When dropped, the object (a) will not rotate, as it is right-left symmetric. Asymmetry can be introduced by removing rectangular portions (marked as black rectangles) as shown in sketches (b) and (c). The copters (b) and (c) are non-superimposable mirror images, and hence they rotate in opposite directions during falling.

Model III: Here a T-shaped figure is cut, and the ends of the transverse arm are bent in opposite directions to form two objects of opposite handedness (a), (b) and (c), (d) as depicted in Figure 7.18.

The two pairs differ in the position of loading with a paper clip marked as the darkened small rectangle. When dropped, the first pair will fall without rotating and will maintain the longitudinal arm in a vertical orientation, whereas the second pair will fall a little distance, turn around into a flat position with the plane of the T-shaped area nearly horizontal, and rotate around the position of the load in opposite directions as shown in the sketches (e) and (f) of Figure 7.18. In the first pair, air flows similarly on both sides of the wings, and therefore a torque is not exerted to impart rotation. The models (c) and (d) simulate the ingenious seed dispersal mechanism evolved in of *Hiptage benghalensis* shown in Figure 7.8.

Model IV: This model simulates the autogyration of the tulip tree samaras. Figure 7.19 depicts four rectangles (~2.5 × 10 cm^2) cut out of ordinary thin

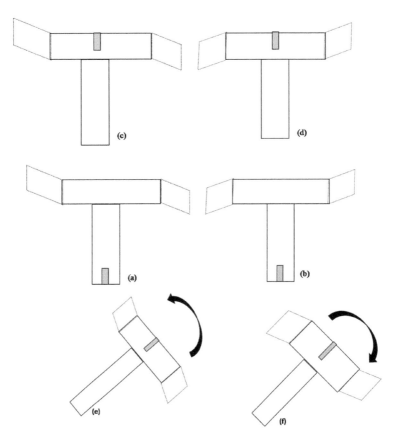

FIGURE 7.18 Copters (a) and (b) loaded at the bottom do not rotate when dropped, whereas (c) and (d) loaded at the top rotate in opposite senses almost flat with and around the position of the load with the longitudinal flap slightly inclined above, as shown in (e) and (f).

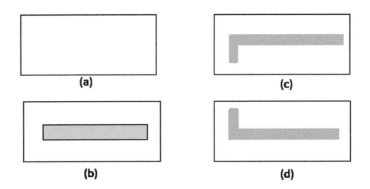

FIGURE 7.19 Rectangular sheets of paper: (a) blank; (b) rectangular strip pasted symmetrically; (c) and (d) letter L pasted; and (d) inverted letter L pasted to one another.

writing paper. In (b), a thin rectangular strip of paper is pasted symmetrically, and in (c), (d) with letter L cut from same paper as shown. Because of the thickness of the letter L, two configurations (c) and (d) are distinct objects possessing handedness. When dropped from a height, all four glide and rotate with the length-wise side remaining nearly horizontal but with following differences. The symmetric ones rotate in either sense when viewed from a fixed direction relative to the rectangle. However, the one with the middle ridge is much more stable. Ones pasted with letters L rotate in opposite directions. A simple experiment sometimes gets complicated owing twist developed when the strips are pasted.

7.12   CONCLUSION

The introductory section of this article defined chirality as the character possessed by objects when the object and image are non-superimposable. Sometimes both object form and image form are equally possible and exist in nature. This situation corresponds to right-left symmetry, each object taken separately being right-left asymmetric. The usage of terms right or left is a matter of convention. When right defined, the mirror image form would be left. In many instances, global right-left asymmetry gets broken so that there is a preponderance of one form (right or left) or behavioral difference of the two forms in a given environment.

Matter manifests right-left asymmetries at all levels of its organization. The amazing feature of the most successful fundamental theory of matter—The Standard Model in Elementary Particle Physics—is that the asymmetry between right and left owing to the chiral structure of the weak interaction is unified with the electromagnetic force. It is not absolutely clear whether a deeper theory would preserve this symmetry or not. Recent experiments suggest that there might be a difference in masses of neutrinos and antineutrinos. Confirmation of this idea would lead to the conclusion that not only matter but matter and anti-matter together will not conserve parity, pointing to an even more universal difference between right and left.

At the next level, there is an intrinsic right-left biochemical asymmetry in all living systems on earth whether extant or extinct. Living matter is based on preferred molecular handedness of amino acids, sugars, and other biomolecules. As discussed in this work, the cause and origin of this asymmetry continues to remain elusive. It is hard to ascertain whether this asymmetry happens to be an amplification of the intrinsic right-left asymmetry as expressed in matter, spontaneously developed, or incidental in origin.

Just as chiral objects can be crafted artificially, plants and animals also possess macroscopic right-left distinguishing attributes. Studies on such asymmetries in animals are more extensive compared to those in the case of plants. In animals, asymmetry seems to have evolved, initially as a morphogenetic requirement, followed by a developmental optimization need and a favorable character for locomotion and sensing in the environment. Apparently, in the case of animals, the right-left asymmetry characters have gotten more prominent with advancement, the extreme seen in humans, and among them it is an integral part of social organization.

The initial cause of origin of right-left asymmetry in plants seems to be the same as that of animals that it has a morphogenetic origin. However, in subsequent development, such asymmetry aspects have not been amplified to the same degree as in the case of animals.

The present study classifies right-left asymmetries observed in plants into four groups. In the group classified as *Type I*, all the individuals in species, genera, or family can be assigned right- or left-handedness determined by the chirality of an organ or organs. Both types occur with equal probability, and the handedness has no genetic or external influence. The nature of the physical mechanism involved seems to be a spontaneous breaking of right-left symmetry during the early stages of morphogenesis in the seedling. It is not clear whether instability is auxin induced or a result of cellular stress. Palms are prototype plants belonging to this category. Despite few reports to the contrary, there is no convincing evidence that the symmetry between right and left, even weakly breaks in this class of plants.

In the *Type II*, individual plants in species have organs of both handedness in near equal probabilities. This class seems to be a transition of the *Type I*, because when seeds of species are germinated the seeding has either right or left symmetry. Later on, both types of branches are developed, washing out the initial symmetry. This phenomenon manifests to varying degrees in different species belonging to *Type II*.

Type III plants possess no chiral organs or organ arrangements. The majority of plants, notably those evolved, later steps up the phylogenetic ladder, belong to this class. Possibly *Type III* was evolved independently as well as a transition from *Type II*.

Type IV is a special group where the entire species possesses an organ(s) of one chirality (right or left) inherited genetically.

Contrary to animals, plants in general have lost the degree of right-left asymmetry with evolutionary advancement. Natural trends are seen in non-living matter, where initial symmetries of the constituents disappear as the complicity of the organization continues to increase. Unlike for animals, the possession of a handedness is considered to be only rarely advantageous.

As evolution progresses, organisms retain or amplify the advantageous characters and eliminate traits that are not meaningful for the survival. Despite the general trend of plants losing symmetries as evolution proceeds towards advancement, there exists few instances where they are retained, because of the advantageous nature of these attributes. The most notable is value of handedness in wind dispersal of seeds. Reduction of the terminal velocity of seeds falling from a tree via autogyration greatly increases the chance of them being carried farther away by wind. Right-left asymmetry ensures initiation of rotation. Clear evidence has been presented to show that plants belonging to widely varying families have amplified the existing asymmetry to optimize the wind-dispersing potential via autogyration.

REFERENCES

Allard, H.A. 1946. Clockwise and counterclockwise spirality in the phyllotaxy of Tobacco. *Journal of Agricultural Research*, 73, 237–242.

Anderson, P.W. 1972. More is different. *Science*, 117(4047), 393–396. doi:10.1126/science.177.4047.393

Azuma, A. and Yasuda, K., 1989. Flight performance of rotary seeds. *Journal of Theoretical Biology*, 138(1), 23–53. doi:10.1016/S0022-5193(89)80176-6.

Bailey, J., 2000. Chirality and the origin of life. *Acta Astronautica*, 46(10–12), 627–631. doi:10.1016/S0094-5765(00)00024-2.

Bible, B.B. 1976. Non-equivalence of left-handed and right-handed phyllotaxis in tomato and pepper. *Hortscience*, 11, 601–602.

Blackmond, D.G. 2010. The origin of biological homochirality. *Cold Spring Harbor Perspectives in Biology*, 2(5), a002147. doi:10.1101/cshperspect.a002147.

Blum, M., Feistel, K., Thumberger, T. and Schweickert, A., 2014a. The evolution and conservation of left-right patterning mechanisms. *Development*, 141(8), 1603–1613. doi:10.1242/dev.100560.

Blum, M., Schweickert, A., Vick, P., Wright, C.V. and Danilchik, M.V., 2014b. Symmetry breakage in the vertebrate embryo: When does it happen and how does it work? *Developmental Biology*, 393(1), 109–123. doi:10.1016/j.ydbio.2014.06.014.

Boyd, R.N., Kajino, T. and Onaka, T., 2011. Supernovae, neutrinos and the chirality of amino acids. *International Journal of Molecular Sciences*, 12(6), 3432–3444. doi:10.1088/1742-6596/403/1/012032.

Brown, N.A. and Wolpert, L.E.W.I.S., 1990. The development of handedness in left/right asymmetry. *Development*, 109(1), 1–9.

Camposano, A.V.C., Virtudes, N.C., Otadoy, R.E.S. and Violanda, R., 2015. Vertical and rotational motion of mahogany seed. In *IOP Conference Series: Materials Science and Engineering*, 79(1), 012006. doi:10.1088/1757-899X/79/1/012006.

Capdevila, J., Vogan, K.J., Tabin, C.J. and Belmonte, J.C.I., 2000. Mechanisms of left–right determination in vertebrates. *Cell*, 101(1), 9–21. doi:10.1016/S0092-8674(00)80619-4.

Davis, T.A. and Davis, B., 1987. Association of coconut foliar spirality with latitude. *Mathematical Modelling*, 8, 730–733. doi:10.1016/0270-0255(87)90680-4.

Davis, T.A., 1962. The non-inheritance of asymmetry in Cocos nucifera. *Journal of Genetics*, 58(1), 42–50. doi:10.1007/BF02986118.

Davis, T.A., 1963. The dependence of yield on asymmetry in coconut palms. *Journal of Genetics*, 58(2), 186–215. doi:10.1007/BF02986139.

Davis, T.A., 1974. Enantiomorphic structures in plants. *Proceedings of the Indian National Academy of Sciences*, 40B, 424–429.

Denis, B. and Jean, R.V., 1998. *Symmetry in Plants*. World Scientific, Singapore.

Dixon, R., 1989. Spiral phyllotaxis, computers mathematics applications. In: *Symmetry 2: Unifying Human Understanding*. Hargittai, I (Ed.). Pergamon Press, Oxford, UK.

Douady, S. and Couder, Y., 1992. Phyllotaxis as a physical self-organized growth process. *Physical Review Letters*, 68(13), 2098. doi:10.1103/PhysRevLett.68.2098.

Dreiling, J.M. and Gay, T.J., 2014. Chirally sensitive electron-induced molecular breakup and the Vester-Ulbricht hypothesis. *Physical Review Letters*, 113(11), 118103. doi:10.1103/PhysRevLett.113.118103.

Fang, R., Zhang, Y. and Liu, Y., 2017. Aerodynamics and flight dynamics of free-falling ash seeds. *World Journal of Engineering and Technology*, 5(4), 105. doi:10.4236/wjet.2017.54B012.

Flack, H.D., 2009. Louis Pasteur's discovery of molecular chirality and spontaneous resolution in 1848, together with a complete review of his crystallographic and chemical work. *Acta Crystallographica Section A: Foundations of Crystallography*, 65(5), 371–389. doi:10.1107/S0108767309024088.

Flanders M. and Swan D., 1950. "Misalliance" on the album at the drop of hat available at https://www.flashlyrics.com/lyrics/flanders-and-swann/misalliance, accessed 2-4-2018.

Fredeen, A.L., Horning, J.A. and Madill, R.W., 2002. Spiral phyllotaxis of needle fascicles on branches and scales on cones in *Pinus contorta* var. latifolia: Are they influenced by wood-grain spiral? *Canadian Journal of Botany*, 80(2), 166–175. doi:10.1139/b02-002.

Grimes, D.T. and Burdine, R.D., 2017. Left–right patterning: Breaking symmetry to asymmetric morphogenesis. *Trends in Genetics*, 33(9), 616–628. doi:10.1016/j.tig.2017.06.004.

Hein, J.E. and Blackmond, D.G., 2012. On the origin of single chirality of amino acids and sugars in biogenesis. *Accounts of Chemical Research*, 45(12), 2045–2054. doi:10.1021/ar200316n.

Imai, Y., 1927. The right- and left-handedness of phyllotaxis. *Botanical Magazine*, 41, 592–596.

Inaki, M., Liu, J. and Matsuno, K., 2016. Cell chirality: Its origin and roles in left–right asymmetric development. *Philosophical Transactions of the Royal Society B*, 371(1710), e20150403. doi:10.1098/rstb.2015.0403.

Jean, R.V., 1994. *Phyllotaxis: A Systemic Study in Plant Morphogenesis*. Cambridge University Press, Cambridge, UK.

Jesson, L.K. and Barret, S.C., 2002. Solving the puzzle of mirror image flowers. *Nature*, 417, 707. doi:10.1038/417707a.

Kanahama, K., Saito, T., and Qu, Y., 1989. Right and left handedness of phyllotaxis and flower arrangement in inflorescences of Solanaceae plants. *Journal of the Japanese Society for Horticultural Science*, 57, 642–647. doi:10.2503/jjshs.57.642.

King, S., Beck, F. and Lüttge, U., 2004. On the mystery of the golden angle in phyllotaxis. *Plant, Cell & Environment*, 27(6), 685–695. doi:10.1111/j.1365-3040.2004.01185.x.

Klingenberg, C.P., McIntyre, G.S., Zaklan, S.D., 1998. Left-right asymmetry of fly wings and the evolution of body axes. *Proceedings Royal Society London B*, 265, 1255–1259.

Komatsu, Y. and Mishina, Y., 2013. Establishment of left–right asymmetry in vertebrate development: The node in mouse embryos. *Cellular and Molecular Life Sciences*, 70(24), 4659–4666. doi:10.1007/s00018-013.

Kondepudi, D.K. and Nelson, G.W., 1985. Weak neutral currents and the origin of biomolecular chirality. *Nature*, 314(6010), 438–441. doi:10.1038/314438a0.

Konnen, G.P., 1985. *Polarized Light in Nature*. Cambridge University Press.

Korn, R.W., 2008. Phyllotaxis theories and evaluation. *International Journal of Plant Developmental Biology*, 2, 1–12.

Kurita, Y. and Wada, H., 2011. Evidence that gastropod torsion is driven by asymmetric cell proliferation activated by TGF-β signaling. *Biology Letters*, 7(5), 759–762. doi:10.1098/rsbl.2011.0263.

Lentink, D., Dickson, W.B., Van Leeuwen, J.L. and Dickinson, M.H., 2009. Leading-edge vortices elevate lift of autorotating plant seeds. *Science*, 324(5933), 1438–1440. doi:10.1126/science.1174196.

Levin, M., 2005. Left–right asymmetry in embryonic development: A comprehensive review. *Mechanisms of Development*, 122(1), 3–25. doi:10.1016/j.mod.2004.08.006.

Levin, M., Johnson, R.L., Sterna, C.D., Kuehn, M. and Tabin, C., 1995. A molecular pathway determining left-right asymmetry in chick embryogenesis. *Cell*, 82(5), 803–814. doi:10.1016/0092-8674(95)90477-8.

Ligoxygakis, P., Strigini, M. and Averof, M., 2001. Specification of left-right asymmetry in the embryonic gut of Drosophila. *Development*, 128(7), 1171–1174.

Looy, C.V. and Stevenson, R.A., 2014. Earliest occurrence of autorotating seeds in conifers: The Permian (Kungurian-Roadian) *Manifera talaris* gen. et sp. nov. *International Journal of Plant Sciences*, 175(7), 841–854. doi:10.1086/676973.

Louis, I.H. and Chidambarm, A., 1976. Inheritance studies on phyllotaxy of coconut palm. *Ceylon Coconut Quarterly*, 27, 22–24.

Lubensky, T.C., Harris, A.B., Kamien, R.D. and Yan, G., 1998. Chirality in liquid crystals: From microscopic origins to macroscopic structure. *Ferroelectrics*, 212(1), 1–20. doi:10.1080/00150199808217346.

Martinez, C.C., Chitwood, D.H., Smith, R.S. and Sinha, N.R., 2016. Left–right leaf asymmetry in decussate and distichous phyllotactic systems. *Phil. Trans. R. Soc. B*, 371(1710), 20150412. doi:10.1098/rstb.2015.0412.

McCutchen, C.W., 1977. The spinning rotation of ash and tulip tree samaras. *Science*, 197(4304), 691–692. doi:10.1126/science.197.4304.691.

Minorsky, P.V. and Bronstein, N.B., 2006. Natural experiments indicate that geomagnetic variations cause spatial and temporal variations in coconut palm asymmetry. *Plant Physiology*, 142(1), 40–44. doi:10.1104/pp.106.086835.

Minorsky, P.V., 1998. Latitudinal differences in coconut foliar spiral direction: A re-evaluation and hypothesis. *Annals of Botany*, 82(1), 133–140. doi:10.1006/anbo.1998.0651.

Mirabet, V., Besnard, F., Vernoux, T. and Boudaoud, A., 2012. Noise and robustness in phyllotaxis. *PLoS Computational Biology*, 8(2), e1002389. doi:10.1371/journal.pcbi.1002389.

Mönchengladbach, U.K., 2010. Fruit availability and dispersal processes in a highly fragmented landscape in the northeastern Brazilian Atlantic Forest region, PhD Dissertation, FakultätfürNaturwissenschaften der Universität Ulm.

Nayani, K., Chang, R., Fu, J., Ellis, P.W., Fernandez-Nieves, A., Park, J.O. and Srinivasarao, M., 2015. Spontaneous emergence of chirality in achiral lyotropic chromonic liquid crystals confined to cylinders. *Nature Communications*, 6, 8067. doi:10.1038/ncomms9067.

Okabe, T., 2015. Biophysical optimality of the golden angle in phyllotaxis. *Scientific Reports*, 5, 15358. doi:10.1038/srep15358.

Ong, M., Bulmer, M., Groening, J. and Srinivasan, M.V., 2017. Obstacle traversal and route choice in flying honeybees: Evidence for individual handedness. *PLoS One*, 12(11), e0184343. doi:10.1371/journal.pone.0184343.

Palmer, A.R., 1996. From symmetry to asymmetry: Phylogenetic patterns of asymmetry variation in animals and their evolutionary significance. *Proceedings of the National Academy of Sciences*, 93(25), 14279–14286. doi:10.1073/pnas.93.25.14279.

Pennybacker, M.F., Shipman, P.D. and Newell, A.C., 2015. Phyllotaxis: Some progress, but a story far from over. *Physica D: Nonlinear Phenomena*, 306, 48–81. doi:10.1016/j.physd.2015.05.003.

Petch, T. 1911. Right and left handed coconut trees. *Annals of the Royal Botanical Gardens, Peradeniya*, 5, 538.

Rutishauser, H. 2001. *Phyllotaxy, Encyclopedia of Life Sciences*. Macmillan, pp. 6.

Schilthuizen, M. and Gravendeel, B., 2012. Editorial left-right asymmetry in plants and animals a gold mine for research. *Contributions to Zoology*, 81(2).

Schouts, J.C., 1921. On whorls and phyllotaxis. Available at http://natuurtijdschriften.nl/download?type=document;docid=552505.

Sherry, A.W. and Levin, M., 2008. What's left in asymmetry? *Developmental Dynamics*, 237(12), 3453–3463. doi:10.1002/dvdy.21560.

Skutch, A.F., 1927. Anatomy of leaf of banana, *Musa sapien* L. var. Hort. Gros Michel. *Botanical Gazette*, 84: 337–391.

Smith, R.S., 2008. The role of auxin transport in plant patterning mechanisms. *PLoS Biology*, 6(12), e323. doi:10.1371/journal.pbio.0060323.

Snow, M. and Snow, R., 1934. The interpretation of phyllotaxis. *Biological Reviews*, 9(1), 132–137. doi:10.1111/j.1469-185X.1934.tb00876.x.

Stevenson, R.A., Evangelista, D. and Looy, C.V., 2015. When conifers took flight: A biomechanical evaluation of an imperfect evolutionary takeoff. *Paleobiology*, 41(2), 205–225. doi:10.1017/pab.2014.18.

Stewart, G.T., 1966. Liquid crystals in biological systems. *Molecular Crystals*, 1: 566–580. doi:10.1080/15421406608083293.

Tennakone, K., 1984. Biochemical, L-D stereo selection by weak neutral currents: A mathematical model. *Chemical Physics Letters*, 105:444–446. doi:10.1016/0009-2614(84)80060-3.

Tennakone, K. M.P., Ariyasinghe, W.M., Ariyaratne S., 1978. Right-left symetry of flowers. *Vidyodaya Journal Arts and Science Letters*, 6:80–84.

Tennakone, K., 1991. The kinetics of the growth of an accidentally created chiral biomolecule in the racemic prebiotic medium. *Applied Mathematics and Computation*, 43(2), 139–144. doi:10.1016/0096-3003(91)90029-M.

Tennakone, K., 2017. Aerodynamics and right-left symmetry in wind dispersal of maple, dipterocarps, conifers and some genera of apocyanaceae and magnoliaceae. *Journal of the National Science Foundation of Sri Lanka*, 45(3), 201–217. doi:10.4038/jnsfsr.v45i3.8184.

Tennakone, K., Dayatilaka, R.K.D., Ariyaratne, S., 1982. Right-left symmetry in phyllotaxy and imbrication of flowers in *Hibiscus furcartus* L. *Annals of Botany*, 50, 397–400. doi:10.1093/oxfordjournals.aob.a086379.

Tennakone, K., Peiris, M.G.C., Divigalpitiya, W.M.R. and Ariyaratne, S., 1979. Right-left symmetry preservation by flowers of malvaceae family. *Journal of Biological Physics*, 7(1), 26–38. doi:10.1007/BF02349937.

van Eeden, F. and St Johnston, D., 1999. The polarisation of the anterior-posterior and dorsal-ventral axes during Drosophila oogenesis. *Current Opinions in Genetics Development*, 9, 396–404. doi:10.1016/S0959-437X(99)80060-4.

Vandenberg, L. and Levin, M., 2013. A unified model for left–right asymmetry? Comparison and synthesis of molecular models of embryonic laterality. *Developmental Biology*, 379, 1–15. doi:10.1016/j.ydbio.2013.03.021.

Venkateswarlu, T., 1982. Non-inheritance of isomerism in cocoyams. *Proceedings: Plant Science*, 91, 17–23.

Wandelt, J., and Nagy, L.M., 2004. Left-right asymmetry: More than one way to coil a shell. *Current Biology*, 14, R654–R656. doi:10.1016/j.cub.2004.08.010.

Watson, A.R., 2017. The golden relationships: An exploration of Fibonacci numbers and phi, Master of arts thesis, Duke University.

Zaidel, E., 2001. Brain asymmetry. *International Encyclopedia of Science and Behavioral Sciences*, 1321–1329. doi:10.1016/B0-08-043076-7/03548-8.

8 Molecular Bases and Genetic Aspects of Handedness in Plants

Andrey A. Sinjushin and Bir Bahadur

CONTENTS

8.1 INTRODUCTION

Handedness, or chirality, in living things is manifested as a structural or behavioral feature that can be realized in two alternative ways that cannot be superimposed by a rotation, which is most obvious in left versus right hands. In plants, the best-known alternatives of such kind are connected with the direction of the spiral of the shoot of a climbing plant, such as honeysuckle (*Lonicera* spp.: Caprifoliaceae). When viewed from above the shoot, if the coiling is to the right, this may be called "clockwise" and when it is to the left "counterclockwise." Chirality is more often illustrated and discussed in animals, including humans, than in plants. Examples of chirality such as the coiling of snail shells or preferential usage of one hand in humans have been under investigation for a long time (Schilthuizen and Davison, 2005; Medland et al., 2009). In some cases, genetic bases of chirality are known together with molecular factors, defining the choice between left and right alternatives. In plants, chirality became an object of systematic surveys later than in animals (Darwin, 1875, and references cited therein; Schmucker, 1924).

Generally, two scenarios are possible. Left- and right-handed forms may occur by chance, in which case they would be expected in equal frequency within or among plants. On the other hand, chirality might be genetically determined or predisposed, in which case the alternative forms would occur with unequal frequency.

Frequencies of different forms depend on history of every population. Some groups can be fully uniform with respect to this trait. Moreover, if this trait is governed by certain genetic factor(s), it would readily be heritable. For instance, that is the case of handedness in humans that has been studied for long (reviewed in Warren, 1980), but one may also conclude that this trait is affected by both environmental and genetic factors. The number of genes controlling this feature remains unknown (Medland et al., 2009). Alternatively, if chirality is defined by factors other than genetic, it is hardly expected to be heritable. A population would be either uniform (if environment is constant) or exhibit a 1:1 ratio (if chirality is established by chance). For example, in one population of *Impatiens parviflora* DC. (Balsaminaceae) in the Moscow region, the ratio between individuals with right and left leaf spirals comprised 497:501 (in 2015; chi-square 0.016, $p = 0.899$) and 553:497 (in 2017; chi-square 2.99, $p = 0.084$), which is not significantly different from a 1:1 ratio (original unpublished data). In a study of legume and cereal seedlings there was, for most species, an equal frequency of left- or right-handed folding of the first pair of leaves (Compton, 1912; Rao and Bahadur, 1980; Bahadur and Rao, 1981). Numerous examples of plant species exhibiting either complete uniformity or 1:1 ratio with respect to handedness are listed in a review of Kihara (1972).

For the genetic determination of handedness, it is of interest whether a gene elaborates a pre-existing asymmetry or determines the handedness directly (Palmer, 2016). It is hard to imagine how a specific handedness can emerge *ex nihilo*, as this would break some fundamental physical principles, but it should be recalled that there is an underlying chirality in many biological molecules.

In some cases, mechanisms of chirality formation in animals are known in detail. In snails, body chirality is established early in embryogenesis through geometry of the cleavage process, which is itself defined by spatial cytoskeleton dynamics (Shibazaki et al., 2004), and in *Drosophila*, a mutation in gene encoding myosin reverses the internal asymmetry of the body plan (Spéder et al., 2006). In mammals, internal body asymmetry is generally defined by asymmetric distribution of some laterality signal molecules. This asymmetry derives from a vortex created by movements of cilia in the primitive node (Nonaka et al., 2002). Defects of protein dynein participating in movements of cilia result in their immobility or abnormal motility, and this explains why, in humans, the rare congenital variation *situs inversus* (inverted position of visceral organs, such as dextrocardia, right position of spleen, etc.) is associated with mutations in the dynein gene (Chodhary et al., 2004).

Compared with animals, many plants comprise a complicated model for studies on chirality. This complexity is connected with whole mode of plant ontogeny with its iterative production of serially homologous structures, such as flowers, inflorescences, leaves, and axillary shoots. Plants develop structures with a wider range of possible positions than most of animals. For instance, a flower can be terminal/lateral on the main stem or terminal/lateral on the second-order axis. Chirality of different structures can be order-dependent. For example, handedness of lateral cincinni (a class of inflorescence) appeared correlated with chirality of the inflorescence axis in *Dioscorea tokoro* Makino (Dioscoreaceae) (Remizowa et al., 2010). Similarly, asymmetric features of the leaf are dependent on its position in the leaf spiral, that is, whether a certain side of the leaf lamina faces the next or previous leaf primordium (Troll, 1935; cited from Meyen, 1973). The latter phenomenon was termed "anodic asymmetry" by Korn (2006). That is why in left- and right-handed leaf spirals "anodic" (oriented in the direction up the leaf spiral) and "cathodic" (running down the spiral) sides of leaf lamina may be different with respect to more easily under-stood terms, such as "left side" or "right side." Generally, one may summarize that chirality of shoot of order $n + 1$ is opposite to chirality of shoots having order n and $n + 2$. Some irregularities can be found. For example, in *Cajanus cajan* (L.) Millsp. (Leguminosae), this rule is not followed rigorously (Table 8.1). More correlations between chirality of different structures within a single plant were illustrated for the palm *Ptychosperma macarthurii* H. Wendl. (Davis, 1974) (Table 8.1). Frequencies of different associations between chirality of the main (first order) stem and axillary (second order) branches in *C. cajan* (original unpublished data). Distribution is sig-nificantly different from equiprobable (chi-square 30.000, $p < 0.001$), but none of the combinations are prohibited in ontogeny.

One of the special causes of "anodic asymmetry" is the so-called pendulum sym-metry, which takes place in distichous stems (reviewed by Charlton, 1998). In the legume tribe Fabeae, such symmetry is evident in leaves (*Vicia faba* L., *V. sylvatica* Benth.), axillary inflorescences (most of species) (Sinjushin and Belyakova, 2015), and even flowers (e.g., *Lathyrus tuberosus* L.). In the latter case, the style of gynoe-cium bends to the left in all flowers of axillary raceme in node n, to the right in node $n + 1$, and so on. This phenomenon is known as enantiostyly and may have a genetic basis in certain taxa (see Section 8.3). However, in *Lathyrus* this feature, if present, is defined exclusively by the position of flower in inflorescence. These observations lead to the following conclusions:

1. Chirality of an organism is established early in ontogeny, usually by inter-actions on the intracellular level.
2. Different metameric structures of a single plant body may have different chirality, thus evidencing for independence of this feature from genetic factors—at least in some cases. This fact narrows down a range of models for studies on chirality in plants.
3. Chirality of plant structures is position-dependent in many cases studied.

TABLE 8.1

Frequencies of Different Associations Between Chirality of Main Stem and Axillary Branches in *Cajanus*

		Second-Order Stem	
		Right	Left
First-Order Stem	**Right**	7	31
	Left	39	13

The latter phenomenon deals with the concept of positional information introduced by Wolpert (1969) for animal embryogenesis. It proposes that cells acquire information on their differentiation pattern from their position in the developing system. This concept can be applied to plant development.

In this chapter, we will briefly review the present state of knowledge on genetic and molecular mechanisms underlying handedness (chirality) in plants.

8.2 MOLECULAR BASES OF HANDEDNESS IN PLANTS

8.2.1 CYTOSKELETON CONTENT AND DYNAMICS

The development of a chiral pattern on the cellular level is connected with a suite of structural proteins, especially self-assembling components of the cytoskeleton (Tee et al., 2015). Similarly to animals, components of the cytoskeleton can define chirality of axial structures (roots, stems, flowers) in plants. The most precisely dissected case is helical growth in model species *Arabidopsis thaliana* (L.) Heynh. (Furutani et al., 2000; Thitamadee et al., 2002). Normally this plant possesses roots and other axial organs lacking any signs of twisting. Two recessive non-allelic *spiral* mutations (*spr1* and *spr2*, the latter is allelic to *tortifolia1*) caused twisting of epidermal cells along root and hypocotyl. As a result, these structures exhibited right-handed coiling (Furutani et al., 2000). A similar effect was observed in petioles and petals, so the corolla acquired a contorted shape, as normally occurs in families like Apocynaceae, Malvaceae, or Torricelliaceae, but not in Brassicaceae. Phenotypes of both mutants manifested to varying degrees depending on temperature and light regime. Addition of microtubule-interacting drugs, such as propyzamide or taxol, altered the expression of mutant phenotype up to inversion of chirality (Furutani et al., 2000). It was proposed that the genes *SPR1* and *SPR2* act antagonistically with some processes of cytoskeleton assembly and normally prevent elongating cells from twisting. As a result, the normally observed symmetry is acquired and masks an underlying asymmetry. The gene *SPR1* has been isolated and characterized precisely (Nakajima et al., 2004). The *SPR1* protein is associated with cortical microtubules and found in different organs undergoing rapid cellular elongation. The affinity of *SPR1* is specifically to the plus-ends of microtubules (Sedbrook et al., 2004). The *SPR2* gene product is also a microtubule-associated small protein (Shoji et al., 2005).

In a parallel survey, Thitamadee et al. (2002) identified two semidominant mutations, which suppress the effect of the *spr1* and *spr2* mutations. These mutations were designated *lefty1* and *lefty2*, as mutant plants exhibited left-handed skewing of the same structures as those undergoing right-handed affection in *spr1/spr2* plants. The mutations *lefty1* and *lefty2* were shown to correspond to lesions in genes encoding α-tubulin 6 and α-tubulin 4, respectively, again pointing at the key role of the cytoskeleton in directed cell growth, elongation, and orientation. One more mutation in the gene of α-tubulin 4, *tortifolia2* (*tor2*), distorts normal interaction between α- and β-tubulin molecules (Buschmann et al., 2009). As a result, roots, leaf petioles, trichomes, and even isolated cells grown in suspension culture exhibit right-handed coiling. The *tor3* mutant exhibits left-handed coiling but yet remains uncharacterized in detail (Buschmann et al., 2009). Mutation in the gene of α-tubulin 1 causes right-hand twisting of roots and leaves in *Eragrostis tef* (Zuccagni) Trotter (Poaceae) (Jöst et al., 2015), and mutation in the orthologous gene in rice has a similar phenotypic effect (Sunohara et al., 2009). Smith and Mehlenbacher (1996) studied inheritance of contort growth in the hazelnut (*Salix matsudaba* Koidz.: Salicaceae). Curled and twisted emergences also characterize mutant *curly* of a garden pea (*Pisum sativum* L.: Leguminosae) (Sidorova and Uzhintzeva, 1975). These and other mutations in non-model objects remain uncharacterized but probably will be dissected using data obtained from *A. thaliana*.

These results were summarized in a review by Hashimoto (2002) who proposed that they shed light upon origin of helical growth of twining plants (such as lianas and vines) and tendrils, together with processes like circumnutation (the oscillating growth of axial structures) and corolla aestivation. However, it should be noted that *A. thaliana* normally exhibits no features of helical growth, nor do rice or tef. Identification of genes *SPR1/SPR2* and *LEFTY1/LEFTY2* undoubtedly widens our understanding of problem of chirality. No surveys on the association between polymorphism of tubulin gene(s) and normal handedness (e.g., in naturally existing populations) have been conducted to date.

Discovery of the microtubule-associated proteins *SPR1* and *SPR2* and their role in plant morphogenesis was one of the first topics in a number of surveys on the cytoskeleton in plants. To date, a set of microtubule tip-tracking proteins has been identified in living cells (Akhmanova and Steinmetz, 2008). Proteins of the END BINDING (EB) family are found in different eukaryotes (while SPR proteins are plant specific). Protein EB1b interacts with *SPR1* and tubulin dimers and possibly participates in the microtubule-loading mechanism (Galva et al., 2014). Double mutants *eb1b spr1* exhibit even more severe defects than in *spr1* mutants, so gene *EB1b* is expected to regulate organ twisting in plants, either directly or indirectly.

Numerous motor proteins are associated with cytoskeleton fibers. Among them, a kinesin is a key player in moving toward the plus-end of the microtubule and enabling transport functions. In the *Arabidopsis* genome, there are at least three paralogous genes encoding kinesin-related proteins ARMADILLO REPEAT KINESIN (ARK1, ARK2, ARK3) (Sakai et al., 2007). Mutations in these genes, as well as in the gene encoding kinesin-associated protein kinase, NEK6, result in abnormal differentiation of root epidermis, including root twisting (Sakai et al., 2007).

Regulatory components interacting with microtubules may also affect the twisting habit. Among such components, cytoskeleton-specific mitogen-activated protein kinases (MAPK) and their phosphatases can be listed. Plants of *A. thaliana* with impaired function of any of these proteins exhibit root twisting typical for microtubule-defective mutants (Walia et al., 2009).

Root twisting was also observed in *A. thaliana* knockout plants lacking protein *TNO1* (Roy and Bassham, 2017). This protein is involved in intracellular transport of vesicles from the trans-Golgi network to different destinations, which may include cellulose synthase complexes trafficked to the cell membrane, so *TNO1* may interfere with building the cell wall, which is important in connection with organ twisting (see Section 8.2.2). Anomalies in *tno1* mutants become enhanced in case of treatment with some microtubule-interacting agents, so this source of root skewing is also cytoskeleton sensitive (Roy and Bassham, 2017).

In addition to microtubules and tubulins (Hashimoto, 2013), actin microfilaments also seem to be engaged in the prevention of organ twisting in plants. Plants of *A. thaliana* homozygous for mutation in an actin gene *ACT7* are characterized by root waving and twisting (Gilliland et al., 2003). In rice, mutants *villin2* (*vln2*) also exhibit twisting of roots and shoots (Wu et al., 2015). Villin is a protein that binds with actin microfilaments and participates in their dynamics. Protein *VLN2* is not only involved in microtubule assembly but also contributes to polar auxin transport, which is known as one of the main factors defining plant architecture (Reinhardt et al., 2003).

What one may state definitely, is that molecular bases of chirality in plants, if determined mostly by cytoskeleton proteins and interactions between them, are difficult to dissect. Genes encoding cytoskeleton proteins (such as tubulin or actin) are often referred to as "housekeeping genes." This is in keeping with their constitutive and ubiquitous expression together with existence of numerous paralogous copies in genome. Even in *Arabidopsis* with its small genome (ca. 135 Mb), there are at least six expressed α-tubulin genes (Kopczak et al., 1992) and eight actin genes (Meagher et al., 1999), although expression of different paralogous copies is organ-specific. A set of induced mutations in tubulin genes cause anomalous twisting phenotype in *A. thaliana*, which has no twisting normally (Ishida et al., 2007). A normal handedness of a certain plant may result from a complex interaction of alleles at multiple loci.

8.2.2   CELL WALL STRUCTURE

Cytoskeleton-dependent organ shaping in plants is very likely mediated through deposition of cellulose in cell walls (Hashimoto, 2015). This structural polysaccharide is synthesized by a membrane-localized enzyme, cellulose synthase (CESA complex), which seems to be functionally associated with cortical microtubules of the cytoskeleton, that is, microtubules guide the enzyme's distribution (Paredez et al., 2006). The necessity of association between the CESA and the microtubules is, however, debatable (Baskin, 2001). One may expect that altered arrangement of microtubules might affect cellulose deposition. This hypothesis is supported with the fact that mutation in the gene encoding CESA interactive protein 1 (*CSI1*) also leads to twisting of the stem in *Arabidopsis* (Bringmann et al., 2012; Landrein et al., 2013). This protein links the CESA complex and cortical microtubules. Interestingly, inflorescences of *csi1* mutants acquire atypical phyllotaxis with a torsion angle, either 90° or 180°, but floral primordia initiate in a typical helical mode. Thus, some authors (Landrein et al., 2013) concluded that unusual phyllotaxis of *csi1* mutants resulted from "a late postmeristematic event." This observation gives the idea that features, which we refer to as possessing chirality, may be either "meristematic" or "postmeristematic." One more example of "postmeristematic" coiling of the axial structure is a dominant mutation *Screwed spike rachis 1* (*Scr1*) of wheat (*Triticum aestivum* L.), which causes clockwise or counterclockwise coiling of the stem and inflorescence (Amagai et al., 2017). This feature was found to be cytoskeleton independent (at least, not connected with α-tubulin). However, spikes of *Scr1* plants are normal at initiation, so this twisted habit develops "postmeristematically" (Dobrovolskaya et al., 2017).

Although in cereals the inflorescence normally has no features of twisting; this can be found in other monocots, such as orchids (Scopece et al., 2017).

All the phenomena discussed above, such as root or stem twisting, belong to the latter group and develop comparatively late in ontogeny, that is, in the course of cell elongation. Chirality of phyllotaxis, however (whether leaf spiral is clockwise or counterclockwise), is defined by spatial patterning in apical meristem. This process is beyond the scope of this review, but it is realized generally through spatial orientation of polar auxin flows, that is, earlier than a certain primordium becomes structurally discernible (Reinhardt et al., 2003). This type of chirality seems fully independent from peculiarities of cytoskeleton assemblage and dynamics and requires application of concept of positional information (see Introduction) – but yet remains far from understanding.

The chemical composition of the cell wall is also a determinant of organ shape, including twisting. The mutant *rhamnose biosynthesis 1* (*rhm1*) of *Arabidopsis* is deficient in the pectic polysaccharide rhamnogalacturonan-I due to non-functional enzyme UDP-L-rhamnose synthase encoded by gene *RHM1* (Saffer et al., 2017). As a result, petal epidermal cells become helically twisted, thus causing left-handed coiling of petals. A coiling growth is also observed in the roots of mutant plants. These authors especially highlight that cytoskeleton structure remains intact in *rhm1* mutants, that is, this type of growth regulation is independent from the cytoskeleton assembly.

Details of cytoskeleton dynamics and cell wall synthesis, which, when impaired affect organ coiling in *Arabidopsis*, are summarized in Figure 8.1.

8.2.3 OTHER MECHANISMS

Twisting of roots due to deformation of epidermal cells, although without fixed chirality, is characteristic for mutants *twisted dwarf1* (*twd1*) of *A. thaliana* (Wang et al., 2013). Protein *TWD1* is required for proper functioning of B family auxin

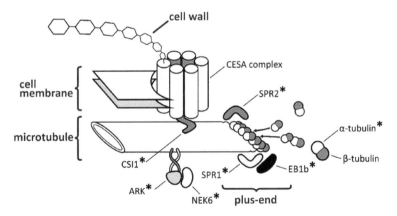

FIGURE 8.1 Interacting systems of cytoskeleton assembly and cell wall synthesis, as revealed from studies on *A. thaliana* (simplified scheme, see text for explanation). Asterisks indicate components that were proved as participating in the prevention of organ coiling.

transporters, the latter performing a polar auxin transport. This represents a way to acquire a twisted habit independent of cytoskeleton dynamics. Interestingly, this phenomenon partly links two aspects of morphogenetic patterning in plants, viz. one connected with cytoskeleton-dependent regulation and positional information conditioned by auxin gradients and polar transport. The *lopped1* (*lop1*) mutant of *A. thaliana* with impaired basipetal auxin transport has stems, roots, leaf petioles, and gynoecia twisted (Carland and McHale, 1996). Later, the mutation *lop1* was localized in gene *TORNADO1* (*TRN1*) (Cnops et al., 2006). Genes *TRN1* and *TRN2* act in the same regulatory pathway and are involved in auxin-dependent patterning in plant body, but their proteins are not similar (though both are transmembrane proteins) (Cnops et al., 2006).

It has been shown previously that polar auxin transport is mediated through special transporters, such as the PIN protein (Křeček et al., 2009). The distribution of PIN in the cell membrane is atypical in *Arabidopsis* mutants having a defective CESA complex (Feraru et al., 2011). This means that there is an interplay between polar auxin transport (and positional information) and cell wall biosynthesis. The latter, in its turn, also appears dependent on cytoskeleton structure and dynamics, so three main aspects of morphogenetic regulations in plants cooperate with each other.

The transmembrane protein QUIRKY (QKY) with several interacting components participates in the interrelation between cells and their anisotropic growth in *A. thaliana* (Trehin et al., 2013). However, its action seems organ specific, as mutants *qky* have anomalously twisted gynoecium and, later, silique. This regulatory pathway does not leave the cytoskeleton fully uninvolved, as cortical microtubules become disorganized in cells of *qky* plants (Trehin et al., 2013).

KNAT1, a wide-range regulator of development and a factor establishing meristematic properties of cells, also appears to be participating in prevention of root twisting. Plants homozygous for some mutant alleles of *KNAT1* have coiled roots (Qi and Zheng, 2013). Polar auxin transport seems to mediate this connection.

The total number of factors (proteins, regulatory RNAs, etc.) involved in the process of root skewing is far from precisely assessed; numerous genes potentially responsible for this habit were identified through transcriptomic studies (Schulz et al., 2017). The exact role of these genes needs further exploration.

Treatment of fronds of *Lemna minor* L. (Araceae) with stereoisomers of two amino acids had no effect on handedness, so this feature was found independent from stereochemical differences between different morphs (Kasinov, 1980). This discards the previously proposed hypothesis on the role of stereoisomers of hormones or other signal molecules in determination of handedness (Bahadur and Reddy, 1975). In *Lemna*, however, X-ray irradiation causes inversion of clone handedness, and this shift is stably reproduced through vegetative reproduction (Posner and Hillman, 1960). Rao and Bahadur (1982a) reported the effect of irradiation and centrifugation on chirality of seedlings in two species of *Vigna* Savi (Leguminosae). However, this example needs deeper examination, as the first pair of leaves acquires left- or right-handed aestivation already in embryogenesis, that is, mutual arrangements of leaves cannot be influenced by factors affecting fully mature seeds. Possibly this case deals with differential viability of different chiral morphs rather than with true changes in their handedness. Kihara (1972) describes an experiment with centrifugation of *Triticum* seedlings, which significantly skewed the left-right ratio as assessed by the direction of coiling in the third leaf. The position of germinating seed with respect to the magnetic field of Earth was also revealed as a factor potentially affecting seedling handedness (Kihara, 1972).

One more example of influence of genotype and external factors on handedness was dissected in a series of papers on *Triticum* spp. (Poaceae) (Kihara et al., 1951; Kojima, 1953; Kojima et al., 1954; Ono et al., 1954). In wheats, left- and right-handed spikelets alternate regularly along the axis of an ear, hence comprising a particular case of pendulum symmetry (see Introduction). The stability of this alternation can be described by a quantitative parameter, which Kojima (1953) called a mean concordance proportion. The average value of this parameter is unequal in different cereal species and can be changed in experiment, for example, by X-ray irradiation (Ono et al., 1954) or different regimes of mineral nutrition (Kojima et al., 1954). The concordance proportion may serve as a quantitative measure of developmental stability, similarly to fluctuating asymmetry in animals (Palmer and Strobeck, 1986); X-ray irradiation or malnutrition may shift stability of left-right alternation.

8.3 GENETICS OF HANDEDNESS IN PLANTS

In addition to the aforementioned genes of *Arabidopsis*, which prevent organ coiling (and, when mutated, cause twisting of different axial structures), genetic factors dealing with handedness are known in several species, but their molecular basis remains to be characterized.

Most species of genus *Medicago* L. (Leguminosae) are remarkable in having a spirally coiled pod. This feature is heritable in three species of this genus (Lilienfeld and Kihara, 1956; Lilienfeld, 1959; Thoquet et al., 2002). In all studied species, the right (clockwise) coiling habit dominates over left (counterclockwise). A single nuclear gene defines this feature. However, Lilienfeld and Kihara (1956) noticed that some pods on "right" plants could be either left or irregular, changing their chirality from right (in proximal portion of pod) to left. This corroborates the observation that chirality can sometimes be unstable (see Table 8.1).

A special case of asymmetry in angiosperms is connected with so-called dimorphic enantiostyly, when all flowers on a plant have styles deflected either to the left or to the right. We have seen that this feature can be fully position dependent (e.g., in some species of *Lathyrus*), a phenomenon referred to as a monomorphic enantiostyly, but at least in one species it is heritable. A single Mendelian locus governs enantiostyly in *Heteranthera multiflora* (Griseb.) C. N. Horn (Pontederiaceae) with the right-styled habit genetically dominant (Jesson and Barrett, 2002a). In natural populations of different species of *Wachendorfia* Burm. (Haemodoraceae), different ratios between left- and right-styled morphs were observed, some populations being monomorphic (Jesson and Barrett, 2002b). If style deflection is heritable in this genus, then one needs to agree with authors that frequencies of phenotypes are affected by reproduction strategy in a given population, that is, whether inbreeding or outbreeding prevails, or population propagates clonally. However, the observed

1:1 ratio in many populations may point at a non-heritable nature of this feature. Jesson and Barrett (2002b) mentioned no intermediate or neutral plants. If style deflection is controlled by a single gene with one allele (say, A for right-styled morph) dominant over another (a for left-styled morph), then a 1:1 ratio in a fully panmictic population can arise only if $p_A = 0.3$ and $p_a = 0.7$. It is hardly probable that this condition is identical in different populations. An equally probable explanation is that in some species handedness is defined by chance or by non-heritable factors. Interestingly, a 1:1 ratio between resupinate and non-resupinate flowers (another case of floral dimorphism) was recorded in two species of *Eplingiella* Harley & J.F.B. Pastore (Lamiaceae) (Harley et al., 2017). One may conclude that decisive evidence for (non-) heritable enantiostyly in any taxon must rely on controlled crossing experiments rather than on population statistics.

The same situation is seen for corolla chirality. In some plants, left- and right-contorted flowers coexist within the same inflorescence, for example, the case of Hypericaceae (Diller and Fenster, 2014). A similar distribution can be found in structures of higher order than flowers, for example, inflorescences (Bahadur and Reddy, 1975). In some angiosperm families, corolla contortion has a fixed chirality, either clockwise or counterclockwise throughout the whole plant, species, or even subfamiliar clade (Endress, 1999) for numerous examples.

The two aforementioned examples of heritable chirality (pod coiling in *Medicago* spp. and enantiostyly in *Heteranthera multiflora*) remain the only known examples of handedness defined by polymorphic genetic factors. In addition, the expression of left- and right-handedness was found to be under the control of a polygenic system in wheat (Kojima et al., 1955). However, in this model such expression reflects the ontogenetic stability rather than a switch between two chiral morphs. In marked contrast, numerous surveys revealed a non-hereditary bases of chirality in different species (Compton, 1912; Davis, 1962; Kasinov, 1969; Kihara, 1972). In *Lemna gibba* L., handedness of a certain frond is maintained through vegetative reproduction, that is, it characterizes the whole clone, but a 1:1 ratio arises in seed progeny (Kasinov, 1969).

Some observational evidence suggests that one of two alternative chiral morphs may be more productive in terms of yield components (e.g., Bible, 1976; see also Kihara, 1972) or possesses some adaptive benefits compared to the other one. For example, Rao and Bahadur (1982b) found that plants emerging from left-handed seedlings of *Vigna* spp. had significantly more root nodules. This would make handedness a desirable for breeding purposes, but there is no evidence that selection for chirality is possible in most species.

The non-heritable nature of the chirality of leaf phyllotaxis was shown for *Cocos nucifera* L. (Arecaceae), in which a ratio of 1:1 between hybrids with clockwise and counterclockwise leaf spirals persisted in progenies of all possible crosses (Davis, 1962). Interestingly, the ratio between two morphs in coconut palms seems to be dependent on magnetic latitude, that is, the position of a certain population with respect to the magnetic field of Earth (Minorsky, 1998).

The distribution and variation of chirality in natural plant populations seems partly similar to that of animals, especially gastropods. In this taxon, most of the species have right-handed (dextral) shells, but gene mutations revert this regularity.

The opposite situation, when left- and right-handed individuals appear equiprobably, is very rare among gastropods (Schilthuizen and Davison, 2005).

8.4 CONCLUSIONS

Our present understanding of mechanisms underlying "handedness" in plants is far from complete. This state of affairs is connected with a few problems. A major challenge is that it is an overgeneralization to refer to "plants" as a uniform group and to "handedness" as a discrete phenomenon. Indeed, when considering humans, it makes no sense to conflate asymmetry of position of visceral organs (resulting from motility of nodal cilia in early embryogenesis) with hand preference (consequence of brain lateralization) or asymmetric deviation of nasal septum (arising comparatively lately in development). Nevertheless, all these comprise different forms of left-right asymmetry.

In plants, different forms of chirality may be roughly classified either as "meristematic," that is, arising together with organ initiation via meristem activity, or as "postmeristematic," resulting in the course of differentiation of organ. In plants with their open growth and long-lasting activity of meristems, such separation is more meaningful than "inborn" versus "acquired."

Most of the available data on the molecular mechanisms of chirality in plants were obtained through studies on *Arabidopsis*, which normally lacks (almost) any signs of such chirality. A suite of mutants with root coiling or epicotyl twisting was described. A precise analysis of these mutants demonstrated that helical growth is a kind of "postmeristematic" event resulting from different malfunctioning of either cytoskeleton and associated proteins, or cell wall biosynthesis. These mechanisms may successfully explain numerous naturally occurring phenomena, such as helical twining of vines and lianas, coiling of tendrils, circumnutation or helical aestivation of corolla, ptyxis in floral, and vegetative buds. However, this deals only with cases when clockwise or counterclockwise coiling of any structure is heritable. Numerous examples of coexistence of both types within a single plant body seem to be explained by positional information. This is probably the case of position-dependent chirality of phyllotaxis, "anodic" (including pendulum) asymmetry, and other non-heritable symptoms. We are still far from a complete understanding of how molecular mechanisms influence chirality on the cellular level and how the spatial structure of a single cell influences chirality of the whole plant body. The ratio between molecules and biomechanical events is still obscure (Weizbauer et al., 2011).

In wild-type versus mutant plants of *A. thaliana*, there is a choice between two alternatives, viz. to have no chirality (normal) or to have any kind of coiling, twisting, etc. (abnormal). We still know nothing about mechanisms that govern a choice between **normal** right and **normal** left. However, this particular case of ontogenetic switch is realized in very many plants and seems likely dependent on non-heritable factors. Regardless of the nature of these factors, they remain anonymous.

To date, one may confidently state that structural proteins (especially those of the cytoskeleton) play an important role in defining body asymmetry in both plants in animals, although not all forms of asymmetry and chirality can be explained by these mechanisms.

ACKNOWLEDGMENT

A.S. expresses his cordial gratitude to Dr. T.H.N. Ellis (The University of Auckland, New Zealand) for language correction and helpful comments on the manuscript.

REFERENCES

Akhmanova, A., Steinmetz, M.O. 2008. Tracking the ends: A dynamic protein network controls the fate of microtubule tips. *Nat. Rev. Mol. Cell Biol.* 9: 309–322. doi:10.1038/nrm2369.

Amagai, Y., Martinek, P., Kuboyama, T., Watanabe, N. 2017. Microsatellite mapping of the *Scr1* locus conferring screwed spike rachis to modify the spatial orientation of spikelets in *Triticum aestivum* L. *Genet. Resour. Crop Evol.* 64: 1569–1579. doi:10.1007/s10722-016-0455-3.

Bahadur, B., Rao, M.M. 1981. Seedling handedness in Fabaceae. *Proc. Indian Acad. Sci. (Plant Sci.)* 90: 231–236.

Bahadur, B., Reddy, N.P. 1975. Types of vernation in the cyathia of *Euphorbia milli* Des Moulins. *New Phytol.* 75: 131–134. doi:10.1111/j.1469-8137.1975.tb01379.x.

Baskin, T.I. 2001. On the alignment of cellulose microfibrils by cortical microtubules: A review and a model. *Protoplasma.* 215: 150–171. doi:10.1007/BF01280311.

Bible, B.B. 1976. Non-equivalence of left-handed and right-handed phyllotaxy in tomato and pepper. *HortScience.* 11: 601–602.

Bringmann, M., Landrein, B., Schudoma, C., Hamant, O., Hauser, M.-T., Persson, S. 2012. Cracking the elusive alignment hypothesis: The microtubule-cellulose synthase nexus unraveled. *Trends Plant Sci.* 17: 666–674. doi:10.1016/j.tplants.2012.06.003.

Buschmann, H., Hauptmann, M., Niessing, D., Lloyd, C.W., Schäffner, A.R. 2009. Helical growth of the *Arabidopsis* mutant *tortifolia2* does not depend on cell division patterns but involves handed twisting of isolated cells. *Plant Cell* 21: 2090–2106. doi:10.1105/tpc.108.061242.

Carland, F.M., McHale, N.A. 1996. *LOP1*: A gene involved in auxin transport and vascular patterning in *Arabidopsis. Development.* 122: 1811–1819.

Charlton, W.A. 1998. Pendulum symmetry. In: *Symmetry in Plants.* Eds. Jean, R.V., Barabe, D. World Scientific Publishing Ltd. pp. 61–87. doi:10.1142/9789814261074_0003.

Chodhary, R., Mitchinson, H.M., Meeks, M. 2004. Cilia, primary ciliary dyskinesia and molecular genetics. *Paediatr. Respir. Rev.* 5: 69–76. doi:10.1016/j.prrv.2003.09.005.

Cnops, G., Neyt, P., Raes, J., Petrarulo, M., Nelissen, H., Malenica, N., Luschnig, C. et al. 2006. The *TORNADO1* and *TORNADO2* genes function in several patterning processes during early leaf development in *Arabidopsis thaliana. Plant Cell* 18: 852–866. doi:10.1105/tpc.105.040568.

Compton, R.H. 1912. A further contribution to the study of right- and left-handedness. *J. Genet.* 2: 53–70. doi:10.1007/BF02981547.

Darwin, C. 1875. On the movements and habits of climbing plants. *Bot. J. Linn. Soc.* 9: 1–118. doi:10.1111/j.1095-8339.1865.tb00011.x.

Davis, T.A. 1962. The non-inheritance of asymmetry in *Cocos nucifera. J. Genet.* 58: 42–50. doi:10.1007/BF02986118.

Davis, T.A. 1974. Enantiomorphic structures in the ornamental palm *Ptychosperma macarthurii* (H. Wendland) Nicholson (Arecaceae). *J. Plant. Crops.* 2: 9–14.

Diller, C., Fenster, C.B. 2014. Corolla chirality in *Hypericum irazuense* and *H. costaricense* (Hypericaceae): Parallels with monomorphic enantiostyly. *J. Torrey Bot. Soc.* 141: 109–114. doi:10.3159/TORREY-D-13-00026.1.

Dobrovolskaya, O.B., Krasnikov, A.A., Popova, K.I., Martinek, P., Watanabe, N. 2017. Study on early inflorescence development in bread wheat (*T. aestivum* L.) lines with non-standard SCR-morphotype. *Vavilov J. Genet. Breed.* 21: 222–226 (In Russian). doi:10.18699/VJ17.240.

Endress, P.K. 1999. Symmetry in flowers: Diversity and evolution. *Int. J. Plant Sci.* 160(6 Suppl.): S3–S23. doi:10.1086/314211.

Feraru, E., Feraru, M.I., Kleine-Vehn, J., Martinière, A., Mouille, G., Vanneste, S., Vernhettes, S., Runions, J., Friml, J. 2011. PIN polarity maintenance by the cell wall in *Arabidopsis*. *Curr. Biol.* 21: 338–343. doi:10.1016/j.cub.2011.01.036.

Furutani, I., Watanabe, Y., Prieto, R., Masukawa, M., Suzuki, K., Naoi, K., Thitamadee, S., Shikanai, T., Hashimoto, T. 2000. The *SPIRAL* genes are required for directional control of cell elongation in *Arabidopsis thaliana*. *Development*. 127: 4443–4453.

Galva, C., Kirik, V., Lindeboom, J.J., Kaloriti, D., Rancour, D.M., Hussey, P.J., Bednarek, S.Y., Ehrhardt, D.W., Sedbrook, J.C. 2014. The microtubule plus-end tracking proteins SPR1 and EB1b interact to maintain polar cell elongation and directional organ growth in *Arabidopsis*. *Plant Cell* 26: 4409–4425. doi:10.1105/tpc.114.131482.

Gilliland, L.U., Pawloski, L.C., Kandasami, M.K., Meagher, R.B. 2003. *Arabidopsis* actin gene *ACT7* plays an essential role in germination and root growth. *Plant J.* 33: 319–328. doi:10.1046/j.1365-313X.2003.01626.x.

Harley, R.M., Giulietti, A.M., Abreu, I.S., Bitencourt, C., de Oliveira, F.F., Endress, P.K. 2017. Resupinate dimorphy, a novel pollination strategy in two-lipped flowers of *Eplingiella* (Lamiaceae). *Acta Bot. Bras.* 31: 102–107. doi:10.1590/0102-33062016abb0381.

Hashimoto, T. 2002. Molecular genetic analysis of left-right handedness in plants. *Phil. Trans. R. Soc. Lond. B*. 357: 799–808. doi:10.1098/rstb.2002.1088.

Hashimoto, T. 2013. Dissecting the cellular functions of plant microtubules using mutant tubulins. *Cytoskeleton*. 70: 191–200. doi:10.1002/cm.21099.

Hashimoto, T. 2015. Microtubules in plants. *Arabidopsis Book*. 13: e0179. doi:10.1199/tab.0179.

Ishida, T., Kaneko, Y., Iwano, M., Hashimoto, T. 2007. Helical microtubule arrays in a collection of twisting tubulin mutants of *Arabidopsis thaliana*. *Proc. Natl. Acad. Sci. USA*. 104: 8544–8549. doi:10.1073pnas.0701224104.

Jesson, L.K., Barrett, S.C.H. 2002a. The genetics of mirror-image flowers. *Proc. Roy. Soc. Lond. B*. 269: 1835–1839. doi:10.1098/rspb.2002.2068.

Jesson, L.K., Barrett, S.C.H. 2002b. Enantiostyly in *Wachendorfia* (Haemodoraceae): The influence of reproductive system on the maintenance of polymorphism. *Am. J. Bot.* 89: 253–262. doi:10.3732/ajb.89.2.253.

Jöst, M., Esfeld, K., Burian, A., Cannarozzi, G., Chanyalew, S., Kuhlemeier, C., Assefa, K., Tadele, Z. 2015. Semi-dwarfism and lodging tolerance in tef (*Eragrostis tef*) is linked to a mutation in the α-*Tubulin 1* gene. *J. Exp. Bot.* 66: 933–944. doi:10.1093/jxb/eru452.

Kasinov, V.B. 1969. On the inheritance of the left- and right-handedness in *Lemnaceae* and other organisms. *Genetika*. 5: 22–29.

Kasinov, V.B. 1980. The action of arginine, asparagine and atebrine stereomers upon the left and right *Lemna minor* plants. *Biol. Plant.* (*Prague.*) 22: 321–326. doi:10.1007/BF02908974.

Kihara, H. 1972. Right- and left-handedness in plants. A review. *Seiken Ziho*. 23: 1–37.

Kihara, H., Kimura, M., Ono, H. 1951. Studies on the right-and left-handedness of spikelets in Einkorn wheats. *1. Proc. Jpn. Acad. Ser. B*. 27: 678–683.

Kojima, K.I. 1953. Studies on the right- and left-handedness of spikelets in einkorn wheats. III. Concordance proportion, mean concordance proportion and their relation. *Proc. Jpn. Acad. Ser. B*. 29: 576–580.

Kojima, K.I., Mukai, T., Suemoto, H., Sueoka, N., Ono, H. 1955. Studies on the right- and left-handedness of spikelets in einkorn wheats. VI. Polygenic analysis in a species cross. *Proc. Jpn. Acad. Ser. B*. 31: 228–233.

Kojima, K.I., Suemoto, H., Ono, H., Sueoka, N. 1954. Studies on the right- and left-handedness of spikelets in einkorn wheats. IV. Effect of nutritional conditioning. *Proc. Jpn. Acad. Ser. B*. 30: 214–220.

Kopczak, S.D., Haas, N.A., Hussey, P.J. 1992. The small genome of Arabidopsis contains at least six expressed α-tubulin genes. *Plant Cell* 4: 539–547. doi:10.1105/tpc.4.5.539.

Korn, R.W. 2006. Anodic asymmetry of leaves and flowers and its relationship to phyllotaxis. *Ann. Bot.* 97: 1011–1015. doi:10.1093/aob/mcl047.

Křeček, P., Skůpa, P., Libus, J., Naramoto, S., Tejos, R., Friml, J., Zažímalová, E. 2009. The PIN-FORMED (PIN) protein family of auxin transporters. *Genome Biol.* 10: 249. doi:10.1186/gb-2009-10-12-249.

Landrein, B., Lathe, R., Bringmann, M., Vouillot, C., Ivakov, A., Boudaoud, A., Persson, S., Hamant, O. 2013. Impaired cellulose synthase guidance leads to stem torsion and twists phyllotactic patterns in *Arabidopsis. Curr. Biol.* 23: 895–900. doi:10.1016/j.cub.2013.04.013.

Lilienfeld, F.A. 1959. Dextrality and sinistrality in plants. III. *Medicago tuberculata* Willd. and *M. litoralis* Rohde. *Proc. Jpn. Acad. Ser. B.* 35: 476–481.

Lilienfeld, F.A., Kihara, H. 1956. Dextrality and sinistrality in plants. II. *Medicago litoralis* Rohde. *Proc. Jpn. Acad. Ser. B.* 32: 626–632.

Meagher, R.B., McKinney, E.C., Vitale, A.V. 1999. The evolution of new structures: Clues from plant cytoskeletal genes. *Trends Genet.* 15: 278–284. doi:10.1016/S0168-9525(99)01759-X.

Medland, S.E., Duffy, D.L., Wright, M.J., Geffen, G.M., Hay, D.A., Levy, F., van-Beijsterveldt, C.E.M. et al. 2009. Genetic influences of handedness: Data from 25,732 Australian and Dutch twin families. *Neuropsychologia.* 47: 330–337. doi:10.1016/j.neuropsychologia.2008.09.005.

Meyen, S.V. 1973. Plant morphology in its nomothetical aspects. *Bot. Rev.* 39: 205–260. doi:10.1007/BF02860118.

Minorsky, P.V. 1998. Latitudinal differences in coconut foliar spiral direction: A re-evaluation and hypothesis. *Ann. Bot.* 82: 133–140. doi:10.1006/anbo.1998.0651.

Nakajima, K., Furutani, I., Tachimoto, H., Matsubara, H., Hashimoto, T. 2004. *SPIRAL1* encodes a plant-specific microtubule-localized protein required for directional control of rapidly expanding arabidopsis cells. *Plant Cell* 16: 1178–1190. doi:10.1105/tpc.017830.

Nonaka, S., Shiratori, H., Saijoh, Y., Hamada, H. 2002. Determination of left-right patterning of the mouse embryo by artificial nodal flow. *Nature* 418: 96–99. doi:10.1038/nature00849.

Ono, H., Sueoka, N., Suemoto, H., Kojima, K.I. 1954. Studies on the right- and left-handedness of spikelets in einkorn wheats. V. Effect of X-ray irradiation. *Proc. Jpn. Acad. Ser. B.* 30: 221–225.

Palmer, A.R. 2016. What determines direction of asymmetry: Genes, environment or chance? *Phil. Trans. R. Soc. Lond. B.* 371: 20150417. doi:10.1098/rstb.2015.0417.

Palmer, A.R., Strobeck, C. 1986. Fluctuating asymmetry: Measurement, analysis, patterns. *Ann. Rev. Ecol. Syst.* 17: 391–421. doi:10.1146/annurev.es.17.110186.002135.

Paredez, A.R., Somerville, C.R., Ehrhardt, D.W. 2006. Visualization of cellulose synthase demonstrates functional association with microtubules. *Science.* 312: 1491–1495. doi:10.1126/science.1126551.

Posner, H.B., Hillman, W.S. 1960. Effect of X irradiation on *Lemna perpusilla. Am. J. Bot.* 47: 506–511. doi:10.2307/2439567.

Qi, B., Zheng, H. 2013. Modulation of root skewing responses by *KNAT1* in *Arabidopsis thaliana. Plant J.* 76: 380–392. doi:10.1111/tpj.12295.

Rao, M.M., Bahadur, B. 1980. Seedling handedness in *Phaseolus vulgaris* L. *Curr. Sci.* 49: 61–62.

Rao, M.M., Bahadur, B. 1982a. Effect of irradiation and centrifugation on seedling enantiomorphism in green gram and black gram. *Curr. Sci.* 51: 242–243.

Rao, M.M., Bahadur, B. 1982b. Note on the relationship between seedling-handedness and root and nodule characters in green gram and black gram. *Indian J. Agric. Sci.* 52: 706–708.

Reinhardt, D., Pesce, E.-R., Stieger, P., Mandel, T., Baltensperger, K., Bennett, M., Traas, J., Friml, J., Kuhlemeier, C. 2003. Regulation of phyllotaxis by polar auxin transport. *Nature* 426: 255–260. doi:10.1038/nature02081.

Remizowa, M.V., Sokoloff, D.D., Kondo, K. 2010. Early flower and inflorescence development in *Dioscorea tokoro* (Dioscoreales): Shoot chirality, handedness of cincinni and common tepal-stamen primordia. *Wulfenia.* 17: 77–97.

Roy, R., Bassham, D.C. 2017. TNO1, a TGN-localized SNARE-interacting protein, modulates root skewing in *Arabidopsis thaliana. BMC Plant Biol.* 17: 73. doi:10.1186/s12870-017-1024-4.

Saffer, A.M., Carpita, N.C., Irish, V.F. 2017. Rhamnose-containing cell wall polymers suppress helical plant growth independently of microtubule orientation. *Curr. Biol.* 27: 1–12. doi:10.1016/j.cub.2017.06.032.

Sakai, T., van der Honing, H., Nishioka, M., Uehara, Y., Takahashi, M., Fujisawa, N., Saji, K. et al. 2007. Armadillo repeat-containing kinesins and a NIMA-related kinase are required for epidermal-cell morphogenesis in *Arabidopsis. Plant J.* 53: 157–171. doi:10.1111/j.1365-313X.2007.03327.x.

Schilthuizen, M., Davison, A. 2005. The convoluted evolution of snail chirality. *Naturwissenschaften.* 92: 504–515. doi:10.1007/s00114-005-0045-2.

Schmucker, T. 1924. Rechts- und Linkstendenz bei Pflanzen. *Beih. Bot. Zbl.* 41: 51–81.

Schulz, E.R., Zupanska, A.K., Sng, N.J., Paul, A.-L., Ferl, R.J. 2017. Skewing in Arabidopsis roots involves disparate environmental signaling pathways. *BMC Plant Biol.* 17: 31. doi:10.1186/s12870-017-0975-9.

Scopece, G., Gravendeel, B., Cozzolino, S. 2017. The effect of different chiral morphs on visitation rates and fruit set in the orchid *Spiranthes spiralis. Plant Ecol. Divers.* 10: 97–104. doi:10.1080/17550874.2017.1354093.

Sedbrook, J.C., Ehrhardt, D.W., Fisher, S.E., Scheible, W.-R., Somerville, C.R. 2004. The *Arabidopsis SKU6/SPIRAL1* gene encodes a plus end-localized microtubule-interacting protein involved in directional cell expansion. *Plant Cell* 16: 1506–1520. doi:10.1105/tpc.020644.

Shibazaki, Y., Shimizu, M., Kuroda, R. 2004. Body handedness is directed by genetically determined cytoskeletal dynamics in the early embryo. *Curr. Biol.* 14: 1462–1467. doi:10.1016/j.cub.2004.08.018.

Shoji, T., Narita, N.N., Hayashi, K., Asada, J., Hamada, T., Sonobe, S., Nakajima, K., Hashimoto, T. 2005. Plant-specific microtubule associated protein SPIRAL2 is required for anisotropic growth in Arabidopsis. *Plant Physiol.* 136: 3933–3944. doi:10.1104/pp.104.051748.

Sidorova, K.K., Uzhintzeva, L.P. 1975. Induced pea mutant with curled leaves and stipules. *Pisum Newsl.* 7: 56.

Sinjushin, A.A., Belyakova, A.S. 2015. Ontogeny, variation and evolution of inflorescence in tribe Fabeae (Fabaceae) with special reference to genera *Lathyrus*, *Pisum* and *Vavilovia. Flora.* 211: 11–17.doi:10.1016/j.flora.2014.12.003.

Smith, D.C., Mehlenbacher, S.A. 1996. Inheritance of contort growth in hazelnut. *Euphtica.* 89: 211–213. doi:10.1007/BF00034607.

Spéder, P., Ádám, G., Noselli, S. 2006. Type ID unconventional myosin controls left-right asymmetry in *Drosophila. Nature* 440: 803–807. doi:10.1038/nature04623.

Sunohara, H., Kawai, T., Shimizu-Sato, S., Sato, Y., Sato, K., Kitano, H. 2009. A dominant mutation of *TWISTED DWARF 1* encoding an α-tubulin protein causes severe dwarfism and right helical growth in rice. *Genes Genet. Syst.* 84: 209–218. doi:10.1266/ggs.84.209.

Tee, Y.H., Shemesh, T., Thiagarajan, V., Hariadi, R.F., Anderson, K.L., Page, C., Volkmann, N. et al. 2015. Cellular chirality arising from the self-organization of the actin cytoskeleton. *Nat. Cell Biol.* 17: 445–457. doi:10.1038/ncb3137.

Thitamadee, S., Tuchihara, K., Hashimoto, T. 2002. Microtubule basis for left-handed helical growth in *Arabidopsis. Nature* 417: 193–196. doi:10.1038/417193a.

Thoquet, P., Ghérardi, M., Journet, E.-P., Kereszt, A., Ané, J.-M., Prosperi, J.-M., Huguet, T. 2002. The molecular genetic linkage map of the model legume *Medicago truncatula*: An essential tool for comparative legume genomics and the isolation of agronomically important genes. *BMC Plant Biol.* 2: 1. doi:10.1186/1471-2229-2-1.

Trehin, C., Schrempp, S., Chauvet, A., Berne-Dedieu, A., Thierry, A.-M., Faure, J.-E., Negrutiu, I., Morel, P. 2013. *QUIRKY* interacts with *STRUBBELIG* and *PAL OF QUIRKY* to regulate cell growth anisotropy during *Arabidopsis* gynoecium development. *Development.* 140: 4807–4817. doi:10.1242/dev.091868.

Troll, W. 1935. Vergleichende Morphologie der Fiederblätter. *Nova Acta Leopold.* 2: 315–455.

Walia, A., Lee, J.S., Wasteneys, G., Ellis, B. 2009. Arabidopsis mitogen-activated protein kinase MPK18 mediates cortical microtubule functions in plant cells. *Plant J.* 59: 565–575. doi:10.1006/anbo.1998.0651.

Wang, B., Bailly, A., Zwiewka, M., Henrichs, S., Azzarello, E., Mancuso, S., Maeshima, M., Friml, J., Schulz, A., Geisler, M. 2013. *Arabidopsis* TWISTED DWARF1 functionally interacts with auxin transporter ABCB1 on the root plasma membrane. *Plant Cell* 25: 202–214. doi:10.1105/tpc.112.105999.

Warren, J.M. 1980. Handedness and laterality in humans and other animals. *Psychobiology.* 8: 351–359. doi:10.3758/BF03337470.

Weizbauer, R., Peters, W.S., Schulz, B. 2011. Geometric constraints and the anatomical interpretation of twisted plant organ phenotypes. *Front. Plant Sci.* 2: 62. doi:10.3389/fpls.2011.00062.

Wolpert, L. 1969. Positional information and the spatial pattern of cellular differentiation. *J. Theor. Biol.* 25: 1–47. doi:10.1016/S0022-5193(69)80016-0.

Wu, S., Xie, Y., Zhang, J., Ren, Y., Zhang, X., Wang, J., Guo, X. et al. 2015. *VLN2* regulates plant architecture by affecting microfilament dynamics and polar auxin transport in rice. *Plant Cell* 27: 2829–2845. doi:10.1105/tpc.15.00581.

9 Embryo Handedness Caused by Embryo Rotations in Angiosperms

K. V. Krishnamurthy, Thiruppathi Senthil Kumar, and S. John Adams

CONTENTS

9.1 INTRODUCTION

The angiosperm embryo is a miniature sporophyte that is produced as the end-product of zygotic ontogeny (Krishnamurthy 1994, 2015a, 2015b). Several studies on angiosperm embryogenesis with inputs from diverse disciplines, such as morphology, developmental anatomy, embryology, biochemistry, genetics, cytology, physiology, and cell and tissue culture, have revealed that the seemingly simple process of embryogeny has turned out to be very complex and full of exciting problems (Evenari 1984; Krishnamurthy 2015a). The zygote is the starting point for the development of the embryo, and with its division an array of developmental episodes occurs in an ordered sequence, where cell multiplication is progressively associated with morphogenetic processes. Several morphogenetically distinguishable contours, such as filamentous, globular, cordate, torpedo, and mature embryos, form at different stages of embryogeny, but all these developmental stages are to be looked upon as a continuous process in which any given stage is intimately related to the previous stage as well as to the stage that follows it (Krishnamurthy 1994). During different developmental stages, diverse chemical and structural components are synthesized, and the cells of the developing embryo are grouped and organized into distinctive patterned regions of significant morphogenetic importance. Along with these features, the developing embryo also seems to undergo rotation along its vertical axis, which results in embryo handedness.

9.2   DISCOVERY OF EMBRYO ROTATION

In 1992, we accidently discovered embryo rotation in some species of *Polygala* (Krishnamurthy and Senthil Kumar 1992). This embryo rotation is a clear instance of circumnutation that starts in the embryo itself and continues until the death of the plant (see Chapter 10). The embryos of these species, during their development, exhibit rotational movements along their vertical axis. At that time, we could not find out whether the filamentous or globular embryos rotated or not since we did not have the sufficient techniques to detect the rotation at these stages, and because at these stages, the embryos have a radial symmetry. However, we found positive evidence for rotation from the cordate stage of the embryo onward until its maturation (Figure 9.1).

The ovule of these species is anatropous, with a distinct raphe traversed by a prominent vascular trace. The ovule does not undergo any post-fertilization change in configuration from its anatropy. We examined transverse sections of the ovule/seed along with its developing embryo at various stages of ontogeny of the seed, paying special attention to the position of the two developing cotyledons and epicotyl in relation to the position of the raphe vascular trace (marked in black in the figures), which was fixed in its location and was taken as an unchanged marker of the direction.

In the sections we examined at the heart-shaped stage of the embryo, the imaginary line connecting the raphe vascular strand, the centers of the two cotyledons, and the embryonic shoot apical meristem region (= epiphysis region, see Krishnamurthy 2015a, 2015b) in the transectional view of the ovule/developing seed is straight. At subsequent stages of embryo development, the lines connecting the three points make a triangle with the three angles varying at different developmental stages depending upon the disposition of the two developing cotyledons (in relation

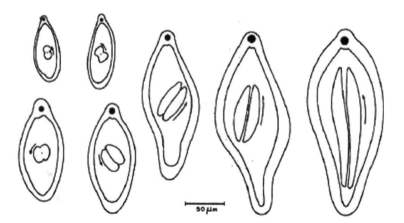

FIGURE 9.1   Transverse section of the developing seeds of the *Polygala* species at successive stages to show rotation of the embryo (arrows). The position of the raphe vascular bundle (dark dot) is used as the marker to find out the direction of rotation of embryo. (From Krishnamurthy, K.V., and Senthil Kumar, T., *Beitr. Biol. Pflanzen.*, 67, 55–58, 1992.)

to the vertical axis of the embryo) of the embryo. At the end of seed development, when rotation stops, the figure formed by the imaginary lines connecting these three points becomes a right-angled triangle. In other words, in the initial stages, the two cotyledons are parallel to the raphe bundle, while at the mature seed stage the two cotyledons are at right angles to the raphe bundle. Thus, the embryo shows a distinct rotational movement. We could not estimate the speed of rotation; neither was it possible to ascertain the number of rotations. Most likely, there were two rotations between the cordate and mature stages of the embryo. We did find that the time taken between pollination and maturation of seed (and embryo) in this species of *Polygala* was 18–24 hours, and hence the rotation(s) must have been completed within this time period.

Although the possibility of an embryo rotation was not thought of by earlier investigators, some illustrations provided by Rao and Mary Roy (1981) for certain other species of *Polygala* indicate the possibility of occurrence of this phenomenon, at least in this genus.

9.3 ADDITIONAL EVIDENCES FOR EMBRYO ROTATION

The first author of the present chapter examined hundreds of illustrations of embryogeny worked out by other investigators in other taxa of angiosperms to find out whether any of them showed evidences for rotation of the embryo. Since 95% of the illustrations show longitudinal sections of the developing embryo (and most of them only embryos and not the whole developing seeds), rotations could not be identified with certainty. However, some figures illustrated in Netolitzky (1926) and Corner (1976) indicated the probable occurrence of embryo rotations. Also, a number of illustrations showed indirect evidence for embryo rotation. This relates to the presence of an irregular empty space, bordered by degenerating cells, in the endosperm around the free end of the developing embryo, from the proembryo stage onward (Figure 9.2). This space was interpreted by all the investigators who

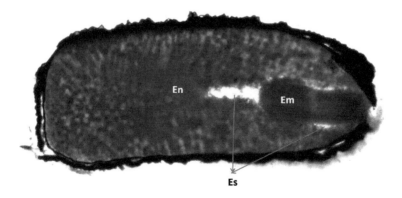

FIGURE 9.2   A fresh longitudinal cut in a developing seed of black cumin showing empty space in the endosperm caused by embryo rotation. Em: embryo; En: endosperm; Es: empty space in the endosperm.

spoke about them, as an effect of the growing embryo utilizing the endosperm tissue around it. This is not the correct interpretation since such degenerating cells could not be seen on the chalazal end of the developing embryos, especially near the suspensor, which is believed to be largely involved in the nutrition of the embryo. We, therefore, feel that the disintegration of the endosperm cells is caused only by the rotation of the embryo. Other evidence is provided by the disposition of the mature embryo when the seeds of the same species are sectioned in two quite different longitudinal planes (along the micropylar axis), which are at right angles to each other (Figure 9.3).

If the sectioned embryo looks similar in all seeds (i.e., sectioned only para-cotyledonarily or anticotyledonarily) in these types of two longitudinal planes of sectioning, it can be assumed that the embryos have not undergone rotation. On the other hand, if in both these types of longitudinal sections the embryos are cut both paracotyledonarily and anticotyledonarily, then it can be assumed that embryo rotation has taken place. These are indicated in transections of seeds in Figure 9.4. In the monocots, rotation can also be verified by the occurrence of by the relative left-right positions of the embryonic shoot apex in relation to the single cotyledon (Figure 9.5).

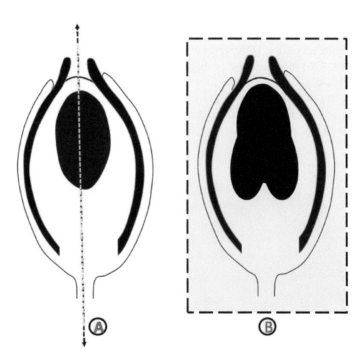

FIGURE 9.3 Diagrammatic section of a developing seed showing the two planes (perpendicular to each other) in which longitudinal sections of the seed can be taken. The two planes of sectioning are shown by broken lines (in A) and dotted lines (in B).

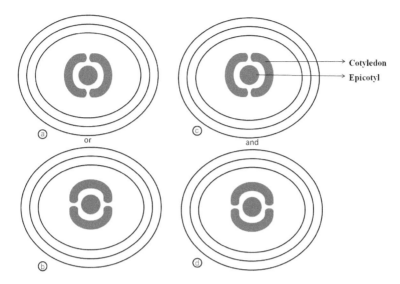

FIGURE 9.4 Transections of developing older seeds. If the transverse sections of all seed examples show only either (a) or (b), then we can say no embryo rotation has happened. And if the condition of all seeds examined shows both (c) and (d), then we can say that the embryo rotation has happened.

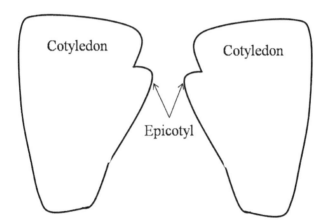

FIGURE 9.5 Mirror image of mature seeds of corn formed by the stoppage of embryo rotation at different times during seed development.

9.4 EMBRYO ROTATION DUE TO OVULE ROTATION

In a number of plants, the ovule undergoes conformational changes immediately after fertilization; this is particularly evident in many legumes. According to a study made by Endo and Ohashi (2009), there are two patterns of seed arrangements in legume fruits, which have a row of seeds one below the other: transverse and longitudinal. The latter type of arrangement is more common, while the former is less common. Their studies show that during the early stages of seed development, the ovules were obliquely orientated or transversely arranged but later on, due to restrictions imposed by the ovary wall or neighboring ovules, the obliquely orientated ovules rotated inward or outward and resulted in the final orientation characteristic of each taxon. Since the ovules are mostly campylotropous, when they rotate the embryo also rotates.

9.5   SIGNIFICANCE OF EMBRYO ROTATION

Embryo rotation is significant in two respects: (1) it indicates that circumnutation has a very early origin in the plants, that is, from the zygote stage onward; and (2) it is responsible for the so-commonly observed seedling handedness in plants. The position of first leaf in the seedlings of epicotyls of plants with alternate phyllotaxy, in relation to the cotyledon (s), is decided by the position of the two cotyledons in dicots and one cotyledon in monocots, when embryo rotation stops in the mature seed.

REFERENCES

Corner, E.J.H. 1976. *The Seeds of Dicotyledons*. Vol. 1 & II. Cambridge University Press, Cambridge.

Endo, Y. and Ohashi, H. 2009. Diversification of seed arrangement induced by ovule rotation and septum formation in Leguminosae. *J. Plant Res.* 122: 541–550. doi:10.1007/s10265-009-0242-8.

Evenari, M. 1984. Seed physiology: From ovule to maturing seed. *Bot. Rev.* 50: 143–170. doi:10.1007/BF02861091.

Krishnamurthy, K.V. 1994. The angiosperm embryo: Correlative controls in development, differentiation, and maturation. In: *Growth Patterns in Vascular Plants* (ed.) M. Iqbal, Portland, OR: Dioscorides Press, 372–404 pages.

Krishnamurthy, K.V. 2015a. *Growth and Development in Plants*. Scientific Publishers. Jodhpur, India.

Krishnamurthy, K.V. 2015b. Post-fertilization growth and development. In: *Plant Biology and Biotechnology* (Vol. 1) B. Bahadur, Rajam, M.V., Leela Sahijram and Krishnamurthy, K.V., Springer, New Delhi, India.

Krishnamurthy, K.V. and Senthil Kumar, T. 1992. Embryo rotation in polygala arvensis willd. *Beitr. Biol. Pflanzen.* 67: 55–58.

Netolitzky, F. 1926. Anatomie der Angiospermen Samen. *Linsbauer Handb. Pfl. Anat.* 10.

Rao, P.N. and Mary Roy, M.J.A. 1981. From embryo sac to seed in polygala linn. *Indian J. Bot.* 4 (2): 115–121.

10 Circumnutation

*K. V. Krishnamurthy, A. V. P. Karthikeyan,
and S. John Adams*

CONTENTS

10.1 WHAT IS CIRCUMNUTATION?

Rapidly elongating plant organs do not grow in a straight line, although their mean growth direction may be maintained in a straight line. The organs' instantaneous growth, however, usually oscillates about this mean straight line. Viewed from a distal point, the organ tip (shoot and root tips, leaf tips, floral tips, etc.) or the elongating cylindrical organ itself (e.g., a tendril) describes an ellipse or a circular pendulum-like movement about a plumb line; this movement can alternate between a clockwise (right-handed) and counterclockwise (left-handed) direction. The ellipse axis can vary, with the ellipse approximating, at one extreme, a line and on the other extreme, a circle. Sometimes it forms an irregular zigzag shape (Stolarz, 2009). As the organ grows and its tip advances further, a series of single oscillatory movements produce an irregular three-dimensional helix. This type of oscillatory movement was very well-known to many 19th century botanists under the name "revolving nutation," a term coined by Sachs, but was subsequently changed into "circumnutation" (CN) by the Darwins (Darwin and Darwin, 1880); this revised usage persists even today.

The Darwins also considered CNs as universal kinds of plant movements and that the origins of all other movements are from CNs. Today, these movements are known to occur in angiosperms, gymnosperms (in all plant organs of both groups) (Baillaud, 1962), fungi (especially, Basidiomycetes), bryophytes, algae, bacteria, and cyanobacteria (Brown, 1993; Kern et al., 2005; Kim et al., 2016; Hoiczyk, 2000). CN is a clear case of asymmetry/handedness in plants.

10.2 PARAMETERS OF CIRCUMNUTATION

Amplitude, period, shape, and direction (Figure 10.1) are the most important parameters of CN (Stolarz, 2009). These are reported to vary with plants, plant parts, and in the same plant part, under different environmental conditions such as light intensity, photoperiod, mechanical stress, and temperature (Anderson-Bernadas et al., 1997; Buda et al., 2003).

10.2.1 AMPLITUDE

Amplitude represents the distance between the plumb line and one end of the oscillation, that is, half of the width of the ellipse or straight line created by a CN. Amplitude does not normally depend on changes in the length of the organ that circumnutates. However, changes in amplitude are induced by various stimuli, such as gibberellins/AMO1618 treatment (in *Phaseolus vulgaris*, Millet and Badot, 1996), microgravity in space lab (in *Helianthus annuus*, Brown et al., 1990), gravity increase (in *H. annuus*, Zachariassen et al., 1987), and light/dark transitions (Stolarz, 2009). In *Arabidopsis*, CNs, which occurred less frequently in green light, has a smaller amplitude. What are the mechanisms by which these treatments cause

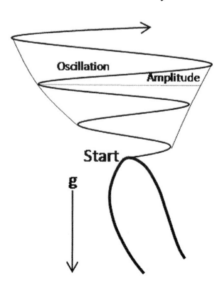

FIGURE 10.1 Diagrammatic representation of circumnutational pattern of the shoot tip and its parameters.

amplitude variations in CN? This has not been answered satisfactorily. A similar unanswered question is, what causes amplitude changes in the same plant organ when all other external conditions are similar? Age of the plant, transition to reproductive phase, plastochron-interplastochron changes in the shoot apical meristem, width of the oscillating apex, resistance from underlying maturing/mature tissues, etc. are likely to control the amplitude of circumnutational movement. More studies need to be done on these aspects.

10.2.2 PERIOD

The period of CN refers to the actual time taken for a full circumnutational movement to get completed. This may range from a few minutes to several hours, mostly between 10 and 480 min, depending on plants. In tulip petioles, it is approximately 4 hr (Hejnowicz and Sievers, 1995), in *Arabidopsis thaliana* seedlings it is 15 min to 24 hr, in adzuki bean hypocotyls it is 60 min, and in wild Alaska pea roots it is 140–190 min (Kim et al., 2016). It is evident from the above that the period may vary from plant to plant, in the same plant between different organs, and even in the same organ at different times. The period also depends on various environmental conditions, particularly temperature, light/dark conditions, gravity including microgravity, mechanical stimuli like touch (especially in tendrils and roots), chemicals (such as ethylene, gibberellins, morphactins, lithium, aluminum, and auxins) etc. The period, however, returns to its initial value once the causative external stimuli/environmental conditions are removed.

There are two periods in CN in some plants, long (LPN) and short (SPN). For example, in the climbers like *Passiflora* and *Sicyos*, the tendrils exhibit two periods in CNs in a mutual ratio of 1:5 (LPN:SPN), one period when CN occurred in the same plane and the other period when CN occurred in a plane that was perpendicular to the first plane (Johnsson and Heathcote, 1973). Two CNs based on period also exist in *Zea mays cv.* Aussie Gold (Shabala and Newman, 1997a, 1997b) with a period respectively of 90 min and 7 min, in *Triticum* coleoptiles with 150 min and 70–80 min periods (Johnsson, 1979), and in *Arabidopsis* with periods respectively of 700 min and 30 min (Schuster and Engelmann, 1997). In *Phaseolus* epicotyles (Johnsson and Healthcote, 1973), there are SPNs (with low-amplitude oscillations) in 27 min at 15°C and in 12 min at 27°C.

Several internal and environmental factors are shown to affect the period of CN. In the same plant, period length increase can be correlated with amplitude increase, although, unlike amplitude, period may be dependent on the length of the circumnutating organ. For example, in the etiolated rice coleoptiles, it increases from 160 min to 200 min (Yoshihara and Iino, 2005), as etiolated ones elongate more than normal coleoptiles. The second possible factor that controls period is circadian rhythm. For instance, the period of CN in *Arabidopsis* inflorescence axis, under continuous white light, was shown to fluctuate in a circadian manner and that this was disturbed in the *toc1* and *elf3* mutants that are deficient in the functioning of internal clock mechanism (Niinuma et al., 2005). Someya et al. (2006) obtained contradictory results: *Arabidopsis* raised under continuous light (as above) did not show fluctuation in the period of CN.

The other question that remains unanswered is why some plants shows two CNs in terms of period (LPN and SPN). Although this has been correlated with H⁺ fluxes in decapitated maizem roots, pH changes, temperature variations (shorter when temperature increases), light color (slower in red light), plant age (longer with increasing age), presence of two separate oscillator mechanisms, etc., definite and convincing proofs have not yet been provided. It is also not clear from earlier literature whether LPNs and SPNs alternate in a continuous manner or whether LPNs follow SPNs or vice versa after some specific intervals of time or whether the two happen without any obvious pattern. Data on the last point raised is very vital to investigate the actual factor(s) involved in the production of LPNs and SPNs.

10.2.3 SHAPE

As already mentioned in this article, the shape of the distal view circumnutational trajectory of the growing organ is most frequently an ellipse; however, again as stated earlier, the shape may be circular, a line, or irregular zigzag (Figure 10.2). In most plants, ellipsoidal and/or circular shapes predominate. The same plant/organ may show various shapes at different times. In a few climbing plants and herbaceous taxa, the shape may be rosette like (see Baillaud 1962; Hejnowicz and Sievers, 1995). Like amplitude and period, we do not clearly know why the shape of the circumnutational trajectory of the growing (during CN) organ varies. It is likely that, if that part of the stem tip above the bending zone of CN or that part of the root tip below the bending

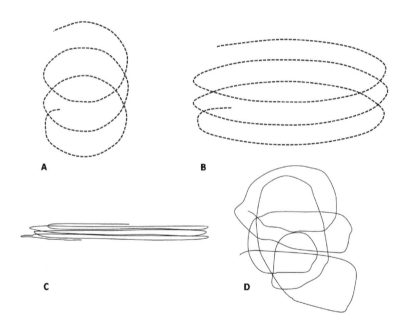

FIGURE 10.2 Diagrammatic representation of circumnutational shapes: (a) circular; (b) elliptical; (c) pendulum-like; and (d) irregular.

zone of CN remains without any longitudinal growth, the shape of the trajectory tends to be a line; an ellipse will be formed when this region shows some degree of longitudinal growth (during CN), and a circle will be formed when this region shows a greater degree of longitudinal growth, while the bending zone remains the same during this growth.

10.2.4 DIRECTION

The direction in which growing plant organs circumnutate may be clockwise or counterclockwise (i.e., right-handed or left-handed) depending on the plant. For example, in *P. vulgaris* (Millet et al., 1988; Millet and Badot, 1996), *Glycine* (Adolfson et al., 1998), the movement is left-handed. In adzuki bean epicotyls, the nutation is clockwise (Iida et al., 2018). In *Arabidopsis*, SPNs are usually counterclockwise while the LPNs are clockwise (Schuster and Engelmann, 1997). In the same plant the direction may change either spontaneously or on specific inductions, such as gravitational or thigmotropic (Johnsson, 1997; Okada and Shimura, 1990). That a directional change in CN has taken place is often indicated by a figure of eight traced by the growing organ. In space-grown sunflower plants, the directional changes are most frequent (under microgravity conditions) than in earth (Brown et al. 1990), which may be accompanied by changes in amplitude and frequency of CNs (Johnsson, 1997). In rice coleoptiles, the ratio of counterclockwise to clockwise CN was approximately 3:1 (Yoshihara and Iino, 2005).

In the shoots the movements are irregular (Orbovic and Poff, 1997) and both right- and left-handed. In roots, by contrast, at least in *Arabidopsis* ecotypes, the movement is helical and right-handed (as viewed from shoot apex) (Simmons et al., 1995); however, Linnaeus and others (Hashimoto, 2002) considered the above movement to be left-handed (pictured from its interior). In wild-type *Arabidopsis*, root movements are not random at all but rather show a clear right-handedness (Okada and Shimura, 1990), that is, they appear to be animated by a process that could be named chiral CN. Thus, the waving/coiling phenomenon is likely to be governed by CN patterns.

By using wild-type ecotypes and different gravitropic mutants (auxin transport mutants such as *aux1*, *eir1*; auxin physiology mutants such as *axr1*; handedness mutants such as *1-6c*), Piconese et al. (2003) observed that wild-type *Arabidopsis* roots made large movements by CN only to the right-hand but auxinic mutants showed a lack of regular chiral CN.

A question that is naturally asked is, why change in direction of CN? Theoretically, one can better understand the presence of left and right directions of CN in different plants than a change in CN in the same plant, but it is very difficult to understand the reason for both types of occurrences (Iida et al., 2018). Smyth (2016) explained that the handedness of some types of helical growth is quite variable and that its direction might reflect on an early developmental asymmetry in the shoot apical meristem. We do not know anything about how early developmental asymmetry happens in the shoot apical meristem, or even if it exists, its mechanism in controlling CN is not clear.

10.2.5 DIURNAL, CIRCADIAN, AND INFRADIAN VARIABILITY OF PARAMETERS

In spite of great variations noticed in the four parameters of CN, there is noticed a certain degree of regularity in its diurnal, circadian infradian and ultradian rhythms. In the sunflower, the amplitude of CN exhibits diurnal (24 h) and infradian (many days long) variations in its length (Buda et al., 2003; Charzewska, 2006). Also, in this plant, changes in trajectory length, period, and shape of CN show a circadian rhythm (close to 24 h) (Charzewska and Zawaszki, 2006). The hypocotyls of the seedling of *Arabidopsis* also show a circadian character of CN (Schuster and Engelmann, 1997). In fact, these seedlings showed a very wide range of rhythms. Niinuma et al. (2005) did experiments with two loss-of-function mutants of *Arabidopsis* TOC1 (mutant that shortens the period for all circadian processes analysed to date) and ELF3 (mutant that causes arrhythmic circadian outputs under constant white light conditions, with an almost constant mutation speed) demonstrated genetically that the circadian clock controls CN speed. These results strongly confirm the hypothesis that rhythmical membrane transport processes play a key role in plant CN, showing a genetic-based control. However, the basic rhythm of CN is ultradian involving 2–3 hr period for one nutational trajectory.

A question arises whether plants are able to sense very weak (i.e., around 10^{-8} g) gravity fluctuations, often called plumb-line variations, which cause tides (Charzewska et al., 2010). These authors applied the model wavelength coherences and spectio-temporal coherences to detect common oscillations in sunflower CNs and variations of the direction of the gravity vector. They found that the mean coherence between time series of sunflower coordinates and the theoretical plumb-line model was in the order of 0.6–0.8 for oscillations with periods greater than two days. They also detected weak common oscillations with periods of 0.5 days and 1.0 days, which may correspond to semidiurnal and diurnal tides, respectively. When they computed the common signal in the time series of sunflowers used in the experiment, they noticed a low, but statistically significant coherence for semidiurnal and diurnal oscillations. They also showed that there is a significant correlation between CN and plumb-line variations. A very recent study by Stolarz and Dziubińska (2017) in the sunflower shows that there is an ultradian spontaneous action potential (SAP) rhythm besides the regular (CNs) and that the number of SAPs and CN trajectory changes depends on light conditions. In continuous light, no SAPs under vigorous regular CN are observed. In continuous very low light conditions, the irregularity of the CN trajectory is accompanied by SAPs. These authors' results show that SAP appearance and CN changes play a role in adaptation to very low light conditions in sunflower plants.

10.3 MODELS OF CIRCUMNUTATION

Ever since CN became an important topic for discussion from the second half of the 19th century, four models have been proposed to explain the mechanism that generates circumnutational movements. These three mechanisms have been reportedly discussed by many researchers (Johnsson, 1979, 1997; Brown et al., 1990; Johnsson et al., 1999; Brown, 1993; Mugnai et al., 2007; Stolarz, 2009).

10.3.1 INTERNAL OSCILLATOR MODEL

This model was first proposed by the Darwin (Darwin and Darwin, 1880), according to whom an internal oscillator mediates circumnutational movements. This model further emphasized that CN was not only universal but also a fundamental process that would "be modified for the good of the plant" to accomplish other types of growth movements and responses. However, the internal oscillator of the Darwins is more a concept than a model (Thain et al., 2002). In this model, earth's gravity does not play a crucial and decisive role in inducing circumnutational movements. This means that the apparatus that drives and regulates CN is exclusively an internal oscillator. Space lab studies on CN in the sunflower have renewed the fact that an internal oscillator is indispensable for generating CN.

10.3.2 EXTERNAL "GRAVITROPIC OVERSHOOT" MODEL

Although first proposed by Baranetzky in 1880 (Baillaud, 1962), this model was later revived by Israelsson and Johnsson (1967) (see also Johnsson and Healthcote, 1973; Brown, 1991; Johnsson et al., 1999; Mugnai et al., 2015; Kim et al., 2016). According to this model, CN is dependent on gravity, and the oscillations are driven and controlled significantly by gravity. This model also emphasizes that CN is a very special type of gravitropic behavior. The key features of this model are the constant ratio of the period of oscillation to the gravitropic lag phase that equals 4, and the overshoot phenomenon that facilitates constant gravitropic response (see Stolarz, 2009). This model is also consistent with the modern version of the Cholodny-Went theory for gravitropic responses, according to which both statocytes (gravity detectors of plants) and IAA-producing sites (see further discussion on IAA's role later in this chapter). The gravity-sensing cells in shoots are stated to be the so-called endodermal cells. However, typical endodermal cells, with their characteristic casparian wall thickenings are absent in stems and what are referred to here are the innermost layer of cortical cells that have starch contents; in fact, in many plants, even such a specialized starch-containing cell layer is totally absent. These cell layers are reported to be located in the elongating zone of the primary internodes; this region is claimed to also be the gravitropic response zone (Kim et al., 2016). In roots, the gravity-sensing cells (i.e., statocytes) are located in the root cap while gravitropic response occurs in the root elongation zone (Spurný 1966, 1968, 1974; Kim et al., 2016), and thus, the two regions are spatially separated (in roots of all plants a typical endodermal layer is present but neither does it have starch grains nor a gravisensing or graviresponding role). Wherever, starch-containing statocytes are present, an asymmetric growth and bending of the organ with positional changes in statioliths are resulted, which, in turn, stimulate a new gravitropic response to cause a repetition of the CN cycle (Johnsson and Heathcote, 1973).

In order to prove the role of gravitropism in CN, studies were undertaken under microgravity conditions in space laboratories and klinostat and centrifuge-based experiments on Earth. These experiments revealed that changes in gravity affected all the parameters' CN mentioned earlier (Israelsson and Johnsson, 1967; Brown et al., 1990; Zachariassen et al., 1987). It has been reported, for example, that sunflower hypocotyls circumnated 40% of the time in space, while on Earth it occurred 100% of the time; in the ground, with klinostats, nutations averaged 21% of the time (i.e., almost half of what was observed in space). That is, the klinostat does not perfectly simulate the μ gravity conditions. The fact that space-lab experiments, although have "microgravity" conditions, did not altogether stop CN is good evidence for the importance of gravity in CN.

The studies of Kitazawa et al. (2005) showed that in *Pharbitis nil* "endodermal cells" with starch were involved in CN indirectly through their control over gravitropism. They identified a gene in this plant, *PnSCR*, that regulates these cells in graviperception. To find out whether "endodermis"-mediated gravisensing is the sole prerequisite for CN, these authors studied CN in the *wel* mutant of this plant that has both abnormal gravitropism and defective CN, which are attributed to a loss of *PnSCR* function in this mutant. Two agetropic mutants of *Arabidopsis*, *sgr2*, and *zig/sgr4* (Kato et al., 2002) and the *scr* mutant (Fukaki et al., 1998) also have abnormal starch sedimentation/defective "endodermal" layers and suggested that these mutants were also defective in CN. Earlier, another *Arabidopsis* mutant, *pgm*, showed reduced gravitropism due to loss of starch grains (or with extremely small grains) and had smaller oscillations than in wild types. The agravitropic *ageotropum* mutant pea roots grew in random directions, exhibited irregular movements, and did not show clear evidence of CN (Kim et al., 2016). These studies are considered as clearly demonstrating that gravitropism controls are essential for root CN.

10.3.3 Two-Oscillator Model

Johnsson et al. (1999) combined the above two models to propose a two-oscillator model, also called the "mediating model," to explain CN. As per this model, CN involves a gravitropic reaction that acts as an externally controlled feedback oscillator along with an internal oscillator that provides a rhythmic signal to the feedback oscillator (see Hejnowicz and Sievers, 1995). The internal oscillator seems to operate independently of the external, that is, gravitropism (Johnsson, 1997). The inner oscillator seems to play a key role, while gravity seems to merely moderate CN. Some researchers feel that the problem still remains and that there has been no direct evidence yet for the involvement of gravitropism as an external oscillator in CN. The so-far proposed evidence for this is still rather controversial. In this connection the work of Yoshihara and Iino (2005) on the dark-grown rice coleoptiles, on the one hand, supported the existence of a close relationship between gravitropism and CN, and on the other mentioned cases in which gravitropism and CN could be separated. Regarding the first point, they showed that CN was interrupted by a gravitropic response and reinitiated at a definable phase after gravitropic curvature. Regarding the second point, they reported that the non-circumnutating *lazy* mutant rice coleoptiles showed almost the same levels of gravitropic responses in their upper half as in wild coleoptiles, where the same upper half is responsible for both CN and gravitropism. Hence, these two authors suggested that gravitropism influenced, but was not directly involved in, CN; they also observed that a gravity signal contributed to the maintenance of CN.

Recently, *Arabidopsis* root movements were interpreted as combined effects of essentially three processes: CN, gravitropism, and negative thigmotropism (Migliaccio and Piconese, 2001), although there was some difficulty in discriminating between these in time and space. Piconese et al. (2003) using a random positioning machine, which provides approximate space lab conditions, showed that the observed root pattern depended only on the circumnutating movement, since both gravitropism and negative thigmotropism has been excluded. They have further shown, based on mutant *Arabidopsis* roots, that gravitropism and CN seem to have a common basis at the level of signal transduction, thus lending support to the two-oscillator model. Ney and Pilet (1981) concluded that CN and gravitropism had a common basis, because when the roots were responding to gravitropism, they stopped circumnutating and then resumed the movement at the end of gravitropic response.

What is the actual relationship between CN and gravitropism? Why have people proposed a two-oscillator model to explain CN? While trying to find an answer to these two questions, it is important to keep in mind that, in the 1-g conditions found on Earth, once a phototropic curvature is initiated in a shoot (or root), induction of gravitropism is an inevitable consequence of the change in the orientation of the curving organ (Hubert and Funke, 1937; Okada and Shimura, 1992; Mullen et al., 2000; see also Millar et al., 2010). In such a situation, phototropism causes a competing gravitropic response, limiting the extent of curvature. We feel that this is true of the CN phenomenon also, where the growing plant organ curves during its helical growth from its upright position (i.e., from its plumb line) at various angles depending on the plant and depending on other prevailing conditions. Once this curvature is affected, as in phototropic curvature, there is an inevitable induction of gravitropism. Consequently, the underside of the bent portion of curved organ is subjected to compressive forces while its upper portion to tensional forces. This is similar to the events preceding reaction wood formation [compression wood (CW) formation in conifers and tension wood (TW) formation in dicots] in leaning main stems and branches (Hariharan and Krishnamurthy, 1995; Karthikeyan, 2000; Krishnamurthy, 2007). The formation of these woods eventually helps the stem to become upright through the shrinking of TW fibers (gelatinous fibers or G-fibres) in dicots and the elongation of CW tracheids in conifers. In this connection, the work done by Gerbode et al. (2012) on circumnutating wild cucumber tendrils is to be mentioned. They found G-fibers as a bilayer of cells closer to the inner, concave edge of the tendril coil. These fibers undergo differential contraction, help in tendril coiling, and have a perversion point (see also Bowling and Vaughn, 2009).

10.3.4 GROWTH-AFFECTED SYMPLASTIC MODEL

Growth-affected symplastic model was proposed in 1991 by Brown (1993). It is based on "local control" of tropistic growth instead of the Cholodny-Went theory. According to this author, it is not incompatible with the gravity-dependent model, but it has a very different way of explaining nutational oscillations. Two observations are stated to support this model: (1) according to the gravity-dependant model, CN would be expected to be more responsive to increased *g*-force (denotes here as *g*), but it is not greatly increased (in a range of 1–20 g). It appeared that above 1 g CN was not very responsive to an increase in g; (2) in simulation tests, the vigour of CN was lowest at simulated 0 g. It increased very rapidly as the simulated level was raised gradually to a few tenths of a g. then it began to level off toward 1 g, after which it increased only very slowly up to about 15 or 10 g. Taken alone, these results would appear to support the gravity-dependent model, but when considered together with the results of space-lab experiments, where some sunflower CNs were initiated and continued under µg, they would show that gravity is not the exclusive explanation for CN.

The vigor of oscillations observed in true µ gravity conditions was shown to be intermediate between that observed at 1 g and that observed at clinostat stimulated 0 g. These results show that gravity does influence both the amplitude and period of CN, but that it is not a mandatory requirement for CN. Accordingly, Brown (1993) argued that accelerating forces enhance the oscillation whether on Earth, on a centrifuge, or in µ gravity conditions, but even without a significant g-force, CNs proceeded. The gravity-dependent model that explained how gravity could influence circumnutational parameters was not thereby disproved, but it was shown to be only a partial explanation for CN because it can be internally driven, without any help from gravity.

Brown (1993) went on to emphasize the following aspects related to CN: (1) consistent with the local-control theory, transport of growth regulators is confined to the symplast and not to apoplast; (2) during cell elongation many existing plasmodesmata are functionally disrupted and new ones are formed in the newly formed walls; this repair took 10 minutes to several hours; (3) if growth just happens to be exactly equal on the flanks of the elongating organ, the pattern of disruption and restoration of growth-controlling plasmodesmata will be radially symmetrical and growth should be exactly straight. However, as a result of a small asymmetry, the organ bends, peripheral tissues stretch, and the concave side of the bent organ gets compressed; (4) compression of the concave side stretching of the convex side (or both these processes) may introduce a functional asymmetry in symplastic transport exactly in the zone of most rapid organ growth. If, by chance, some asymmetry develops, the organ will bend, which will enhance the transport asymmetry and initiate a series of self-sustaining oscillations; and (5) in the relatively quiet, weightless space laboratory, CN gets started but stops capriciously, and even a slight bending of the growing organ could more easily set of a sequence of oscillations in the 1-g "control" plants than in weightless plants. Brown (1993) also emphasized that none of the above observations is incompatible with the growth-affected symplastic communication model.

10.4   CELLULAR BASIS FOR CIRCUMNUTATION

The process underlying CN is based primarily on unequal rates of growth on different sides of an organ; the side with the most rapid growth pushes the apex over bending it toward the least active side. The active area then proceeds around the apex, typically along a helical path (right or left). Thus, CN is the result of helical growth (Brown, 1993) and reversible volume changes occurring in the cells of the growing organ (i.e., in the bending zone below the apex) (Caré et al., 1998). These variations are believed to be due to difference in water content between the cells of the "convex and concave sides of the bending zone," associated with turgor and ion concentration differences between opposite sides of the growing organ (Lubkin, 1994). Possibly a turgor wave rotating around the growing organ during CN drives helical growth. The helical growth is hypothesized to be a mechanism that increases the stability of the hypocotyls (Schuster and Engelmann, 1997) or similar longitudinally growing organs during cell wall loosening (due to the so-called acidic growth) (Cosgrove, 2000) accompanying elongation. As a consequence, oscillatory growth and movement is generated (Van den Driessche, 2000).

Cells in the circumnutating and growing organ must operate in a coordinated and synchronized manner to bring about the dual processes of CN and growth in length. To know about it one must first identify the actual cells that are involved in CN. Epidermal cells are reported to release protons into the cell-external matrix (Hejnowicz and Sievers, 1995; Gruszecki et al., 2002) indicating the possibility of an acid-growth mechanism (mentioned in the last paragraph) for cell growth operating in cell wall plasticity and flexibility (see also Millet et al., 1988). It is the epidermal cells that show reversible length changes during bending in the circumnutating *Phaseolus* (Caré et al., 1998). As already indicated, the gravisensing, "endodermal" (starch-containing cells) cells are reported to play an important role in CN in *Arabidopsis* and *Pharbitis nil*, through an analysis of their mutants (Kitazawa et al., 2005; Hatakeda et al., 2003). Plasmodesmal connections are also shown to be important as indicated in Brown's (1993) model of CN (as already indicated); also important are ion channels (Badot et al., 1990) and aquaporins (Comparot et al., 2000). These are a strong correction between CN and rhythmical patterns of ion fluxes in the elongation region of corn roots (Shabala and Newman, 1997a, and 1997b; Shabala 2003). The authors noted that, when maize roots showed rhythmical movements, H^+ and Ca^{2+} fluxes also change rhythmically, with the same average period and amplitude; when root movement was periodic, so were ion fluxes. Moreover, when root growth was absent or very slow, no oscillations in ion fluxes occurred, and no CN was observed. Shabala (2003) started that correlations between ion flux oscillations and root CN could also be extended to include K^+. Since K^+ is a major osmotic agent in plants and since it is a main factor responsible for differential growth of root cells, an efflux of K^+ from cells results in a loss of turgor within the cell and a consequent "slumping" of the cell. The non-turgid cells cause asymmetric rigidity in the root, which consequently bends to the side that has cells with less turgor (Shabala and Knowless, 2002). This was further supported by the presence of K^+ flux oscillations that are closely associated with root CN (Shabala, 2003). When measured from opposite sides of a vertically growing root, these fluxes are found in reverse phases. Turgor changes are also implicated by Millet et al. (1988), Millet and Badot (1996), and Zachariassen et al. (1987) in CN.

The sequence of cellular events during CNs, as per Stolarz (2009), can be summarized as follows: H^+ *ATPase* continuously acts and hyperpolarizes the cell membrane, which results in opening-inward, voltage-gated K^+ channels; then Cl^- ions are transported inwardly through a co-transport with H^+ ions. In influx of K^+ and Cl^- ions then causes an increase in the osmotic potential, which facilitates water influx into the cell, for instance through aquaporins. The inward water movement results in cell enlargement and cell membrane stretching, which may cause an opening of mechanosensitive Ca^{2+} channels and the resultant Ca^{2+} ion influx. This leads to cell membrane depolarization, which may cause the opening of outward K^+ ion channels that lowers the osmotic potential inside the cell and induces water reflux. The increase in the cell volume would then be inhibited. By releasing H^+ ions into the cell wall, the proton pump acidifies the environment, thus activating expansins, which trigger off "crawling" of the cell wall. Moreover, the positive charge outside the cell membrane enhances cellulose synthase activity, thus promoting cell wall growth. Increased turgor, in turn, facilitates the exocytosis processes, namely, the enlargement of plasma membrane surface and externalization of hemicelluloses. When cell turgor is the maximum Ca^{2+} ion influx or may activate a number of Ca^{2+}-dependent protein kinases, it may also stimulate the activity of cytoskeleton elements, which may contribute to the perpendicular arrangement of cellulose fibrils against the cell long axis as in the case of fast-growing cells. It may also be essential in the polar auxin transport. The effect of gravity on CN may proceed through a system of graviperception based both on the statolith–cytoskeleton model of Sievers et al. (1996) and on the tension- and pressure-sensitive cell membrane-extracellular matrix and model proposed by Wayne and Staves (1996). Both models take into account the possibility of opening of ion channels induced by directional changes related to gravity changes. The hypothetical electrophysiological model of CN by Stolarz (2009) (see also Stolarz and Dziubińska, 2017) is an attempt to include a larger number of factors operating in CN. In a study on the sunflower (Stolarz et al., 2010), a glutamate solution (200, 50, 5 mm) was injected into the plants, and the electrical potential changes were then measured, simultaneously with the measurement of CN, through the use of time-lapse images. The injection provoked a series of action potentials (APs) at the site of the injection as well as in different parts of the plant. This injection also resulted in a transient, approximately five-hour-long decrease in the stem CN rate offering further support to electrophysiological model.

Auxin has to be involved in CN, as it is vital for differential growth (Brown 1991, 1993; Johnsson, 1997; Kim et al., 2016). This is supported by the fact that strong circumnutational movements seen in shoots are completely blocked by auxin-transport inhibitors like NPA (Hatakeda et al., 2003; Britz and Galston, 1983; Kim et al., 2016) and TIBA (Kim et al., 2016) and restored by application of exogenous IAA (Britz and Galston, 1983). The necessity of auxin in gravitropism has been proved for both shoot, and root and in the latter, it is mediated by PIN genes (Perrin et al., 2005; Friml et al., 2002; Kleine-Vetin et al., 2010). Iida et al. (2018) demonstrated the asymmetric distribution of both IAA and gibberellinsA1 in the bending zone of the circumnutating epicotyls of *Vigna angularis*.

10.5 CIRCUMNUTATION IN WOODY ADULT TREES

Douglass (1940) published three photographs of transections of conifer tree logs that showed spiral compression (instead of regular CW only on the lower side) woods; the spiral direction was clockwise (Figure 10.3) in one and anti-clockwise in the other two. He, however, did not explain the importance of these spiral CWs.

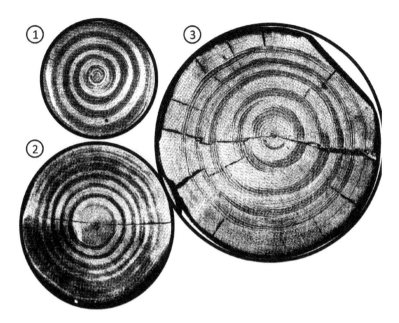

FIGURE 10.3 Transections of woods showing spiral compression wood 1 and 2, *Picea abies*; and 3 *Picea* sp. (After Douglass, A.E., *Tree-Ring Bulletin*, 6, 21–22, 1940. Reproduced with the kind permission of Tree-Ring Society.)

Nearly 50 years later, Telewski (1988) demonstrated for the first time that CN also occurs in the main stems of woody adult plants throughout their life and that it is not restricted to herbaceous taxa or seedlings of woody plants. He found that in two members of Coniferae, *Pinus taeda* and *Abies concolor*, CW was formed in a spiral manner in the main trunk and this, he states, represents an annual record of a slower, circumnutational movement exhibited by the main stem itself. Wilson and Archer (1977) recorded patches of the characteristic TW fibers (called G-fibers) amongst the normal wood regions in the main stems of some dicots, but they did not relate them to CN; they explained this by saying that these fibers are not unique to TW. Fisher and Stevenson (1981) stated that the presence of TW in vertical stems was presumably due to internal growth stress or loading stress because of crown imbalance. A similar opinion had already been expressed by Kaeiser (1955) and Trenard and Gueneau (1975). These authors could have examined already circumnutated vertical stems that might have gotten corrected from leaning positions through the stretching of G-fibers.

Karthikeyan (2000) studied five tree taxa (*Hardwickia binata, Acacia holosericea, Leucaena leucocephala, Pongamia pinnata, and Eucalyptus tereticornis*) and recorded TW and its G-fibers, which are definite products of gravitational effects, not only on the upper side (away from ground) but also on the lower side (facing ground) of leader branches throughout their length. This, he stated, is related to continuous circumnutational growth of the leader branches throughout their life

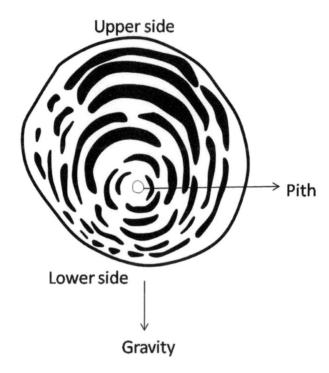

FIGURE 10.4 Transections of central region of the oldest branch of Leucaena leucocephala. Note the discontinuous anticlockwise oriented spirals of tension wood regions (shaded). (After Karthikeyan, A.V.P., Tension wood and maintenance of branch angle and crown architecture in some trees, PhD thesis, Bharathidasan University, Tiruchirappalli, India, 2000.)

from the time of their origin. As a result, the physically upper side and the physically lower side facing the ground periodically changes. He has observed that the oldest leader branch in *H. binata* appears to have changed its physically lower side to the upper side at least 12 times, *A. holosericea* 14 times, *L. lencocephala* 16 times, *P. pinnata* 10 times, and *E. tereticornis* 11 times. Although perfectly spiralled TW is not exhibited by any taxon studied by Karthikeyan (2000), there is an overall spiral disposition of patches of it from the youngest to the oldest growth rings (Figure 10.4). This spiral is anti-clockwise in contrast to the clockwise spiral reported by Telewski (1988). In light of the above, Karthikeyan (2000) also stated that spiral grains in a wood may also be due to circumnutational growth (see Chapter 4 for more information on spiral grains).

10.6 IMPORTANCE OF CIRCUMNUTATION TO THE PLANT

In view of the discussion in the earlier sections of this chapter, the following questions can naturally be raised: what is the importance of CN? Is it true, as claimed by some investigators, that CN is patently advantageous only in a small majority of cases? What is the actual relationship between CN and longitudinal growth? Is there something fundamental about the growth process that endows growing plant organs with the ability to circumnutate, an ability commonly displayed? Or as some researchers consider, is it an oddity of plant growth or, as yet others regard it, is an outward manifestation of some important internal processes that are involved or associated with elongation growth of plant organs? Is it correct that some researchers are not inclined to consider that CN has endured only because it confers some evolutionarily significant advantages to the plant?

CN seems to be very vital for longitudinal growth of plant organs, such as the stem, root, leaf, and tendril. The major evidence for this is that whatever factor that interferes with longitudinal growth inhibits or considerably reduces CN, for example, when tissues derived from longitudinal growth mature or when the elongation of the organ ceases CN, as in the case of a maturing leaf. Mugnai et al. (2007), however, have stated that although these movements are of obvious use to twining plants that seek mechanical support, in other cases they appear to have no useful purpose. A study of CN in relation to growth rate was studied in shoots of *Periploca graeca* (Johnsson, 1979). Here, CN happened above the threshold value of growth at approximately 0.5 mm per hour. In *Arabidopsis* seedlings showing short-period CN, the required minimal growth was 0.05 mm/hr but longer periods of CN occur when the growth is slow, but for this plant this is a general observation. However, there is no strict, proportional dependence between changes in growth rate and changes in the CN period (Johnsson, 1997). The circadian character of CN in *Arabidopsis* hypocotyls is strictly related to the circadian hypocotyl growth. No CN occurred at the low growth rate. Berg and Peacock (1992) also showed dependence between the growth zone and bending zone (during CN) in sunflower hypocotyls. Stolarz et al. (2008) showed, in sunflower, that any disturbance in the growth rate is accompanied by changes in the parameters of CN, but that this is not a simple quantitative relation between the rates of growth and CN.

There are also reports of the possibilities of separation of growth from CNs. Applications of lithium chloride in *Phaseolus* did not stop growth, but CN was not observed (Millet and Badot 1996). Supporting evidence have come from the work of Johnsson (1979) and Tepper and Yang (1996). However, this claim needs to be further investigated in view of the research done by Mulkey (2007).

There are only occasional suggestions in earlier literature that CN may aid underground organs like roots and rhizomes in soil penetration (Fisher, 1964). This idea has gained experimental confirmation in a study of rice varieties (Inoue et al., 1999) through a spectrum analysis of root-tip rotations of various varieties of rice seedlings. Seedlings with a higher CN frequency of 2.0–3.4 cycles per day had the greatest penetration and seedling establishment. Del Dottore et al. (2016, 2017), based on hypothesizing that CN movements could help roots in penetrating soil, had validated their hypothesis by testing an artificial probe at three distinct soil densities and using various combination of CN amplitudes and periods for each soil. They did this comparison basing on the total work done by the system while circumnutating at its tip level with respect to that shown by the same system in straight penetration. They found that the total energy evaluation confirmed an improvement obtained by CN up to 33%. They further showed that there is a trade-off among penetration velocity, CN period, and amplitude toward an energy consumption optimization, expressed by the leading angle of the helical path that should stay in the range between 46° and 65°. They also found that CNs with appropriate amplitude (~10°) and period (~80s) value were more efficient than straight penetration at different dimensions of the root tip, up to a threshold diameter of 2 mm to 55 mm. Based on their results, they suggested that CNs could represent a strategy used by plant roots to reduce the pressure and energy needed to penetrate soil.

In a PhD thesis work of the second author guided by the senior author of this paper (Karthikeyan, 2000), it was suggested that in a growing aerial plant organ like stem tip or leaf, CN would help the organ to spend less energy to overcome the air resistance in its growth through it than when the tip grows straight through it. We argued that a spiral growth would provide a mechanical advantage rather than straight growth. We used the analogy that hitting a nail through wood would require more applied pressure and energy than screwing the same, as the latter takes less force expended over a greater period of time. The amount of air resistance that a growing aerial plant organ would experience (viscous drag) depends on its speed of growth, its cross-sectional area, its shape, its mass, and above all the density of air. It is well-known that air densities vary with altitude, temperature, humidity, wind velocity, pollution, etc., although a mean volume of 1.29 kg/m^3 is a very reasonable value. The parameters of CNs like amplitude, period, shape, and even direction all will then depend on this important factor of air resistance. Hence, we suggest that CN is an important adaptation of a stationary plant's growing organs to facilitate their easier longitudinal growth while confronting soil resistance in the case of roots and air resistance in the case of aerial plant organs like stems, leaves, tendrils, etc. We also suggest that CN is a mechanism by which an organ's longitudinal growth is facilitated with a reduced expenditure of energy and with less exertion of pressure. We further suggest that CN is patently advantageous not merely for a small majority of cases but for all plants with longitudinal growth. CN definitely confers some evolutionarily significant advantages to the plant to facilitate its apical meristem to retain its potential for longitudinal growth till the death of the plant, particularly in view of the plant's evolution as a stationary organism. CN is definitely not an oddity of plant growth, as some researchers opine, but it is definitely an outward manifestation inevitable of some important internal processes that are involved or associated with elongation growth of plant organs. We must also take into consideration the relationship between CN on the one hand, but on the other hand, episodic or continuous longitudinal growth of growing plant organ duration of episodic growth is also important.

10.7   GENETICS OF CIRCUMNUTATION

The first attempt to study CN from a genetic perspective was undertaken by Okada and Shimura (1990). They identified the genes primarily involved in circumnutational growth using roots of *A. thaliana* as the experimental plant organ. They found six *WAVY GROWTH* genes, which on mutation showed disrupted patterns. Although these two authors first attributed the waves to a cyclic response to thigmotropism, subsequently Simmons et al. (1995) showed that the waves represented a flattened helix, which was the result of interactions between helical CN of the root tip and the impenetrable surface of the plate on which the plant roots were growing. Using this experimental approach, Furutani et al. (2000) identified *SPIRAL1* (also known as *SKU6*) and *SPIRAL2* genes from the mutant root of *Arabidopsis*, which slanted strongly in the right-handed (instead of left-handed) directions. The *spiral1* mutant gene probably interfered with CN through the imposition of an abnormal helical growth as a result of which the straight cell files showed a consistent right-handed twist. This changed cellular pattern even extended to the hypocotyls and stem and in *spiral2* mutants, even into the leaf petioles and petals. These genes probably acted through their effect on microtubules, as there are many evidences (although indirect) that exist for this surmise.

Subsequently, as many as 32 root-slanting mutations have been recorded (Ishida et al., 2007); this study showed that point mutations in tubulins could result in left- or right-handed helical phenotypes. Several members of the tubulin gene family (5 of the 6 α-tubulins and 4 of the 9 β-tubulins of *Arabidopsis*) were involved in right- and left-hand mutants present in each subfamily.

A third gene called *GCP2*, with a right-handed helical growth mutant *spiral3*, codes one component of the γ-tubulin-containing complex that is required for microtubule (MT) nucleation (Nakamura and Hashimoto, 2009).

10.8   CONCLUSIONS

Based on the above account, the following conclusions can be made: (1) CN is very vital for any plant organ that undergoes longitudinal growth, and hence CN and longitudinal growth are interrelated. So, CN occurs in all plants with longitudinal growth; (2) circumnutating growing tips will spend less energy and less applied pressure than straight-growing tips, as the tips experience resistance from air (in the case of aerial parts), soil (in the case of underground parts), and water (in the case of aquatic plants); (3) CN involves bending away from the plumb line of the growing tip, and this aspect of CN is probably controlled by an internal oscillator; this bending involves instantaneous changes in the cell wall, cell membrane ion fluxes, turgor, growth regulators, etc; (4) CNs definitely have a geotropic component, which operate when the axis of the growing organ bends away from its plumb line. This change produces a compressive force on the underside (facing the ground) of the bent tip and a tensional force on the upper side (away from the ground). These forces will then help the growing tip to reach the plumb line again and are facilitated by tissue elements such as G-fibers; (5) variations in the various parameters of CN (such as amplitude, period, shape, direction, and response to circadian and other changes) reported are probably related to changes in the environmental conditions within and outside the plant operating at the time of CN and/ or longitudinal growth. However, it is difficult at present to explain the reports of change in the direction of CN (left, right direction); (6) there are evidences (although indirect) that indicate that for CN is also seen in woody adult trees. This aspect needs to be verified by more research work; and (7) genes are involved in controlling CN, and mention has been made on some of these genes.

REFERENCES

Adolfson, K.A., Sothern, R.B. and Koukkari, W.L., 1998. Ultradian movements of shoots of two species of soybeans Glycine soja (Sieb. and Zucc.) and Glycine max (L.) Merr. *Chronobiology International*, 15(1):1–11. doi:10.3109/07420529808998664.

Anderson-Bernadas, C., Cornelissen, G., Turner, C.M. and Koukkari, W.L., 1997. Rhythmic nature of thigmomorphogenesis and thermal stress of *Phaseolus vulgaris* L. shoots. *Journal of Plant Physiology*, 151(5):575–580. doi:10.1016/S0176-1617(97)80233-7.

Badot, P.M., Melin, D. and Garrec, J.P., 1990. Circumnutation in *Phaseolus vulgaris*. II. Potassium content in the free-moving part of the shoot. *Plant Physiology and Biochemistry (Paris)*, 28(1):123–130.

Baillaud, L., 1962. Mouvements autonomes des tiges, vrilles et autres organes à l'exception des organes volubiles et des feuilles. In *Physiology of Movements/ Physiologie der Bewegungen*. Springer, Berlin, Germany, pp. 562–634. doi:10.1007/ 978-3-642-94852-7_17.

Berg, A.R. and Peacock, K., 1992. Growth patterns in nutating and nonnutating sunflower (*Helianthus annuus*) hypocotyls. *American Journal of Botany*, 77–85. http://www.jstor. org/stable/2445200.

Bowling, A.J. and Vaughn, K.C., 2009. Gelatinous fibers are widespread in coiling tendrils and twining vines. *American Journal of Botany*, 96(4):719–727. doi:10.3732/ajb.0800373.

Britz, S.J. and Galston, A.W., 1983. Physiology of movements in the stems of seedling Pisum sativum L. cv Alaska: III. Phototropism in relation to gravitropism, nutation, and growth. *Plant Physiology*, 71(2):313–318. doi:10.1104/pp.71.2.313.

Brown, A.H., 1991. Gravity perception and circumnutation in plants. In *Advances in Space Biology and Medicine*. Elsevier, Vol. 1, pp. 129–153. doi:10.1016/S1569-2574(08)60123-9.

Brown, A.H., 1993. Circumnutations: From Darwin to space flights. *Plant Physiology*, 101(2):345.

Brown, A.H., Chapman, D.K., Lewis, R.F. and Venditti, A.L., 1990. Circumnutations of sunflower hypocotyls in satellite orbit. *Plant Physiology*, 94(1):233–238.

Buda, A., Zawadzki, T., Krupa, M., Stolarz, M., and Okulski, W. 2003. Daily and Infradian rhythms of circumnutation intensity in Helianthus annuus. *Physiol. Plant.* 119: 582–589. doi: 10.1046/j.1399.3054.2003.00198.x

Caré, A.F., Nefed'ev, L., Bonnet, B., Millet, B. and Badot, P.M., 1998. Cell elongation and revolving movement in *Phaseolus vulgaris* L. twining shoots. *Plant and Cell Physiology*, 39(9):914–921. doi:10.1093/oxfordjournals.pcp.a029454.

Charzewska, A., 2006. *Floriculture, Ornamental and Plant Biotechnology: Advances and Topical Issues*. The rhythms of circumnutation in higher plants. Global Science Books, UK, Vol. 1, pp. 268–275.

Charzewska, A. and Zawadzki, T., 2006. Circadian modulation of circumnutation length, period, and shape in *Helianthus annuus. Journal of Plant Growth Regulation*, 25(4):324–331. doi:10.1007/s00344-006-0042-5.

Charzewska, A., Kosek, W. and Zawadzki, T., 2010. Does circumnutation follow tidal plumb line variations? *Biological Rhythm Research*, 41(6):449–455. doi:10.1080/ 09291010903411468.

Comparot, S., Morillon, R. and Badot, P.M., 2000. Water permeability and revolving movement in *Phaseolus vulgaris* L. twining shoots. *Plant and Cell Physiology*, 41(1):114–118. doi:10.1093/pcp/41.1.114.

Cosgrove, D.J., 2000. Expansive growth of plant cell walls. *Plant Physiology and Biochemistry*, 38(1–2):109–124. doi:10.1016/S0981-9428(00)00164-9.

Darwin, C. and Darwin, F., 1880. *The Power of Movement in Plants*. John Murray, London, UK.

Del Dottore, E., Mondini, A., Sadeghi, A., Mattoli, V. and Mazzolai, B., 2016, May. Circumnutations as a penetration strategy in a plant-root-inspired robot. In *Robotics and Automation (ICRA), 2016 IEEE International Conference on.* IEEE, pp. 4722–4728. doi:10.1109/ICRA.2016.7487673.

Del Dottore, E., Mondini, A., Sadeghi, A., Mattoli, V. and Mazzolai, B., 2017. An efficient soil penetration strategy for explorative robots inspired by plant root circumnutation movements. *Bioinspiration & Biomimetics*, 13(1):015003.

Douglass, A.E., 1940. Examples of spiral compression wood. *Tree-Ring Bulletin*, 6(3):21–22.

Fisher, J.B. and Stevenson, J.W., 1981. Occurrence of reaction wood in branches of dicotyledons and its role in tree architecture. *Botanical Gazette*, 142(1):82–95. doi:10.1086/337199.

Fisher, J.E., 1964. Evidence of circumnutational growth movements of rhizomes of *Poa pratensis* L. that aid in soil penetration. *Canadian Journal of Botany*, 42(3):293–299. doi:10.1139/b64-024.

Friml, J., Wiśniewska, J., Benková, E., Mendgen, K. and Palme, K., 2002. Lateral relocation of auxin efflux regulator PIN3 mediates tropism in Arabidopsis. *Nature*, 415(6873):806–809. doi:10.1038/415806a.

Fukaki, H., Wysocka-Diller, J., Kato, T., Fujisawa, H., Benfey, P.N. and Tasaka, M., 1998. Genetic evidence that the endodermis is essential for shoot gravitropism in *Arabidopsis thaliana*. *The Plant Journal*, 14(4):425–430. doi:10.1046/j.1365-313X.1998.00137.x.

Furutani, I., Watanabe, Y., Prieto, R., Masukawa, M., Suzuki, K., Naoi, K., Thitamadee, S., Shikanai, T. and Hashimoto, T., 2000. The SPIRAL genes are required for directional control of cell elongation in *Arabidopsis thaliana*. *Development*, 127(20):4443–4453.

Gerbode, S.J., Puzey, J.R., McCormick, A.G. and Mahadevan, L., 2012. How the cucumber tendril coils and overwinds. *Science*, 337(6098):1087–1091. doi:10.1126/science.1223304.

Gruszecki, W.I., Trebacz, K. and Iwaszko, E., 2002. Application of very small force measurements in monitoring the response of sunflower to weak blue light. *Journal of Photochemistry and Photobiology B: Biology*, 66(2):141–147. doi:10.1016/S1011-1344(02)00234-8.

Hariharan, Y. and Krishnamurthy, K.V., 1995. A cytochemical study of cambium and its xylary derivatives on the normal and tension wood sides of the stems of Prosopis juliflora (SW) DC. *Beitrage Zur Biologie Der Pflanzen*, 69:459–492.

Hashimoto, T., 2002. Molecular genetic analysis of left–right handedness in plants. *Philosophical Transactions of the Royal Society B: Biological Sciences*, 357(1422):799–808. doi:10.1098/rstb.2002.1088.

Hatakeda, Y., Kamada, M., Goto, N., Fukaki, H., Tasaka, M., Suge, H. and Takahashi, H., 2003. Gravitropic response plays an important role in the nutational movements of the shoots of Pharbitis nil and *Arabidopsis thaliana*. *Physiologia Plantarum*, 118(3):464–473. doi:10.1034/j.1399-3054.2003.00080.x.

Hejnowicz, Z. and Sievers, A., 1995. Proton efflux from the outer layer of the peduncle of tulip in gravitropism and circumnutation. *Plant Biology*, 108(1):7–13. doi:10.1111/j.1438-8677.1995.tb00825.x.

Hoiczyk, E., 2000. Gliding motility in cyanobacteria: Observations and possible explanations. *Archives of Microbiology*, 174(1–2):11–17. doi:10.1007/s002030000187.

Hubert, B., and Funke, G.L., 1937. The phototropism of terrestrial roots. *Biologisch Jaarboek* 4:286–315.

Iida, M., Takano, T., Matsuura, T., Mori, I.C. and Takagi, S., 2018. Circumnutation and distribution of phytohormones in *Vigna angularis* epicotyls. *Journal of Plant Research*, 131(1):165–178. doi:10.1007/s10265-017-0972-y.

Inoue, N., Arase, T., Hagiwara, M., Amano, T., Hayashi, T. and Ikeda, R., 1999. Ecological significance of root tip rotation for seedling establishment of *Oryza sativa* L. *Ecological Research*, 14(1):31–38. doi:10.1046/j.1440-1703.1999.141282.x.

Ishida, T., Kaneko, Y., Iwano, M. and Hashimoto, T., 2007. Helical microtubule arrays in a collection of twisting tubulin mutants of *Arabidopsis thaliana*. *Proceedings of the National Academy of Sciences*, 104(20):8544–8549. doi:10.1073/pnas.0701224104.

Israelsson, D. and Johnsson, A., 1967. A theory for circumnutations in *Helianthus annuus*. *Physiologia Plantarum*, 20(4):957–976. doi:10.1111/j.1399-3054.1967.tb08383.x.

Johnsson, A., 1997. Circumnutations: Results from recent experiments on Earth and in space. *Plants*, 203:147–158.

Johnsson, A. and Heathcore, D. 1973. Experimental evidence and models on circumnutations. *Zeitschrift für Pflanzenphysiologie*, 70:371–405.

Johnsson, A., 1979. Circumnutation. In: Haupt, W., Feinleib, E. (Eds). *Encyclopedia of Plant Physiology*, N S, Vol. 7, Physiology of Movements. Springer, Berlin, Germany, pp. 627–646.

Johnsson, A., Jansen, C., Engelmann, W., and Schuster, J. 1999. Circumnutations without gravity: A two-oscillator model. *Journal of Gravitational Physiology*, 6:9–12.

Kaeiser, M., 1955. Frequency and distribution of gelatinous fibers in eastern cottonwood. *American Journal of Botany*, 331–334. doi:10.1002/j.1537-2197.1955.tb11127.x.

Karthikeyan, A.V.P. 2000. Tension wood and maintenance of branch angle and crown architecture in some trees. PhD thesis. Bharathidasan University, Tiruchirappalli, India.

Kato, T., Morita, M.T., Fukaki, H., Yamauchi, Y., Uehara, M., Niihama, M. and Tasaka, M., 2002. SGR2, a phospholipase-like protein, and ZIG/SGR4, a SNARE, are involved in the shoot gravitropism of Arabidopsis. *The Plant Cell*, 14(1):33–46. doi:10.1105/tpc.010215.

Kern, V.D., Schwuchow, J.M., Reed, D.W., Nadeau, J.A., Lucas, J., Skripnikov, A. and Sack, F.D., 2005. Gravitropic moss cells default to spiral growth on the clinostat and in microgravity during spaceflight. *Planta*, 221(1):149–157. doi:10.1007/s00425-004-1467.

Kim, H.J., Kobayashi, A., Fujii, N., Miyazawa, Y. and Takahashi, H., 2016. Gravitropic response and circumnutation in pea (*Pisum sativum*) seedling roots. *Physiologia Plantarum*, 157(1):108–118. doi:10.1111/ppl.12406.

Kitazawa, D., Hatakeda, Y., Kamada, M., Fujii, N., Miyazawa, Y., Hoshino, A., Iida, S. et al. 2005. Shoot circumnutation and winding movements require gravisensing cells. *Proceedings of the National Academy of Sciences of the United States of America*, 102: 18742–18747.

Kleine-Vehn, J., Ding, Z., Jones, A.R., Tasaka, M., Morita, M.T. and Friml, J., 2010. Gravity-induced PIN transcytosis for polarization of auxin fluxes in gravity-sensing root cells. *Proceedings of the National Academy of Sciences*, 107(51):22344–22349. doi:10.1073/pnas.1013145107.

Krishnamurthy, K.V. 2007. In pursuit of understanding the tension wood riddle. Prof. V. Puri Award Lecture 2006. *Journal of the Indian Botanical Society*, 86:14–26.

Lubkin, S., 1994. Unidirectional waves on rings: Models for chiral preference of circumnutating plants. *Bulletin of Mathematical Biology*, 56(5):795–810. doi:10.1007/BF02458268.

Migliaccio, F. and Piconese, S., 2001. Spiralizations and tropisms in Arabidopsis roots. *Trends in Plant Science*, 6(12):561–565. doi:10.1016/S1360-1385(01)02152-5.

Millar, K.D., Kumar, P., Correll, M.J., Mullen, J.L., Hangarter, R.P., Edelmann, R.E. and Kiss, J.Z., 2010. A novel phototropic response to red light is revealed in microgravity. *New Phytologist*, 186(3):648–656. doi:10.1111/j.1469-8137.2010.03211.x.

Millet, B. and Badot, P.M., 1996. The revolving movement mechanism in *Phaseolus*; New approaches to old questions. In: Greppin, H., Degli Agosti, R., Bonzon, M. (Eds.) *Vistas on biorhythmicity*, Biorhythmicity. University of Geneva, Geneva, Switzerland, pp. 77–98.

Millet, B., Melin, D. and Badot, P.M., 1988. Circumnutation in *Phaseolus vulgaris*. I. Growth, osmotic potential and cell ultrastructure in the free-moving part of the shoot. *Physiologia Plantarum*, 72(1):133–138. doi:10.1111/j.1399-3054.1988.tb06634.x.

Mugnai, S. Azzarello, E., Masi, E., Pandolfi, C. Mancuso, S., 2007. Nutation in plants. In: Mancuso, S. and Shabala, S. (Eds.), *Rhythms in Plants: Phenomenology, Mechanisms and Adaptive Significance*. Springer Verlag, Berlin, Germany.

Mugnai, S., Azzarello, E., Masi, E., Pandolfi, C., Mancuso, S., 2015. Nutation in plants. In: Mancuso, S. and Shabala, S. (Eds) *Rhythms in Plants*. Springer International Publishing, Cham, Switzerland, pp. 19–34.

Mulkey, T.J., 2007. Alteration of growth and gravitropic response of maize roots by lithium. *Gravitational and Space Research*, 18(2):119–120.

Mullen, J.L., Wolverton, C., Ishikawa, H. and Evans, M.L., 2000. Kinetics of constant gravitropic stimulus responses in Arabidopsis roots using a feedback system. *Plant Physiology*, 123(2):665–670. doi:10.1104/pp.123.2.665.

Nakamura, M. and Hashimoto, T., 2009. A mutation in the Arabidopsis γ-tubulin-containing complex causes helical growth and abnormal microtubule branching. *Journal of Cell Science*, 122(13):2208–2217. doi:10.1242/jcs.044131.

Ney, D. and Pilet, P.E., 1981. Nutation of growing and georsacting roots. *Plant, Cell & Environment*, 4(4):339–343. doi:10.1111/1365-3040.ep11604564.

Niinuma, K., Someya, N., Kimura, M., Yamaguchi, I. and Hamamoto, H., 2005. Circadian rhythm of circumnutation in inflorescence stems of Arabidopsis. *Plant and Cell Physiology*, 46(8):1423–1427. doi:10.1093/pcp/pci127.

Okada, K. and Shimura, Y., 1990. Reversible root tip rotation in Arabidopsis seedlings induced by obstacle-touching stimulus. *Science*, 250(4978):274–276. doi:10.1126/science.250.4978.274.

Okada, K. and Shimura, Y., 1992. Mutational analysis of root gravitropism and phototropism of Arabidopsis thaliana seedlings. *Australian Journal of Plant Physiology*, 19(4):439–448.

Orbovic, V., Poff, K.L., 1997. Interaction of light and gravitropism with nutation of hypocotyls of Arabidopsis thaliana seedlings. *Plant Growth Regulation*, 23:141–146. doi:10.1023/A:1005853128971.

Perrin, R.M., Young, L.S., Narayana Murthy, U.M., Harrison, B.R., Wang, Y.A.N., Will, J.L. and Masson, P.H., 2005. Gravity signal transduction in primary roots. *Annals of Botany*, 96(5):737–743. doi:10.1093/aob/mci227.

Piconese, S., Tronelli, G., Pippia, P. and Migliaccio, F., 2003. Chiral and non-chiral nutations in Arabidopsis roots grown on the random positioning machine. *Journal of Experimental Botany*, 54(389):1909–1918. doi:10.1093/jxb/erg206.

Schuster, J. and Engelmann, W., 1997. Circumnutations of *Arabidopsis thaliana* seedlings. *Biological Rhythm Research*, 28(4):422–440.

Shabala, S., 2003. Physiological implications of ultradian oscillations in plant roots. *Plant Soil*, 255:217–226. doi:10.1023/A:1026198927712.

Shabala, S.N. and Knowless, A. 2002. Rhythmic patterns of nutrient acquisition by wheat root. *Functional Plant Biology*, 29: 595–605. doi:10.1071/PP01130.

Shabala, S.N. and Newman, I.A., 1997a. Proton and calcium flux oscillations in the elongation region correlate with root nutation. *Physiologia Plantarum*, 100 (4):917–926. doi:10.1111/j.1399-3054.1997.tb00018.x.

Shabala, S.N., Newman, I.A.1997b. Root nutation modelled by two ion flux-linked growth waves around the root. *Physiologia Plantarum*, 101(4):770–776. doi:10.1111/j.1399-3054.1997.tb01062.x.

Sievers, A., Buchen, B. and Hodick, D., 1996. Gravity sensing in tip-growing cells. *Trends in Plant Science*, 1(8):249–250. doi:10.1016/1360-1385(96)10028-5.

Simmons, C., Söll, D. and Migliaccio, F., 1995. Circumnutation and gravitropism cause root waving in *Arabidopsis thaliana*. *Journal of Experimental Botany*, 46(1):143–150. doi:10.1093/jxb/46.1.143.

Smyth, D.R., 2016. Helical growth in plant organs: Mechanisms and significance. *Development*, 143(18):3272–3282. doi:10.1242/dev.134064.

Someya, N., Niinuma, K., Kimura, M., Yamaguchi, I. and Hamamoto, H., 2006. Circumnutation of *Arabidopsis thaliana* inflorescence stems. *Biologia Plantarum*, 50(2):287–290. doi:10.1007/s10535-006-0022-4.

Spurný, M., 1966. Spiral feedback oscillations of growing hypocotyl with radicle in *Pisum sativum* L. *Biologia Plantarum*, 8(5):381–392. doi:10.1007/BF02930674.

Spurný, M., 1968. Effect of root tip amputation on spiral oscillations of the growing hypocotyl with radicle of the pea (*Pisum sativum* L.). *Biologia Plantarum*, 10(2):98. doi:10.1007/BF02921024.

Spurný, M., 1974. Interactions of photo-and geotropism with periodical oscillations of growing pea root (*Pisum sativum* L.). *Biologia Plantarum*, 16(1):43–49. doi:10.1007/BF02920819.

Stolarz, M. and Dziubińska, H., 2017. Spontaneous action potentials and circumnutation in *Helianthus annuus*. *Acta Physiologiae Plantarum*, 39(10):234. doi:10.1007/s11738-017-2528-0.

Stolarz, M., 2009. Circumnutation as a visible plant action and reaction: Physiological, cellular and molecular basis for circumnutations. *Plant Signaling & Behavior*, 4(5):380–387. doi:10.4161/psb.4.5.8293.

Stolarz, M., Król, E., Dziubińska, H. and Kurenda, A., 2010. Glutamate induces series of action potentials and a decrease in circumnutation rate in *Helianthus annuus*. *Physiologia Plantarum*, 138(3):329–338. doi:10.1111/j.1399-3054.2009.01330.x.

Stolarz, M., Krol, E., Dziubinska, H., and Zawadzki,T. 2008.Complex relationship between growth and circumnutations in Helianthus annuus stem. *Plant Signaling & Behavior*, 3:376–380. doi.:10.4161/psb.3.6.5714.

Telewski, F.W., 1988. Intra-annual spiral compression wood: A record of low-frequency gravitropic circumnutational movement in trees. *IAWA Journal*, 9(3):269–274. doi:10.1163/22941932-90001076.

Tepper, H.B. and Yang, R.L., 1996. Influence of the shoot tip and leaves on circumnutation in green pea seedlings. *Plant Biology*, 109(6):502–505. doi:10.1111/j.1438-8677.1996.tb00603.x.

Thain, S.C., Murtas, G., Lynn, J.R., McGrath, R.B. and Millar, A.J., 2002. The circadian clock that controls gene expression in Arabidopsis is tissue specific. *Plant Physiology*, 130(1):102–110. doi:10.1104/pp.005405.

Trenard, Y, and P. Gueneau, 1975. Relations entre contraintes de croissance longitudinales et bois de tension dans le Metres (*Fagus sylvatic* L.) *Holzforschung*, 29:217–223.

Van den Driessche, T. 2000. Nutations in shoots and in Desmodium lateral leaflets, nyctinastism and seismonastism in Mimosa pudica. Comparison and evolution of morphology and mechanism. *Biological Rhythm Research*, 31:451–468.

Wayne, R. and Staves, M.P., 1996. A down to earth model of gravisensing or Newton's law of gravitation from the apple's perspective. *Physiologia Plantarum*, 98(4):917–921. doi:10.1111/j.1399-3054.1996.tb06703.x.

Wilson, B.F. and R.R. Archer, 1977. Reaction wood: induction and mechanical action. *Annual Review of Plant Physiology*, 28:23–43. doi:10.1146/annurev.pp.28.060177.000323.

Yoshihara, T. and Iino, M., 2005. Circumnutation of rice coleoptiles: Its occurrence, regulation by phytochrome, and relationship with gravitropism. *Plant, Cell & Environment*, 28(2):134–146. doi:10.1111/j.1365-3040.2004.01249.x.

Zachariassen, E., Johnsson, A., Brown, A.H., Chapman, D.K. and Johnson Glebe, C., 1987. Influence of the g-force on the circumnutations of sunflower hypocotyls. *Physiologia Plantarum*, 70(3):447–452. doi:10.1111/j.1399-3054.1987.tb02841.x.

11 Seedling Handedness in Some Angiosperms

Bir Bahadur, N. Rama Swamy, and T. Pullaiah

CONTENTS

11.1 INTRODUCTION

The presence of several morphologically well-defined organs/characters together with the complexity attained to exploit the maximum of the existing environment make phylogenetic studies of angiosperms difficult. Plant taxonomy is largely based on floral characters and not much emphasis has been given to the vegetative morphological characters, which can usefully serve in delimiting taxa. Seedling characters in general have received scant attention in comparison to floral and other characters, despite several studies during the last 150 years on seedlings of various monocots and the application of embryo in systematics and seedling characters for taxonomic purposes (Clos, 1870; Van Tiegham, 1897; Avery, 1930; Aldulov, 1931; Boyd, 1932; Arber, 1934, 1961; Reeder, 1957; Duke, 1969; Guignard and Mestre, 1971; Burger, 1972; Muller, 1978; de Vogel, 1980). Younger and Mckell (1972) have edited a good informative volume on "*Biology and utilization of grasses*" and devoted three chapters on seedling vigor, seedling establishment, breeding for seedling vigor, and environmental modifications for seedling establishment by various authors, but there is no mention of seedling handedness. Crow (1973) in his book emphasized the importance of seedling characters in teaching genetics. Klebs (1885) appears to be first to define seven seedling types, while Boyd (1932) distinguished three types of seedlings. Tillich (2000) studied seedling diversity and the homologies of seedling organs in the order Poales and commented "The morphological diversity of seedlings in monocots is as yet incompletely known, and the phylogenetic analysis is more diverse than other angiosperms." While Wilson and Morrison (2000) commented, "The seedling characters provide an important character at the family level and serve as a key character to detect phylogenetic relationships."

Remarkably little information is available on studies related to seedling handedness and its application in taxonomy vis-à-vis other related features. Cullen (1978) extensively studied 2000 species of angiosperms for their vernation but has not focused attention on seedling handedness. Incidentally vernation was earlier studied by Kundu (1968) in her PhD work on Indian plants, but there is no mention of seedling handedness. Likewise, Muller (1978) also studied several species but could not observe/detect seedling handedness in dicot and cereals as described by Compton (1910). Revival of interest in the subject is mostly due to Prof. H. Kihara and his associates in Japan (Kihara, 1972). Ono (1956) and Ono and Suemoto (1957) investigated seedling and spikelet handedness in *Aegilops* and *Triticum*, while Suemoto (1961) extensively studied handedness in various species of *Triticum*. In India, the subject was revived by Bahadur and his associates, and they contributed in both seedling handedness in monocot and dicot species (Bahadur et al., 1978) and reviewed the earlier literature (Rao, 1980, 1981; Narsaiah, 1984). Bahadur et al. (1980) reported, for the first time, seedling handedness in 15 cultivars of *Vigna mungo* (Fabaceae). Swamy et al. (1984) studied seedling handedness in Triticale and its parents in relation to yield. In the present chapter, a review of seedling handedness in three angiosperms families is given, and apart from its implications in taxonomy and related aspects, causal factors for the handedness are presented.

11.2   LEAF VERNATION

Leaves have fascinated botanists for ages for their diversity in shape and their highly organized structures that perform vital plant functions. They originate from small groups of cells (primordia) protruding around the shoot apex. The term *vernation* refers to the arrangement of the leaf primordia in the bud that gives them birth. This arrangement is always the same for the species of the same genus, sometimes even to an entire family. To describe the vernation/prefoliation, it is necessary to consider successively the disposition of each leaf, then that of the various leaves of the bud, with respect to each other. The mode of plication as well as the winding mode of individual young leaves within a bud (called ptyxis) can be very varied in the plant kingdom; hence, many distinct terms have been proposed, viz., replical, conduplicated, folded or plicative, crumpled or corrugative, circinate, convolute, involute, etc. Philipson and Philipson (1979) described four distinct vernation patterns in the genus *Nothofagus* (Fagaceae), namely conduplicate, revolute, plicate, and plane. These characterize groups of species that are distinct on other morphological and geographical grounds and are discussed in the importance in infrageneric taxonomy of the genus. Compton (1910) first observed handedness in several members of Poaceae and later Kihara (see review by Kihara, 1972) and publications of his associates in Japan studied the problem in considerable detail.

By seedling handedness, we mean the left or right position of the first leaves and subsequent leaves (or leaf primordia) in relation to the orientation of the two cotyledons. When young, the first two alternate leaf primordia show ptyxis individually and overlapping when they are together. The older leaf primordium of the two may enclose the younger primordium completely inside it through its folding or rolling, or it may partially enclose it on one margin and overlap it in the other margin, or the two leaf primordia may touch each other without being enclosed by one another. Whatever may be the case, when the leaf primordia expand after germination, the older primordium remains either left or right to the cotyledons and also to the next younger leaf.

We have used the terminology of left- and right-handed seedlings depending upon the folding of the first leaf enclosing fully or partially the next younger leaf, above the coleptile either to the left- or to the right-hand side as shown in Figure 11.1, and it is used in the same as described earlier by Compton (1910, 1911, 1912) and later by Kihara et al. (1951) and Bahadur et al. (1978, 1980).

11.3 SEEDLING HANDEDNESS IN FABACEAE

In the literature there is no published account on seedling handedness in any dicot family. Bahadur et al. (1980) first reported seedling handedness in 15 cultivars of *Vigna mungo* (Fabaceae) and counted 7291 seedlings and noted an excess of right-handed seedlings (4448) over left-handed (2755). Incidentally, this condition lasts only for a very short period and needs patience and critical observation. After a short while, maybe 30 seconds to 1 minute, the left- and right-handed seedlings look identical. Neutral seedlings also occur, which do not show overlapping but show an involute seedling condition of two types (Rao et al., 1980) (Figure 11.1). Later on, seedling handedness was studied in *Cajanus cajan* (Rao, 1980; Rao and Bahadur, 1980) and *Psophocarpus tetragonolobus*, which exhibit both seedling and twining handedness (Narsaiah, 1984).

The occurrence of seedling handedness in Fabaceae varies from tribe to tribe, and out of 137 species examined only 78 species exhibit seedling handedness since its presence is either common or sporadic in some tribes while it is completely absent in others. In the tribe Phaseoleae, a majority of the genera examined showed seedling handedness whereas in Hedysareae nine genera and in Dalbergeae three genera showed this character. In the tribes Genisteae and Trifoleae, only one species in each of the genera examined showed seedling handedness; on the contrary, none of the genera examined in tribes Galegeae, Viceae, Sophoreae, and Loteae show seedling handedness. It was noted that for a species to show seedling handedness, the leaves should be alternate as shown in Figure 11.1. However, in *Robinia*, the first seedling leaves are simple, yet overlapping of the margins was found to be absent because these leaves were found to be at two different levels, and this perhaps prevents overlapping. Further, the seedling handedness seems to be associated with a particular seed shape. In general, kidney-shaped seeds show twisting of the first pair

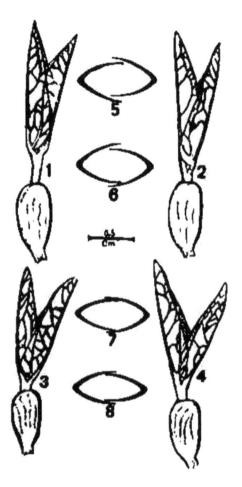

FIGURE 11.1 Seedling handedness of *Vigna mungo*; 1–2. Right- and left-handed seedlings of *V. mungo*; 3–4. Two types of neutral seedlings. 5–8 cross section of right and left-handed and neutral seedlings respectively.

of leaves, with the exception of *Sesbania* and *Tephrosia*, which are not characterized by kidney-shaped seeds. On the other hand, in *Cajanus* and *Glycine*, the shape is broadly oval yet exhibits seedling handedness (Bahadur et al., 1983).

Hence, left- and right-handed seedling handedness may also be used as a taxonomic pointer in delimiting taxa in the family Fabaceae (Bahadur et al., 1983). Further, the seedling handedness in Fabaceae appears to have evolved in a uniform/parallel manner as shown by its presence in most tribes. It may be noted that in the majority of the species examined, the right-handed seedlings were found to be more than the left-handed seedlings. Further, if only *Phaseolus* and *Vigna* species are considered, the American species (*Phaseolus multiflorus* and *P. lunatus*) gave an excess of L seedlings, and the Asian species (*Vigna mungo* and *V. radiata*) showed the reverse condition. Our observations also show that in other species the distribution of

L and R seedlings vary. For instance, in *Psophocarpus tetragonalobus* and *Erythrina stricta*, the L seedlings were found to be more, though statistically not significant. On the other hand, the number of the R seedlings was found to be more in *Glycine max, Dolichos lablab, D. biflorus*, and *Cajanus cajan*. However, what makes the differences in the distribution of the L and R seedlings is presently unknown.

The seedling handedness observed in Fabaceae is a clear case of bioisomerism, as both the L and R seedlings represent stereoisomeric/mirror image forms that have been variously designated in the literature as isomeric and bioisomeric (Meyen, 1973; Bahadur and Reddy, 1974; Bahadur et al., 1977), enantiomorphic (Davis, 1974; Bhaskar et al., 1983) and bio-enantiomorphic objects (Dubrov, 1978). The causal factors for the existence of isomeric forms of seedlings are so far unknown. Bahadur et al. (1977) and Rao (1980) have reviewed the possible reasons for the existence of handedness in plant organs. In their study of *Euporbia milli*, whose cyathia show various types of mirror-image patterns, stated that this may be due to stereoisomerism of a hormone molecule, while Bahadur and Venketeshwalu (1976) opined that some optically active compounds present in plant metabolism determines handedness in biological objects. Thus, molecular chirality is expressed in biological chirality as stated by Onslow is an expression of its chemical constitution (cited by Arber, 1961).

11.4 SEEDLING HANDEDNESS IN POACEAE

The cotyledon of grasses (Poaceae) and other monocotyledons are highly modified leaves composed of a scutellum and a coleoptile. The scutellum is a specialized tissue within the seed that absorbs stored food from the adjacent endosperm tissue. The coleoptile is a protective cap that covers the plumule (precursor to the stem and leaves of the plant). The arrangement of leaves within the seedling bud is called prefoliation/vernation. This condition lasts for just few minutes, and there after the leaves grow, expand, and become angular or horizontal. Depending on the folding of the first leaf, enclosing the second, either to the left or right, a seedling is classified a left- or right-handed (Figure 11.2). This unique observation was first made by Compton (1910) in barley and later in other members of Graminae by Compton (1912). Seedlings that lack folding either to L or R are referred to as neutral, as they are either involute or convolute.

The stereoisomeric dimorphism of the first leaf of most Poaceae is expressed in the direction of L or R in which the first leaf over the coleoptile is folded. During the Pre-Mendelian period, Compton (1910–1912) studied the seedlings of barley and maize, and after studying two-rowed barley concluded that although this character is not inherited, the ratio of L and R seedlings remains constant. In maize he observed equality initially, but different ratios were obtained when the seeds were collected separately from the odd and even rows of the cob, and observed kernels of left rows gave an excess of left-handed seedlings, and those from the odd rows produced an excess of right-handed seedlings. Kihara et al. (1951) have given beautiful illustrations showing the distribution of L and R flag leaves and spikelets, explaining the handedness as mentioned above.

Compton (1910, 1911, 1912) first reported the presence of seedling handedness, that is, left- and right-handedness in seedlings of various cereals. Kihara (1950, 1972) and his associates Ono (1956) and Ono and Suemoto (1957) described handedness

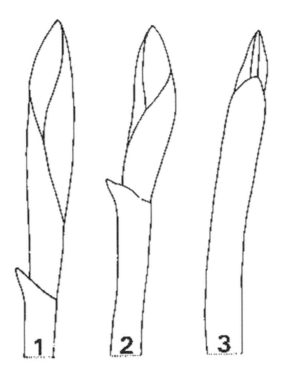

FIGURE 11.2 Right-and left-handed seedlings and neutral seedlings of *Dendrocalamus strictus*. The coleoptiles have round sheaths, and the size varies through which the first leaf emerges. 1–3. Note the folding of the first seedling leaf above the coleoptiles to the right- and left-hand side, while the neutral seedlings are incurved. (From Bahadur, B. et al., *Proc. Indian Acad. Sci. (Plant Sci.)*, 92, 279–283, 1983.)

in seedlings of *Triticum* and *Aegilops*. Suemoto (1961) extensively studied left- and right-handedness in the genus *Triticum*, hence the revival of interest in this interesting area caught up. In India, seedling handedness and some related aspects were first studied by Bahadur et al. (1978) in *Bambusa arundinacea* and dozen varieties of *Sorghum vulgare* (Udaychandra, unpublished). Likewise, the senior author observed seedling handedness in several cereal species and in particular *Oryza sativa* varieties and *Penneistum typhoides* (Bahadur, unpublished).

Bahadur et al. (1983) have made an exhaustive study of seedling handedness in Graminae by examining the first seedling leaf above the coleoptiles in 67 genera and 93 species spread over 11 tribes. They noted seedling handedness to be present in Paniceae, Andropogoneae, Festucae, Hordeae, Avenae, Bambusae, Oryzae, and Choridae but absent in species of tribes Agrostroideae, Phalarideae, and Poaceae examined by them and have discussed the taxonomic significance at the tribal level (Bahadur et al., 1983). Of the eight tribes showing seedling handedness, Paniceae, Andropogoneae, Festuceae, and Hordeae showed a greater number of genera and species than other tribes. Oryzeae, Paniceae, and Chlorideae showed a uniform excess of right-handed seedlings, whereas Hordeae and Andropogoneae showed an excess of left-handed seedlings. Although seedling handedness is present in Andropogoneae, Chlorideae, Festuceae, and Aveneae, surprisingly some of the species under these tribes, viz., *Anthoxanthus ordaratum, Bromus squarrosa, Phragmitis communis, Holcus laciatus*, and *Trisetum flavescens* are characterized exclusively by neutral seedlings only. Thus, of the 6 genera examined in Chlorideae, only 2 show seedling handedness, while in Festuceae out of 13 genera studied only 5 exhibited seedling handedness. The tribes Agrostideae, Phalarideae, and Poaceae didn't show seedling handedness (see Bahadur et al., 1983).

The left- and right-handed seedlings differ in size as shown in Figure 11.2, especially in coleoptiles size. In fact, Bahadur et al. (1978) have earlier described series of differences in *Bambusa arundinacea* between the left- and right-handed seedlings and needs no repetition.

Within the species of eight tribes showing seedling handedness, neutral seedlings also occur, the percentages being: Paniceae (0.61%); Andropogoneae (5.14%); and Chlorideae (31.2%); Bambuseae (1.9%). In Triticale, whose seedling handedness has been described by Bahadur and Swamy (1985), neutral seedlings occur (1.6%). Compton (1911, 1912) and Bahadur and Swamy (1985) have reported a very high incidence of neutral seedlings in *Secale cereale*, but neutral seedlings rarely occur in *Triticum* of its parents. Thus, a relationship exists between the presence or absence of seedling handedness in the various tribes, and this may serve as a useful tool for taxonomic purposes for delimiting the various tribes of Poaceae.

The seedling handedness and handedness in successive leaves, flag leaves, and spikelets have been investigated in Triticale and its parents (Swamy and Bahadur, 1983). The morphological and physiological characters have also been studied in these plants (Swamy and Bahadur, 1984).

11.5 HANDEDNESS IN FIRST LEAF, SUCCESSIVE LEAVES, AND SPIKELETS

Kihara and his associates in a series of papers studied the various aspects of handedness and determined the LH or RH position in the spikelet. The right-handed spikelet is one when viewed from the dorsal side the first floret faces the right hand of the observer, and the vice versa for the left-handed spikelet. The spikelets are successively numbered from the lowest first upward, beginning with spikelet 1 upward, and are designated as odd (O) and even (E). Kihara et al. (1951) terminology has been used in Triticale studies (Bahadur and Swamy 1985) to be discussed later.

Based on these observations of 108 ears of *Triticum monococum*, Kihara (1950) concluded that:

1. The ratio of RH to LH spikelets showed a 1:1 ratio.
2. In the ratio of RH and LH spikelets in the row, RH spikelets were more than LH spikelets and vice versa (i.e., 1.6:0.58, respectively).
3. There is regularity in the alterations of RH and LH spikelets is more apparent than the third spikelet (88%) and becomes less definite in the upper part of the spikelet.

4. Of the seedling raised from the first floret of the RH spikelets, 60% were RH, and from the LH spikelets 60% were LH.
5. The first leaf of the seedling obtained from the second floret tends to hold in the opposite directions to that of the seedling from the first floret.
6. In continuation of their work, Kihara et al. (1951) studied the RH and LH of spikelets in Einkorn wheats. They examined 821 ears—411 OL, 398 OR, and the remaining 12 were with rudimentary spikelets and not taken into account.

From these they concluded the proportion of OR to OL is in equal ratio, and to support that they have given beautiful illustrations showing the distribution of R and L flag leaves and spikelets explaining the handedness as mentioned above (see Kihara et al., 1951).

Kihara et al. (1951) studied antidromy (i.e., controlling the opposite direction) and concordance (i.e., no observed/no expected in a harmonious agreeable manner) in 809 samples of ears and mentioned that the R/L handedness is controlled by the threshold reaction, which is the result of the external influences and the specific potency.

Based on the above, Kihara et al. (1951), "concluded that the determination of RH and L handedness starts in the vicinity of the third spikelet where the potency is at the highest and then proceeds upwards and downwards as the differentiation of the ear goes on." And further commented, "The adequacy of this hypothesis will be judged by a precise examination of developmental processes in the primordial ears."

The left- or right-handedness of spikelets is determined by the position of the first floret in the spikelet; it will be right-handed when viewed from the dorsal side and facing the observer and left-handed when viewed from the left hand. The spikelets are designated by successive numbers from the lowest to the base to top of inflorescence. Kihara examined 108 ears of *T. moncocum* and noted equality of left- and right-handed spikelets in several varieties of Einkorn wheats (Kihara, 1950).

The number of available spikelets per spike varies depending upon on the species. Einkorn wheats have more than 20 spikelets, Emmer and Dinkel wheats have 10 spikelets, while in *Agelops* the spikelets vary from 2 to 14. Furthermore, they found no differences between the results in two different years (Kihara et al., 1956).

Swamy and Bahadur (1983) have described the handedness in successive leaves, flag leaves, and in spikelets (L/R) in a given ear bone on the same spike, on a particular plant developing respectively from the left- or right-handed seedlings (Figure 11.3). They have selected four varieties of *Triticale* viz.: DTS-42-3, DTS-642, DTS-47-1, and DTS-280-7; five varieties of *Triticum* viz., NI-5439, NI-747-19, UP-215, Kalyana Sona, Sonalika, and *S. cereale* for this study. The seedlings were sorted out as described by Bahadur and Swamy (1985). Seedlings (25) of *Triticale*, *Triticum*, and *S. cereale* (neutral seedlings only) representing both left- and right-handed were transplanted in the field for observation of handedness in the subsequent leaves, including the flag leaves and the spikelets.

They noted that the handedness continued up to sixth leaf, both in *Triticale* and *Triticum*. According to their findings, most of the right-handed plants show the sequence R L R L R L with regard to vernation either to their left or right (see Figure 11.3).

Occasionally this sequence gets altered to R L L R L R or the change may be in the third, fourth, or fifth leaf, whereas in the left-handed plants, the sequence was found

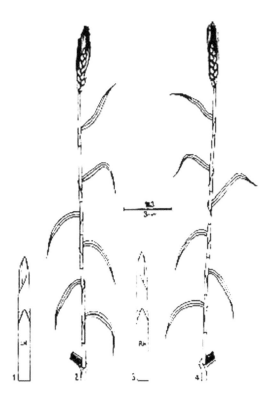

FIGURE 11.3   1 and 3 Drawings of left- and right-handed seedlings of Triticale var. DTS-47-1 showing vernation of the seedling left. 2 and 4 represents semi-diagrammatic representation of mature plants developing from the left- and right-handed seedlings showing the distichous arrangement showing a mirror-image pattern. The vernation can be clearly seen in the petiole region forming a mirror-image pattern. Note the handedness in spikelets at the tip and the flag leaf. (From Swamy, N.R. and Bahadur, B., *Curr. Sci.*, 52, 89–91, 1983.)

reverse, that is, L R L R L R (Figure 11.3). Left-handed plants like the plants occasionally showed altered sequence, that is, L R R L R L. The left-handed plants like the right-handed ones occasionally showed an altered sequence, that is, L R R L R L. It is interesting to observe that although the seedling handedness is generally absent in *S. cereale*, the handedness in the successive leaves was found common as observed in *Triticale* and *Triticum* (Swamy and Bahadur, 1983). With regard to the flag leaves, the left-handed plants showed right-handed flag leaves, and in the right-handed plants this was found reversed (Figure 11.3). The percentage of left-handed flag leaves in the right-handed plants, in general, was higher in *Triticale* (44%) and *Triticum* (29%), except in DTS-642 and in Kalyana sona. Left-handed plants similarly follow the sequence of L R L R L R R. Rarely, however, the sequence gets altered to L R R L R L-L.

In *S. cereale*, out of 25 neutral plants, 10 plants showed left-handedness and 15 plants showed right-handedness in the flag leaves. The folding of the first spikelet in most of the plants with right-handed flag leaves showed right-handed first spikelets while the left-handed produced left-handed first spikelets. The percentage of first spikelets folding to left-handed is higher in *Triticale* (72%) and 65% of left-handed flag leaves in right-handed plants in *Triticum* (Swamy and Bahadur, 1983). They have also observed in left-handed plants the high percentage of right-handed first spikelet in right-handed flag leaves of *Triticale* (78%) and *Triticum* (74%). According to Swamy and Bahadur (1983), the flag leaf and the first spikelet showed tendency for the R/L character. On the other hand, the second spikelet showed a strong tendency in the opposite direction from that of the first one, while the third spikelet repeated the R/L character of the first one. It is interesting to note that similar observations were made by Kihara et al. (1951) in handedness in spikelets of Einkorn wheat.

According to Latting and Aberg (1972), the vegetative and floral meristems are fundamentally alike; hence, the leaves and floral organs have homologous structures. In view of this observation, the handedness in subsequent leaves, flag leaves, and the first spikelets observed by Swamy and Bahadur (1983) represent mirror images, not only at the seedling stage but also at the maturity. Thus, this phenomenon in the flag leaves and spikelets observed in *Triticale* and its parents represent stereoisomeric forms and constitute a case of bioisomerism (Bahadur et al., 1977; Bahadur and Rao, 1981). This is comparable to the work reported earlier by Kihara (1950; 1972) and Kihara et.al. (1951) in their publications.

Morphophysiological considerations with reference to seedling handedness in *Triticale* and its parents were assessed by Swamy and Bahadur (1984). In all those characters, left-handed plants had shown the better results in comparison to the right-handed plants. This may be attributed that the left-handed plants are metabolically superior than the right-handed plants. That was because left-handed plants had contained higher amounts of chlorophylls per gram of tissue, hence a high rate of photosynthesis followed by other metabolic activities. According to Swamy and Bahadur (1984), the stereoisomerism present in the seedlings of left- and right-handedness in *Triticale* and its parents may influence the morphological and physiological characters.

11.6 YIELD IN RELATION TO SEEDLING HANDEDNESS IN TRITICALE AND ITS PARENTS

Swamy et al. (1984) have studied the yield in relation to seedling handedness in two cvs of *Triticale* (DTS-42-3; DTS-208-7), three cvs of *Triticum* (NI-5439, Sonalika, Kalyanasona), and *S. cereale.* Left-handed plants bore longer spikes in all the cvs of *Triticale* and *Triticum* with the exception of cv NI-5439. The left-handed plants of both the cvs of *Triticale* and cv Kalyana Sona of *Triticum* showed more spike weight. The cvs NI-5439 and Sonalika of *Triticum* showed almost the same spike weight in both the left- and right-handed plants. According to their observations, the dry seed weight was found higher in left-handed plants of *Triticale* and *Triticum* with the exception of the cv Kalyanasona.

11.7 HANDEDNESS IN COCONUT SEEDLING: ARECACEAE

Unlike the seedling handedness described above, handedness/spirality in various palms, and in particular the coconut palm, is under Fibonacci control. It is easy to determine the handedness of coconut seedlings by just holding up the coconut with the emerging seedling, by marking the first leaf and then the second leaf. If the second leaf position is nearer to first leaf on the clockwise direction, then it is left-handed seedling. On the other hand, if the direction of the nearer second leaf is anticlockwise, then the seedling is right-handed. Studies have also been done on seedling handedness in *Areca catechu* (see Ghose et al., 1996). Since many palms also show various Fibonacci numbers, they all should exhibit seedling handedness (see Chapter on Fibonacci.....).

11.8 SEEDLING ENANTIOMORPHISM IN BUCKWHEAT (POLYGONACEAE)

Bhaskar et al. (1983) have investigated the seedling handedness in 4 cvs of *Fagopyrum esculentum* (IC 37289, 17369, 16555 & 13411), *F. sagittatum*, 6 cvs of *F. tataricum* (IC 13145, 18889, 13376, 18043, 13374, 1869), and *F. kashmiriyanum*. Out of 730 seedlings, 358 (49%) left-handed and 372 (51%) right-handed seedlings were found. According to these observations, various cvs of *F. tataricum* exhibited more right-handed than left-handed seedlings, except cv IC 13374, in which equality of left- and right-handed seedlings was observed (Figure 11.4). The cv IC 16555 of *F. esculentum* showed an excess of left-handed seedlings (L/R = 2.2), and

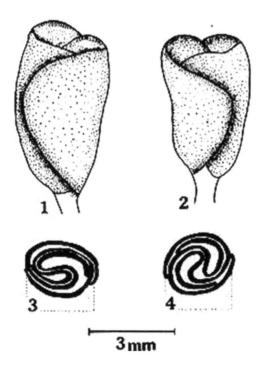

FIGURE 11.4 1 and 2 left- and right-handed seedlings of *Fagopyrum esculentum* showing overlapping of the first pair of leaves towards the left and right hand, respectively. In 3 and 4, the same is shown in cross section showing vernation in the clockwise and counterclockwise direction. (From Bhaskar, K.V. et al., *Curr. Sci.*, 52, 25–26, 1983.)

F. kashmiriyanum also showed a similar feature (L/R = 1.75). But the combined data showed an excess of right-handed seedlings (Bhaskar et al., 1983). This is the first record of seedling handedness in Polygonaceae. Since various *Fagopyrum* species, commonly called buckwheat, are important cereals, we suggest interested researchers take up further work.

11.9 SEEDLING HANDEDNESS IN RELATION TO SEED POSITION

Rao et al. (1983a) observed that the first and second seed position in pods in 4 out of 5 cultivars had an excess of left-handed seedlings. It was only in the third position that the seed tended to show a uniform excess of right-handed seedlings, while in the fourth position seedling handedness was haphazard/wobbled irrespective of the cultivar studied. Thus, seedling isomerism appears to be correlated with mathematical isomerism. Seedling handedness cannot be fixed. This will be described later. What controls seed position vis-a-vis seedling handedness is mystery, and we feel that this may be due to morphogenetical factors that are laid down at the time of ovule initiation, and enantiomer selectivity of certain metabolites (auxins) could be the cause. It may be noted that according to Horovitz et al. (1976), the left and right ovule position in flowers of *Medicago sativa* influences the seed set and out-crossing rates, whether the same holds for *Cajanus cajan* is presently not known. Rao et al. (1983a, 1983b) suggested that the left and right ovule position in flowers of *M. sativa* influences the seed set and out-crossing rates; whether the same holds true of *C. cajan* is presently not known.

11.10 EFFECT OF IRRADIATION AND CENTRIFUGATION

This work was carried out by Rao and Bahadur (1982a, 1982b) in green gram and black gram to verify the yield parameters in relation to handedness. Out of 1600 presoaked seeds, 1193 seedlings were treated with X-irradiation and gamma irradiation at various doses that resulted in right-handed seedlings at 57.7% and the rest left-handed. But at a higher dose of 15kR, the ratio of seedlings was 58.1%. On the contrary, X-irradiation seems to be ineffective, as the results were just the opposite (out of 1024 seedlings only 321 [40%] were LH in control). The effect of gamma irradiation with similar doses also gave a higher percentage for LH seedlings (58.2%) over RH (41.7%). The results of irradiation on seedlings and spikelet handedness in Einkorn wheat by Ono et al. (1954) showed changes in the relation of L and R seedlings and spikelet reduction of the mean concordance proportions and its transfer to subsequent generations. We suggested that higher doses of X-irradiation and gamma irradiation may be prove useful. Rao and Bahadur (1982a, 1982b) also conducted experiments on green and black gram to study the effects for 60 minutes in a refrigerated centrifuge at 15,000 rpm. Out of total of 1500 seedlings, 756 were LH and the rest RH, in contrast to the control almost L-49% and R-50%. Biswas and Bhattacharya (1975) studied the effect of X-irradiation in *Phaseolus vulgaris* and *Glycine max*, while Reddy and Srinivasachar (1972) opined that due to centrifugation a pseudo gravity sets in, thereby altering the ratio as noted by us. Ono et al. (1954) conducted

X-irradiation as an external agent on left- and right-handedness on Einkorn wheat plants at the time of double ridge formation in spike primordia with the objective to find out the cause of the external agent using higher doses ranging from 3200 r to 16000 r. They found no difference, but the sensitivity to irradiation seems to be greater in the lower part of the spike than in upper part of the spikelets. Further work by them revealed the remarkable reduction of mean concordance proportion due to irradiation being partly transmitted to the subsequent generations (i.e., transmissible effect). "They concluded that the mechanism of the radiation effect may throw additional light upon the R and L handedness which is ultimately a fundamental problem of morphogenesis."

11.11 SEEDLING HANDEDNESS IN RELATION TO FIRST LEAF, SUCCESSIVE LEAVES, FLAG LEAVES, AND SPIKELETS

Although this has been discussed under Poaceae, the situation in Triticale studied by Swamy et al. (1984) studied handedness in *Triticale* and its parents in relation to handedness and noted:

1. Left-handed plants had longer spikelet inflorescence in all the cultivars of *Triticale* and *Triticum* with few exceptions.
2. Left-handed plants of two DTS-42-3 and DTS-280-7 and *Triticum* (Kalyansona) showed more spike weight, while *Triticale* cvsNi5439 and *Triticum* (Sonalika) showed the same spike weight in both.
3. The dry seed weight in both was found to be higher in left-handed plants of *Triticale* and *Triticum* with the exception of cv Kalyansona. This suggested that LH plants yielded more than RH plants. It may be mentioned that handedness in *S. cereale* is less pronounced, being only 3%, and the rest of the seedlings were neutral; hence, the above parameters were not recorded.
4. Swamy and Bahadur (1984) noted more roots, more chlorophyll content I leaves, and faster growth rate in LH *Triticale* and its parents. Swamy et al. (1984) concluded that LH plants are superior with respect to spike length, weight and grain weight, and seed yield as in *Vigna radiata* and *V. mungo* (Rao, 1980).

Kihara et al. (1951) have beautiful illustrations showing the distribution of R and L flag leaves and spikelets, explaining the handedness as mentioned above.

11.12 SEEDLING HANDEDNESS IN RELATION TO SELECTION

Rao et al. (1983a, 1983b) studied the effects of selection on seedling handedness for three consecutive years from 1977 to 1980. Ten seedlings each of left- and right-handedness were sown in earthen flats, and the seedlings were transplanted in separate rows in the field. One plant each from left- and right-handed types was randomly selected, and on maturity a total of 89 pods from left-handed and 94 pods from right-handed were obtained. The total seeds produced from these pods were 322 and 352, respectively, and these were used as parents. Left- and right-handed plants showed a 1:1 ratio. The procedure of selecting one plant and selfing 100 flowers was repeated. Seeds obtained were during years 1979–1980. The seeds were sown, and seedlings from three parents were sorted out, the ratio of right to left being 1:1, while the fourth parent showed deviation from equality (Table 11.1). During subsequent generations, seeds from parents of an extreme selection type (L L L and R R R) were used to determine the ration of L and R seedlings. It was observed that the seedlings produced from the parent type L L L showed an excess of the right RH seedling type of 21, but the deviation from equality is statistically insignificant. Thus, the ratio of left- and right-handed seedlings in *C. cajan* derived from the parents selected in successive generations did not appear to reflect the progressive purity of the character with regard seedling handedness (Table 11.1).

Kundu and Sarma (1965) have tried to establish the handedness with regard to the leaf spiral in *Corchorus* species and obtained similar results. To date, there is no satisfactory explanation of handedness available, although this appears hereditary. But in Mendelian monogenic inheritance (RR, RL, and LL), where right dominates over left, reciprocal crosses are identical and segregation is simple. It was shown in twisted pods and L and R spiral seeds of *Medicago tuberculata* and *M. litoralis* species of Fabaceae (Lilianfield and Kihara, 1956a, 1956b).

TABLE 11.1
Results of Continued Selection Experiments in *C. cajan*

Year	Parent	L-Handed Seedlings	R-Handed Seedlings	X^2 1:1 Deviation	P-Value
1977–1978	Unknown	93	96	0.047	>0.75
	L-handed	122 (LL)	137 (LR)	0.868	>0.25
1978–1979	R-handed	131 (RL)	165 (RR)	3.9	<0.05
	LL-handed	117 (LLL)	5 (LLR)	102.8	<0.005
	LR-handed	104 (LRL)	114 (LRR)	0.454	>0.50
1979–1980	RL-handed	115 (RLL)	125 (RLR)	0.416	>0.50
	RR-handed	103 (RRL)	185 (RRR)	23.3	<0.005
	LLL-handed	141	151	0.34	>0.50
	RRR-handed	140	161	1.46	>0.10

Source: Rao, K.L. et al., *Curr. Sci.*, 52, 608–609, 1983a.

Key: LL = Left-handed plants produced from left-handed parents.

LR = Left-handed plants produced from right-handed parents.

LLL = left-handed plants produced in the third generation from left-handed parents in the previous two generations.

11.13 DISCUSSION

From the foregoing account, Bahadur and Reddy (1974) believed that the cause of handedness lies at a chemical level and may be connected with the stereoisomerism of hormone molecules. It may be recalled that Fredga and Aberg (1965) have discussed stereoisomerism in plant growth regulators of the auxin type (see also Eiel and Wilen, 1994). Bahadur and Venkateshwarlu (1976) earlier hypothesized that handedness may be due to the presence of optically active substances, that is, levo and dextro compounds present in the plant metabolism. Thus, according to them, molecular chirality is expressed in biological chirality. However, Dubrov (1978) believes that the geomagnetic force determines the biosymmetric status of living objects, and this probably occurs at the time of untwisting of the chromosomes and replication of DNA. Davis (1974), on the other hand, opines that handedness is influenced by geophysical forces. Sulima (ex. Dubrov, 1978, p. 134) established empirically that the left-handedness prevailed in even years 1960, 1962, 1964, and 1968 when the north-polar geomagnetic conditions predominated, while right-handedness dominated in the odd years 1961, 1963, 1965, 1967, and 1969 during the manifestation of south-polar geomagnetic conditions. Hence, the causal factors for the existence of the L and R forms of seedlings are to be sought, and it is hoped that it might open up a new area in the study of plant biology to which we proposed the term "stereo-botany" (Bahadur and Rao, 1981). According to Rao (1980), plants developing from the left-handed seedlings were found to be metabolically superior with regard to quantity of chlorophyll (l/g) tissue, size, and density of stomata coupled with transpiration rates and some of root characters. See also similar observations in this regard by Ghose et al. (1996). This is possibly responsible for the higher yield in contrast to the plants developing from the right-handed seedlings. Such differences in the left- and right-handed plants was not only observed with regard to seedlings but also with regard to ovule position either to the left or right (Horovitz et al., 1976). And according to them, left/right ovule position influences the seed set and outcrossing rates in *M. sativa* (Fabaceae).

Weatherwax (1948) made an interesting observation that the endosperm in maize shows handedness. It's strange but true that maize shows seedling handedness coupled with endosperm handedness. We believe that most cereals must show this condition, and the physiological and biochemical factors that are involved needs investigation. Since most cereals show this condition, it would be interesting to find out the physiological and biochemical and even genetic factors needs further investigation in relation to handedness.

Kihara (1972) points out the use of R/L characters will enable us to achieve clear theoretical and practical results and consequently fuller understanding of the vital process. He further says, "As there is a possible relationship between the foliar arrangement and yield of crop plants, it is necessary to examine the differences between the R and L strains in their utilization of solar energy."

The left- or right-handedness of spikelets is determined by the position of the first floret in the spikelet, and it will be right-handed when viewed from the dorsal side and facing the observer and left-handed when viewed from ventral side. The spikelets are designated by successive numbers from the lowest to the top of inflorescence. Kihara examined 108 ears of *Triticum moncocum* and noted equality of left- and right-handed spikelets in several varieties of Einkorn wheat (Kihara, 1950).

Kihara et al. (1951) studied antidromy, concordance, and mentioned that the R/L handedness is controlled by the threshold reaction, which is the result of the external influences and the specific potency, on the assumption that the external influences are involved by a random variable that distributes normally. They stated that "it seems that the determination of right- and left-handedness is at the highest and then proceeds upwards and downwards as the differentiation of the ear goes on this is largely based on developmental processes in the primordial leaves." Since then, no new work has been carried out in this direction.

Two important reviews by Kihara (1972), and more recently Hashimoto (2002), have shown the current status on the genetic basis and related aspects of plant handedness and are good eye-openers for the current state of knowledge on the subject. But relatively few publications on genetic basis is an indication that not all handedness cases in plants are under genetic control and Mendelian. However, the molecular basis using *Arabidopsis* mutants and cortical microtubules are new developments in the study of left-right handedness in plants. Recently, Andrey and Bahadur have also reviewed this subject (see Chapter 8 in this book).

Thus, it's obvious that there is a lacuna in our fuller understanding, and we feel that the subject of handedness is still in a nascent stage despite its long-recorded history. Hopefully, concerted efforts at interdisciplinary research will unravel the hidden mystery for better understanding of the mechanism of various types of handedness, including seedling handedness and various kinds of handedness on display by plants.

REFERENCES

Aldulov, N.P. 1931. Karyo-systematicsches untersuchungen der Familis Gramineen. *Bull. Appl. Bot. Suppl.* 44: 428.

Arber, A. 1934. *The Graminae: A Study of Cereals, Bambus and Grasses.* Cambridge University Press, London, UK.

Arber, A. 1961. *Monocotyledons.* Hafner, New York.

Avery, G.S. 1930. Comparative anatomy and morphology of embryos and seedlings of maize, oats and wheat. *Bot. Gaz.* 89: 1–39.

Bahadur, B. and N.P. Reddy. 1974. Types of vernation in the cyathia of *Euphorbia milli* Des moulins. *New Phytologist.* 75: 131–134.

Bahadur, B. and T. Venkateswarlu. 1976. Isomerism in flowers of *Carica papaya. J. Indian bot. Soc.* 55: 89-94.

Bahadur, B., N.P. Reddy, P.V. Kumar, and M.M. Rao. 1977. On the occurrence of alternative Germination in *Phaseolus. Curr. Sci.* 46: 499–501.

Bahadur, B., K.L. Rao and M.M. Rao. 1978. Left and right handedness in seedlings of *Bambusa arundenacea* Willd. *Curr, Sci.* 47(16): 548–586.

Bahadur, B., M.M. Rao, and N.P. Reddy. 1980. The left hand, right-handed, and neutral seedlings in *Vigna mungo* (L.) Hepper. *Indian J. Bot.* 3: 13–17.

Bahadur, B. and M.M. Rao. 1981. Seedling handedness in Fabaceae. *Proc. Indian Acad. Sci. (Plant Sci.)* 90 (3): 231-236

Bahadur, B., M.M. Reddy, N.R. Swamy, and G. Narsaiah. 1983. Seedling handedness in Graminae. *Proc. Indian Acad. Sci. (Plant Sci.)* 92(3): 279–283.

Bahadur, B. and N.R. Swamy. 1985. Seedling handedness in Triticale and its parents. *Acta Agron. Scientiarum Hungariceae* 34:117–122

Bhaskar, K.V., B. Bahadur, G. Narsaiah, and N.R. Swamy. 1983. Seedling enantiomorphism in Buckwheat (Polygonaceae) *Curr. Sci.* 52: 25–26.

Biswas, A.L. and N.K. Bhatacharya. 1975. Studies on the X-irradiated *Phaseolus vulgaris* L. and *Glycine max*. *Geobios*. 2: 68–70.

Boyd, L. 1932. Monocotyledonous seedlings. Morphological studies in the post-seminal development of the embryo. *Trans. Proc. Bot. Soc. Edinburgh.* 31: 1–224.

Burger, H.D. 1972. Seedlings of some tropical trees and shrubs mainly of South East Asia (Wagnengin). Translated from Dutch by Burger, D. and Festings, B. Centre for Agric. Publ and Documentation.

Clos, D. 1870. *Monographie de la préfoliation, dans ses rapports avec les divers degrés de classification*. Rouget frère et Delahaut, Toulouse, France.

Compton, R.H. 1910. On right and left handedness in barley. *Proc. Camb. Phil. Soc.* 15: 495–506.

Compton, R.H. 1911. Right and left handedness in cereals. *4th Int. Conf. Genetics* Paris. pp. 4–7.

Compton, R.H. 1912. A further contribution to the study of right- and left-handedness. *J. Genet.* 2: 53–70.

Crow, L.K. 1973. Seedling characters in teaching genetics. In: *Teaching Genetics in Schools and Colleges*. Oliver and Boyd, London, UK.

Cullen, J. 1978. A preliminary survey of ptyxis (vernation) in the Angiosperms. *Proc. R. Bot. Gdns. (Edinburgh)* 37: 161–214.

Davis, T.A. 1974. Enantiomorphic structures in plants. *Proc. Indian Natl. Sci. Acad.* 40: 424–429.

de Vogel, E.F. 1980. *Seedlings of Dicotyledons. Structure, Development, Types. Descriptions of 150 woody Malesian Taxa*. Centre for Agricultural Publishing and Documentation, Wageningen, the Netherlands. 1–465.

Dubrov, A.P. 1978. *The Geomagnetic Field and Life, Geomagnetic Biology*. Plenum Press, New York.

Duke, J.A. 1969. On tropical tree seedlings. I: Seeds, seedlings systems and systematics. *Ann. MO. Bot. Gard.* 56: 125–161.

Eiel, E.L. and S.H. Wilen. 1994. *Stereoselectivity of Organic Compounds*. John Wiley & Sons, New York.

Fredga, A. and B. Aberg. 1965. Stereoisomerism in plant growth regulators of the auxin type. *Anu Rev. Plant Physiol.* 16: 53–73.

Ghose, M., K. Bhattacharya, S.S. Ghosh, K. Roychoudhury, and A. Datey. (1996). Effect of foliar arrangement on the leaf epidermal structures in Areca palm (*Areca catechu* Linn.). *Acta Bot. Neerl.* 45: 303–308.

Guignard, J.L. and J.C. Mestre. 1971. L'embryon des Graminées. *Phytomorphology.* 20: 190–197.

Hashimoto, T. 2002. Molecular genetics analysis of left-right handedness in plants. *Phil. Trans R. Soc. London. B.* 357: 799–808.

Horovitz, A., L. Meiri, and A. Beiles. 1976. Effects of ovule position in Fabaceous flowers on seed set and outcrossing rates. *Bot. Gaz.* 137: 250–254.

Kihara, H. 1950. Right and left handedness in plants. In: *Problems of Modern Biology*. Ed: Nakamura, Japan. pp. 11–30.

Kihara, H., M. Kimura, and H. Ono. 1951. Studies on the right left handedness of spikelets in Einkorn wheats. III. *Proc. Jpn. Acad.* 27: 678–683.

Kihara, H. 1972. Right and left-handedness in plants: A review. *Seiken Ziho* 23: 1–37.

Klebs, G. 1885. Beiträge zur Morphologie und Biologie der Keimung. *Untersuchungen aus dem Botanischen Institut Tübingen.* 1: 536–635.

Kundu, A. and M.S. Sarma. 1965. Direction of leaf spiral in *Corchorus capsularis. Trans. Bose Res. Inst. (Calcutta.)* 28: 107.

Kundu, A. 1968. Studies in the dextro and levorotatory phyllotaxy and their vernation in relation to fruits production. Ph. D. Thesis, Calcutta University, Calcutta, India.

Latting, A. and B. Aberg. 1972. *Biology and Utilization of Grasses.* Eds. Mckell Academic Press, New York.

Lilianfield, F.A. and H. Kihara. 1956a. Dextrality and sinistrality in plants.1. *Medicago tuberculata. Proc. Jpn. Acad.* 32: 620–632.

Lilianfield, F.A. and H. Kihara. 1956b. Dextrality and sinistrality in plants. II. *Medicago littoralis. Proc. Jpn. Acad.* 32: 639–650.

Meyen, S.V. 1973. Plant morphology in its nomothetical aspects. *Bot. Rev.* 39: 205–260.

Muller, F.M. 1978. *Seedlings of the North-Western European Low Land (A Flora of Seedlings).* Dr W Junk Publishers, the Netherlands.

Narsaiah, G. 1984. Handedness in twining plants with reference to *Psophocarpus tetragonolobus* (L.) DC. and *Dioscorea alata* L. in relation to yield. Ph. D. Thesis, Kakatiya University, Warangal, India.

Ono, H., N. Sueoka, N. Suemoto, and K. Kojima. 1954. Studies on the right and left and right handedness of spikelets in Einkorn wheats. V. Effect of X-rays irradiation. *Proc Jpn. Acad.* 30: 221–225.

Ono, H. 1956. Right and left handedness in *Aegilops.* III. The effect of external factors. *Wheat Inf. Serv.* 3: 17–19.

Ono, H. and H. Suemoto. 1957. The right and left handedness in seedlings of *Triticum. Seiken Ziho.* 8: 60–65.

Philipson, W.R. and M. Philipson. 1979. Leaf vernation in *Nothofagus. New Zealand Jour. Bot.* 17: 417–421.

Rao, K.L. and B. Bahadur. 1980. Seedling handedness to *Cajanus cajan* (L.) Millsp. *Curr. Sci.* 49: 201–202.

Rao, K.L. 1981. Studies on seedling and corolla handedness in *Cajanus cajan* (L.) Millsp. (Fabaceae) and its possible influence on yield. Ph.D. Thesis Kakatiya University, Warangal, India.

Rao, K.L., B. Bahadur, and A. Satyanaryana. 1983a. Effect of selection on seedling handedness in Pigeonpea. *Curr. Sci.* 52: 608–609.

Rao, K.L., B. Bahadur, and A. Satyanarayana. 1983b. Seedling handedness in *Cajanus cajan* (L.) Millsp. in relation to seed position. *Curr. Sci.* 52: 1029–1030.

Rao, M.M. 1980. Studies on seedling and corolla handedness in Papilionaceae with special reference to *Vigna* and *Phaseolus* spp. and its possible influence on yield. Ph. D. Thesis, Kakatiya University, Warangal, India.

Rao, M.M. and B. Bahadur. 1982a. A note on the relationship between seedling handedness and root and nodule characters in greengram and blackgram. *Indian J. Agric. Sci.* 52: 706–708.

Rao, M.M. and B. Bahadur. 1982b. Effect of irradiation and centrifugation on seedling enantiomorphism in green gram and black gram. *Curr. Sci.* 51(5): 242–243.

Rao, M.M., B. Bahadur, and N.P. Reddy 1980. Seedling handedness in *Phaseolus vulgaris. Curr. Sci.* 49: 61–62.

Reddy, T. P. and D. Srinivasachar. 1972. *Curr. Sci.* 40: 163.

Reeder, J.R. 1957. The embryo in grass systematics. *Amer J. Bot.* 44: 756–763.

Suemoto, H. 1961. Morphological studies on the right- and left-handedness in the genus *Triticum.* Ph.D. Thesis, Kyoto University, Japan.

Swamy, N.R. and B. Bahadur. 1983. A preliminary study of the handedness in first leaf and successive leaves, flag leaves and spikelets. I. Triticale and its parents. *Curr. Sci.* 52: 89–91.

Swamy, N.R. and B. Bahadur. 1984. Right and left handed seedlings of Triticale and its parents: Some morphological and physiological considerations. *J. Indian Bot. Soc.* 63: 412–415.

Swamy, N.R., B. Bahadur, and G. Narsaiah. 1984. Seedling handedness in Triticale and its parents ll. Yield in relation to handedness. *Curr. Sci.* 53: 538 –539.

Tillich, H.J. 2000. Seedling diversity and the homologies of seedling organs in the order Poales. *Ann. Bot.* 100(7): 1413–1439.

Van Tiegham, P. 1897. Morphologie de la embryon et dela Plantae chez les Graminae's les Cyperaceae. *Ann. Sci. Nat. Bot.* 8(3): 259–309.

Weatherwax, P. 1948. Right handed and left handedness in corn embryo. *Ann. MO. Bot. Gdn.* 35: 317–321.

Wilson, K.L. and D.A. Morrison. Eds. 2000. *Ancestral and Derived Character States in the Seedlings of Monocots: Systematics and Evolution.* CSIRO, Melbourne, Australia.

Younger, V.B. and C.M. Mckell. Eds. 1972. *Biology and Utilization of Grasses.* Academic Press, New York.

12 A Comparison of Chirality Patterns in Climbing Plants (Lianas) of Moist Tropical Forest in Peru and Brazil

Robyn J. Burnham, Wladimir Hermínio de Almeida, Rainiellen S. Carpanedo, Cristiane Miranda da Cruz, Aline C. S. Dresch, Monique Machiner, Lucinere P. Pinto, Rozangela Cristina Alves de Oliveira, Evandro F. dos Santos, and Bradley Spilka

CONTENTS

12.1   INTRODUCTION

A fascination with twining plants has been present in scientific literature for centuries, with grapevines the focus of an entire book, *The Natural History, Pliny the Elder* (Bostock 1855). The nature of rotation in climbing stems of peas, bindweed, and even woody nightshade (*Solanum*) were observed and painstakingly recorded by Darwin (1865). Even Linnaeus, in the *Philosophia Botanica*, documented the presence of variation in twining orientation (Linné, C. von and S. Freer 2005). Twining mechanisms are even documented in the fossil record of climbing plants (Wang et al. 2013). The mechanical or physiological basis for dextral versus sinistral orientation of circumnutation among apically twining plants has not been determined, despite several papers on the topic in recent years (Thitamadee et al. 2002, Kitazawa et al. 2005, Goriely and Neukirch 2006, Edwards et al. 2007, Mugnai et al. 2007, Edwards and Moles 2009, Isnard and Silk 2009, Stolarz 2009, Bastien and Meroz 2016, Smyth 2016). The majority of the data supporting a general hypothesis that twining is caused by microfibril orientation and their overlying microtubules is based on research on the roots and seedlings of mutant forms of *Arabidopsis thaliana*, a non-twining plant species (Hashimoto 2002, Smyth 2016). The specific orientation of microfibrils in twining plant species has apparently received little attention.

We present data on the chirality (direction of circumnutation) among climbing species from two censuses of local floras in lowland Amazonian Peru and Brazil. Our goal was to determine whether field observations of species in distinct regions of the Amazon basin were consistent in the proportion of dextral and sinistral climbing orientations, particularly at the level of family, genus, and species (rather than at the individual level). Our work differs from the global approach of Edwards et al. (2007) in two aspects: (1) our surveys include identifiable species or morpho-species as they are encountered in field transects, rather than censusing a fixed number of individuals for which taxonomic identities are not identified; and (2) our focus on a single geographic region (the Amazon basin) reduces the possibility that compositional differences at higher taxonomic levels (e.g., families) influence the observations.

We address the following questions via comparison of the two sites sampled:

- Is the proportion of dextral/sinistral climbers different in two disparate parts of the Amazon basin?
- Compared to the larger number of species evaluated in Peru, does the sample from Brazil show bias?
- Given that we anticipate that the same suite of families would be represented in the Amazon flora, are any families restricted to one region, thus influencing the summary data with respect to our hypothesis of similarity?

Finally, we discuss the insights that the distribution of climbing chiralities among species and families might provide to investigations on the basis for overwhelming dextral chirality among stem-twining climber species. We compare our data to those of prior similar surveys in India and to global results presented by Edwards et al. (2007).

12.2   METHODS

12.2.1   Los Amigos, Peru Sampling

In southeastern Peru (Department of Madre de Dios), we sampled climbing plants at the Los Amigos Biological Station (12°34′07″S, 70°05′57″W). Concurrent with a complete census of climber distribution, diversity, and density of two 0.5 ha forest plots, we also made observations on the chirality of stem-twining species (Burnham and Revilla-Minaya 2011). We also surveyed 5.2 km of forested trails in the surrounding area to record twining chirality. No more than eight individuals of the same taxon were included in our tallies. Eight was our maximum observation number per species, based on the probability (1 in 250) of incorrectly coding a taxon as demonstrating a single twining chirality if both dextral and sinistral were actually present in equal numbers in the population. However, most climbing species are sufficiently rare in the Los Amigos area, that we observed most taxa only two to three times. Thus, we maximized phylogenetic diversity and minimized duplicate observations of the same taxon. All individuals grew in densely shaded *terra firme* habitats and were vigorous but sterile adults, that is, without evident fruits or flowers. Herbarium vouchers were made of species from the 0.5 ha plots, but not from all individuals on the trail censuses (see Burnham and Revilla-Minaya 2011). Specimens are deposited at the herbarium of Universidad Agraria La Molina in Lima, Peru, and at the University of Michigan Herbarium, in Ann Arbor, Michigan, USA.

12.2.2 MATO GROSSO, BRAZIL SAMPLING

In Mato Grosso, Brazil, sampling was carried out at the Rio Ronuro Ecological Station, Fazenda São Judas Tadeu, municipality of Nova Ubiratã. This conservation unit covers 102,000 hectares between latitudes 12°46′00″ and 14°07′00″S and longitudes 55°15′00″ to 54°19′00″W. Mean annual temperature is 25°C, and precipitation is about 2000 mm per year with a general Am climate of the Köppen system (Silva et al. 2009). The vegetation of the general area is a transition between semi-deciduous and open evergreen forest and has been preserved because of the high diversity of habitats and species that are present in the forests. Over two days in December 2017, two habitats were sampled: forest interior via trails and a disturbed forest edge via an access road. Methods were similar to those used in Peru; however, sampling was limited to about 1000 m along two major trails. Individual climbing species with an apical twining mechanism were classified to family, identified when known, or assigned an informal morpho-species name and scored as dextral, sinistral, or dual-chirality. Specimens collected in Mato Grosso are housed at the herbarium of Universidade Federal do Mato Grosso (UFMT), Sinop, Mato Grosso.

At both sites, we excluded all taxa whose main mechanism of ascent is by tendrils, hooks, or spines (members of the Bignoniaceae, Passifloraceae, Rhamnaceae, Sapindaceae, etc.), even if their stems twined weakly. We record an individual's chirality only when the stem twined at least three times around its support. To avoid species duplication via individual clonality, a minimum distance of 20 m was established between observations of individuals of the same species.

Individuals were included in the census with identification to family as a minimum, because our intent was to maximize representation of the local floras. In Brazil, one species was included without a family identification. Each morpho-species is a distinct entity in the case of incompletely identified taxa, regardless of taxonomic level. Where distinction between potential morpho-species was not possible, individuals were lumped, creating a minimum estimate of species present. Based on the first author's experience with climbers of the Amazon region, we are confident of our placement of the liana individuals into appropriate taxonomic groups.

Comparison with an earlier census of twining chirality in 26 families from India (Narsaiah et al. 1985) was performed at the family level only.

12.2.3 DATA ANALYSIS

Family, morpho-species, and individual counts were summarized for each site and coded by climbing chirality (dextral, sinistral, both). The proportions of each climbing mechanism were hypothesized to be equal for the two sites and were tested by means of individual chi-squared tests. We evaluated compositional differences using the Sorenson's Index of Similarity at the family level.

12.3 RESULTS

Just over half the number of morpho-species were observed in Brazil as in Peru (33 vs. 60). However, a clear pattern in chirality orientation across climbers emerges from comparison of the two Amazonian sites. Vines and lianas in Brazil and Peru display dextral chirality in 81%–82% of 11 and 21 families observed, respectively. Sinistral chirality was present in 0% and 14% of families, while both chiralities were present in 5% and 18% of families, respectively. At the morpho-species level, we found a similar pattern: Peru had 85%, 7%, and 8% of morpho-species with chirality of dextral, sinistral, and both, respectively. Brazil had 82%, 3%, and 15% of morpho-species with dextral, sinistral, and both, respectively. Neither sinistral nor dextral proportions of climbers were significantly different between the two sites (dextral chi-square = 0.222, P = 0.6372; sinistral chi-square = 0.548, P = 0.4590).

The family with the highest number of morpho-species in Peru was Malpighiaceae with eight morpho-species, followed by Apocynaceae and Menispermaceae, both with seven morpho-species. In Brazil, Apocynaceae included seven morpho-species, Dilleniaceae included six, and Malpighiaceae included five. We observed only 59 individuals in Brazil versus 146 individuals in Peru, and thus included fewer morpho-species in Brazil. However, the relatively low abundance of Menispermaceae in the Brazil census (only two morpho-species) is suggestive of more disturbed conditions and more seasonal rainfall in the Rio Ronuro area. In a broader survey of lianas at Rio Ronuro, we did encounter two additional species of Menispermaceae (Burnham et al. in press). In that broader survey of lianas at Rio Ronuro, Brazil, we also observed seven species each in Apocynaceae and Malpighiaceae, slightly increasing the number of species in those dextral climbing families.

Chirality is not often observed and recorded at the species level, but one study to which we can compare chirality of the Amazonian lianas is the comprehensive survey of 278 species in 26 families of cultivated and wild climbers from India (Narsaiah et al. 1985). In that study, chirality was reported at the family level, and not for genera and species. We updated the taxonomic assignments of the families represented (e.g., merging Asclepiadaceae and Apocynaceae) and calculated the proportion of families with each chirality state. These data from a broad geographic area in India show 84% dextral chirality at the family level, similar to the 81%–82% dextral chirality seen in Peru and Brazil (see Table 12.1 for a list of morpho-species at the Amazonian sites).

The overwhelming majority of families, species, and individuals of lianas in the Amazon region display dextral chirality. Sinistral chirality was represented in only 3 of 21 families in Peru: Asteraceae, Dioscoreaceae, and Piperaceae, but none in Brazil, other than the dual chirality Dilleniaceae. The climbing species in the sinistral families in Peru

TABLE 12.1

(a) Climber Species Evaluated at Los Amigos, Peru

Family	Genus	Species	Number Individuals	Chirality	Number Species per Family
Acanthaceae	Mendoncia	sp. 1	3	D	3
Acanthaceae	Mendoncia	sericea	3	D	
Acanthaceae	Mendoncia	sp. 3	3	D	
Apocynaceae	Forsteronia	amblybasis	7	D	7
Apocynaceae	Forsteronia	acouci	2	D	
Apocynaceae	Odontadenia	puncticulosa	1	D	
Apocynaceae	Odontadenia	verrucosa	1	D	
Apocynaceae	Gonolobus	sp. 1	3	D	
Apocynaceae	Genus indet. 1	sp. 1	2	D	
Apocynaceae	Genus indet. 2	sp. 1	1	D	
Aristolochiaceae	Aristolochia	sp. 1	1	D	3
Aristolochiaceae	Aristolochia	sp. 2	1	D	
Aristolochiaceae	Aristolochia	dalyi	2	D	
Asteraceae	Mikania	guaco	8	S	
Asteraceae	Mikania	sp. 1	1	S	
Combretaceae	Combretum	laxum	7	D	3
Combretaceae	Combretum	assimile	1	D	
Combretaceae	Genus indet. 1		1	D	
Connaraceae	Pseudoconnarus	macrophyllus	1	D	1
Convolvulaceae	Ipomoea	sp. 1	5	D	3
Convolvulaceae	Maripa	sp. 1	1	D	
Convolvulaceae	Dicranostyles	sp. 1	1	D	
Dichapetalaceae	Dichapetalum	sp. 1	2	D	
Dilleniaceae	Genus indet.	sp. indet. 1	2	S, D	5
Dilleniaceae	Genus indet.	sp. indet. 2	4	S, D	
Dilleniaceae	Genus indet.	sp. indet. 3	4	S, D	
Dilleniaceae	Genus indet.	sp. indet. 4	2	S, D	
Dilleniaceae	Genus indet.	sp. indet. 5	2	S, D	
Dioscoreaceae	Dioscorea L.	sp. 1	2	S	2
Euphorbiaceae	Plukenetia	brachybotrya	2	D	1
Fabaceae	Deguelia		3	D	4
Fabaceae	Genus indet. 1		3	D	
Fabaceae	Genus indet. 2		1	D	
Fabaceae	Genus indet. 3		1	D	
Hernandiaceae	Sparattanthelium	tarapotanum	1	D	1
Malpighiaceae	Mezia	sp. indet. 1	4	D	8
Malpighiaceae	Genus indet. 1	sp. indet. 2	3	D	
Malpighiaceae	Genus indet. 1	sp. indet. 3	1	D	
Malpighiaceae	Banisteriopsis	muricata	1	D	
Malpighiaceae	Mascagnia	dissimilis	2	D	

(*Continued*)

TABLE 12.1 (*Continued*)

(a) Climber Species Evaluated at Los Amigos, Peru

Family	Genus	Species	Number Individuals	Chirality	Number Species per Family
Malpighiaceae	Hiraea	fagifolia	4	D	
Malpighiaceae	Hiraea	sp. indet. 2	1	D	
Malpighiaceae	Tetrapterys	sp. indet. 1	1	D	
Malvaceae	Byttneria	benensis	2	D	1
Menispermaceae	Sciadotenia	toxifera	2	D	7
Menispermaceae	Abuta	rufescens	2	D	
Menispermaceae	Anomospermum	grandifolium	4	D	
Menispermaceae	Anomospermum	chloranthum	4	D	
Menispermaceae	Cissampelos	pareira	2	D	
Menispermaceae	Disciphania	sp. indet. 1	3	D	
Menispermaceae	Telitoxicum	minutiflorum	1	D	
Olacaceae	Heisteria	scandens	2	D	1
Piperaceae	Manekia	sydowii	3	S	2
Piperaceae	Piper	sp. indet. 1	2	S	
Polygonaceae	Coccoloba	excelsa	3	D	3
Polygonaceae	Coccoloba	marginata	5	D	
Polygonaceae	Coccoloba	parimensis	2	D	
Rubiaceae	Manettia	cordifolia	1	D	1
Verbenaceae	Petrea	blanchetiana	1	D	2
Verbenaceae	Petrea	maynensis	4	D	

(b) Climber Species Evaluated at Rio Ronuro, Mato Grosso, Brazil

Family	Genus	Species	Number Individuals	Chirality	Number Species per Family
Apocynaceae	"Asclepiadoid"	sp. indet. 1	1	D	7
Apocynaceae	Forsteronia sp. 1	sp. indet. 1	2	D	
Apocynaceae	Odontadenia sp. 1	sp. indet. 1	1	D	
Apocynaceae	Prestoniasp. 1	sp. indet. 1	6	D	
Apocynaceae	Genus indet. 1 RSC 188	sp. indet. 1	3	D	
Apocynaceae	Genus indet. 2	sp. indet. 2	1	D	
Apocynaceae	Genus indet. 3	sp. indet. 3	2	D	
Aristolochiaceae	Aristolochia	sp. indet. 1	1	D	1
Celastraceae	Genus indet. 1 RSC 218	DESTRO	1	D	1
Convolvulaceae	Dicranostyles	densa cf.	1	D	2
Convovulaceae	Dicranostyles/ Maripa	"yellow vein"	1	D	
Dilleniaceae	Genus indet. 1	sp. indet. 1	2	D/S	7
Dilleniaceae	Genus indet. 2	sp. indet. 2	3	S	
Dilleniaceae	Genus indet. 3	sp. indet. 3	1	D/S	
Dilleniaceae	Genus indet. 4	sp. indet. 4	2	D/S	

(Continued)

TABLE 12.1 (*Continued*)

(b) Climber Species Evaluated at Rio Ronuro, Mato Grosso, Brazil

Family	Genus	Species	Number Individuals	Chirality	Number Species per Family
Dilleniaceae	Genus indet. 5	sp. indet. 5	2	D	
Dilleniaceae	Genus indet. 6. BB-9	sp. indet. 6	1	D	
Dilleniaceae	Genus indet. 7	sp. indet. 7	1	D	
Fabaceae	Centrosema	sagittata	2	D/S	3
Fabaceae	Deguelia	rariflora	5	D	
Fabaceae	Mimosa cf.	sp. indet. 1	1	D	
Malpighiaceae	Genus indet. 1	sp. indet. 1	1	D	5
Malpighiaceae	Diplopterys	sp. indet. 1	1	D	
Malpighiaceae	Mascagnia	sp. indet. 1	1	D	
Malpighiaceae	Mascagnia	sp. indet. 2	4	D	
Malpighiaceae	Genus indet. 2	sp. indet. 1	4	D	
Menispermaceae	Cissampelos	sp. indet. 1	2	D	2
Menispermaceae	Disciphania	sp. indet. 1	1	D	
Piperaceae	Sarcohachis	sp. indet. 1	1	D	1
Polygonaceae	Coccoloba	sp. indet. 1	2	D	2
Polygonaceae	Coccoloba	sp. indet. 2	1	D	
Unknown	Genus indet. 1	sp. indet. 1	1	D	1

Chirality: D = dextral, S = sinistral.

are less common among tropical forest lianas, occurring most often along forest edges and in disturbed areas. Both chiralities are present in Dilleniaceae throughout Brazil and Peru, but in India dual chirality was reported only for Asteraceae and Dioscoreaceae.

The liana chiralities from Peru, Brazil, and India were also compared for similarity among the families represented. We hypothesized that if family composition was largely the same across sites and that if chirality was a family-level attribute, that chirality proportions should be similar among sites. Among the three sites, 35 families were scored for chirality, yet only 6 families were present in all three regions. Sorenson's Similarity Coefficient was used to evaluate the compositional similarity of families among sites. Peru and Brazil had a similarity coefficient of 56%, India and Peru followed with 55%, and Brazil and India were the least similar with 36%. In contrast, the proportion of dextral chirality for all sites was very close: 81%–84%, suggesting that family composition is not an overriding factor. The family-level similarity between Peru and India was as high as that between Peru and Brazil, which was not expected, given the geographic distances involved. The small sample size from Brazil (only 11 families were observed) clearly biased the comparison of the family similarity coefficients. However, with respect to chirality, because most species have dextral chirality, the high level of chiral similarity is expected, regardless of family overlap.

12.4   DISCUSSION

We found that the proportion of dextral chirality among climber species in two disparate areas of the Amazon basin is virtually identical at the family level (81% in Peru vs. 82% in Brazil), and at the morpho-species level, the proportions are also similar (83% vs. 82%, respectively). This result is found in spite of different sampling intensities in the censuses. While only 9 of 21 families in Peru are found in Brazil, those 9 families comprise the majority of families in the Brazil sample, thus contributing to a similar distribution of chirality. The census from Brazil is limited because of the small sample, yet it still shows a similar pattern to the Peruvian census. Relatively high numbers of morpho-species in the dual-chirality Dilleniaceae increases the similarity of the two Amazonian samples at the species level. One family observed in Los Amigos, Peru, and not in Rio Ronuro, Brazil, is the Asteraceae. This family has been observed frequently in the western and southern Amazon basin (Yasuní, Ecuador, and Mata Atlantica, Brazil) by the first author and showed sinistral twining among all species observed in the field. Further, *Dioscorea* (Dioscoreaceae) has been reported as including species that are either sinistral or dextral, but not both (Wilkin et al. 2005, Lin et al. 2010). This family was found in Peru but not in Brazil. Combined, these two families contribute three of the five morpho-species that are sinistral climbers in Peru. These morpho-species are relatively uncommon and the families are not diverse in either site, but in areas where a higher richness of Asteraceae climbers are found (e.g., Mata Atlantica of southern Brazil: Araújo and Alves 2010, Oliveira-Gomes et al. 2018), the proportions of sinistral versus dextral climbers may differ from those reported here.

Similar to the summary of climber chirality in India by Narsaiah et al. (1985), we found a high proportion of dextral climbers among families, even within disparate lineages. Their work states that Asteraceae includes at least one dual chirality species (*Mikania micrantha*, reported in Narsaiah and Bahadur 1984), a condition that otherwise is only observed by us to occur in Dilleniaceae. We verified their observation of *M. micrantha* by using online images of herbarium specimens deposited at the Missouri Botanical Garden. There is clear support of dextral chirality in specimen 21493 collected by S. P. Churchill from Bolivia, but clear support of sinistral chirality in specimen 3526 collected by C. Galdames from Panama. Edwards et al. (2007) found, in a global survey of unidentified climbers, that 92% of climbing individuals had dextral chirality and that this pattern was related neither to latitude nor to hemisphere. We emphasize that the ontogenetic or physiological basis for the preponderance of dextral chirality among climber families, genera, species, and individuals is not presently understood. A promising comparison would involve closely related species whose chiralities are opposite (e.g., members of the genus *Dioscorea*) or species whose members can climb in either direction (e.g., *M. micrantha* in the Asteraceae). Anatomical and/or biochemical differences may then be linked to the genetic basis for these differences. In the light of entire families showing sinistral or dual chiralities, we propose that environmental differences will not be found as the ultimate cause for this plant behavior.

REFERENCES

Araújo, D. and M. Alves. 2010. Climbing plants of a fragmented area of lowland Atlantic Forest, Igarassu, Pernambuco (northeastern Brazil). *Phytotaxa* 8: 1–24.

Bastien, R. and Y. Meroz. 2016. The kinematics of plant nutation reveals a simple relation between curvature and the orientation of differential growth. *PLoS Computational Biology* 12(12): e1005238. doi:10.1371/journal.pcbi.1005238.

Bostock, J. 1855. *The Natural History. Pliny the Elder.* London, UK: Taylor & Francis Group. http://www.perseus.tufts.edu/hopper/text?doc=Perseus%3Atext%3A1999.02.0137%3A book%3D14%3Achapter%3D1.

Burnham, R.J. and C. Revilla-Minaya. 2011. Phylogenetic influence on twining chirality in lianas from Amazonian Peru. *Annals of the Missouri Botanical Garden* 98(2): 196–205.

Burnham, R.J., R.S. Carpanedo, and D. de J. Rodrigues. (in press). Species richness and dispersal of lianas of the Rio Ronuro Ecological Reserve, central Mato Grosso, Brazil. *In* D. de J. Rodrigues et al., eds. *Biodiversidade da Reserva Ecologica do Rio Ronuro*.

Darwin, C. 1865. On the movements and habits of climbing plants. *Journal of the Linnean Society of London, Botany* 9: 1–118. doi:10.1111/j.1095-8339.1865.tb00011.x.

Edwards, W. and A.T. Moles. 2009. Re-contemplate an entangled bank: The power of movement in plants revisited. *Botanical Journal of the Linnean Society* 160(2): 111–118. doi:10.1111/j.1095-8339.2009.00972.x.

Edwards, W., A.T. Moles, and P. Franks. 2007. The global trend in plant twining direction. *Global Ecology and Biogeography* 16: 795–800. doi:10.1111/j.1466-8238.2007.00326.x.

Goriely, A. and S. Neukirch. 2006. Mechanics of climbing and attachment in twining plants. *Physical Review Letters* 97: 184302.

Hashimoto, T. 2002. Molecular genetic analysis of left-right handedness in plants. *Philosophical Transactions of the Royal Society, B Biological Sciences* 357: 799–808.

Isnard, S. and W.K. Silk. 2009. Moving with climbing plants from Charles Darwin's time into the 21st century. *American Journal of Botany* 96(7): 1205–1221.

Kitazawa, D., Y. Hatakeda, M. Kamada, N. Fujii, Y. Miyazawa, A. Hoshino, S. Iida et al. 2005. Shoot circumnutation and winding movements require gravisensing cells. *Proceedings of the National Academy of Sciences, USA* 102(51): 18742–18747.

Lin, G., H. Ci, Y. Liu, and J. Su. 2010. Phylogenetic analysis of chirality of climbing plants. Agricultural Science and Technology 11(2): 34–38.

Linné, C. von and S. Freer. 2005. *Linnaeus' Philosophia Botanica.* Oxford, UK: Oxford University Press.

Mugnai, S., E. Azzarello, E. Masi, C. Pandolfi, and S. Mancuso. 2007. Nutation in plants. pp. 77–90. *In* S. Mancuso and S. Shabala, eds., *Rhythms in Plants: Phenomenology, Mechanisms, and Adaptive Significance.* Springer-Verlag, Berlin, Germany.

Narsaiah, G. and B. Bahadur. 1984. Twining handedness in flowering plants. *Indian Journal of Botany* 7(1): 41–49.

Narsaiah, G., N. Ramaswamy, and B. Bahadur. 1985. Foliar and epidermal features in relation to twining habit in flowering plans. *Journal of the Swamy Botanical Club* 2(4): 117–124.

Oliveira-Gomes, L.C., J. Durigon, P.T. Padilha, and V. Citadini-Zanette. 2018. Composiçãoflorística e estrutura da comunidade de trepadeiras da Floresta Atlântica no Sul de Santa Catarina, Brasil. *Iheringia Série Botânica* 73(1): 5–12.

Silva, da, N.M., A.M. Batistella, A.M.M. Coelho, and V.L.N. Kuroyanagi. 2009. Monitoring of environmental impacts in the buffer zone of the Rio Ronuro protected area, Nova Ubiratã, Mato Grosso, Brazil. *Engenharia Ambiental* 6(2): 484–491.

Smyth. D.R. 2016. Helical growth in plant organs: Mechanisms and significance. *Development* 143: 3272–3282. doi:10.1242/dev.134064.

Stolarz, M. 2009. Circumnutation as a visible plant action and reaction: Physiological, cellular and molecular basis for circumnutations. *Plant Signaling & Behavior* 4(5): 380–387.

Thitamadee, S., K. Tuchihara, and T. Hashimoto. 2002. Microtubule basis for left-handed helical growth in Arabidopsis. *Nature* 417: 193–196.

Wang, Q., S. Shen, and Z. Li. 2013. A left-handed, stem-twining plant from the Miocene Shanwang Formation of Eastern China. *American Journal of Plant Sciences* 4: 18–22. doi:10.4236/ajps.2013.45A003.

Wilkin, P., P. Schols, M.W. Chase, K. Chayamarit, C.A. Furness, S, Huysmans, F. Rakotonasolo, E. Smets, and C. Thapyai. 2005. A plastid gene phylogeny of the Yam genus, *Dioscorea*: Roots, fruits and Madagascar. *Systematic Botany* 30(4): 736–749.

13 Handedness in Plant Tendrils

Pedro E. S. Silva, Fernão Vistulo de Abreu, Ana Isabel D. Correia, and Maria Helena Godinho

CONTENTS

13.1 INTRODUCTION

In 1865, the seminal work of Darwin (1865), about climbing plants, described not only the circumnutation movements of tendrils but also how they became twisted. Darwin found that some plants had their tendrils curled forming spiral-like shapes, like those shown in Figure 13.1a and b. However, when the end of the tendrils was attached to a support, so both extremities were constrained, tendrils would not only curl but also twist and acquire helical-like shapes. Examples of tendrils with helical shapes are shown in Figure 13.1c and d. These helices are not regular: left- and right-handed loops are interconnected by segments that reverse the handedness of the structure. This phenomenon is always observed whenever both ends are constrained and in tendrils with a curling behavior.

The inversion of the handedness can also be observed in different everyday life contexts, such as in gift ribbons and telephone cords (Goriely and Tabor, 1998). For instance, gift ribbons acquire a coiled shape when a scissor blade is swept along the dull side of the stripe. Afterwards, if the coiled ribbon is stretched with both ends fixed, loops tighten up until a maximum extension is reached (Figure 13.2a,b). Upon release, the ribbon returns to its original configuration (Figure 13.2c). However, if a straight configuration is imposed by unwinding the helix (Figure 13.2d–f), the release of the ribbon, with both ends clamped, leads to the rising of shapes similar to the ones observed on tendrils of a climbing plant (Figure 13.2g–i). These reversal segments that connect right- and left-handed helices were named as "perversions" by Goriely and Tabor (1998) and since then used by different authors (Domokos and Healey, 2005; Liu et al., 2016; McMillen and Goriely, 2002).

FIGURE 13.1 Tendrils from *Bryonia alba* and *Cucurbita pepo*: (a) and (b) tendril with spiral shapes; (c) and (d) tendril with right- and left-handed helices connected by perversions; (d) dry tendril showing smooth and rough surfaces, on inner and outer parts of the helix, respectively.

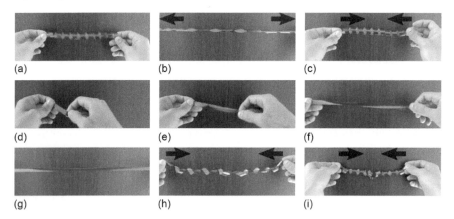

FIGURE 13.2 Formation of perversions in ribbons. Gift ribbons: (a) to (c) pull and release of a helical gift ribbon, with both ends fixed; the loops get tightened and afterwards the ribbon recovers to the original shape; (d) to (f) removal of the loops from the ribbon; and (g) to (i) due to the constraint imposed in the ends of the ribbon, when the stripe is allowed to relax a perversion connecting the two helices with opposite handedness first arises as in tendrils of plants. (Reprinted by permission from Macmillan Publishers Ltd. *Sci. Rep.*, Silva, P.E.S. et al., 2016, Copyright 2016.)

In this work, the formation of helices and perversions in tendrils and the functions that these shapes can provide in climbing plants is first addressed. Then, illustrated by experimental results, we explain how helical filamentary materials with perversions can be produced from the macro to the nanoscale. Additionally, computer simulations of structures containing tendril perversions are provided to complement theoretical and experimental results. Finally, we address current and potential technological applications and provide some current insights on research inspired on the shapes of the tendrils.

13.2   HELICAL CONFORMATIONS

The helical shape is present in many natural systems and synthetic materials. A helix is a smooth curve in a three-dimensional space characterized by its tangent making a constant angle with a fixed line (center line). A simple way to describe the shape of a helix is to think of a circular motion and add a translation motion in the perpendicular direction (to the plane that contains the circular motion). A consequence from the combination of both motions, translational and rotational, is the handedness of a helix. Considering the movement of a closing hand, as represented in Figure 13.3, where the thumb gives the direction of translational motion and the remaining fingers, performing the movement of a closing hand, the direction of rotational motion. If a helical curve follows the action of a right hand, then it is called a right-handed helix (Figure 13.3a). If instead, a curve follows the action of a left hand, then it is denominated a left-handed helix (Figure 13.3b). The handedness of a helix is invariant under rotation. If instead of starting in the bottom end of the curve, in Figure 13.2a, we consider the upper end, then both directions of translational and rotational motions are reversed. The thumb now points down, and the remaining fingers close in the clockwise direction. However, handedness remains unchanged. The same hand is still used to describe the evolution of the curve path.

Mathematically, a helical curve, $\mathbf{r}(s)$, can be written parametrically by:

$$x = R\cos\alpha,$$
$$y = R\sin\alpha, \qquad\qquad (13.1)$$
$$z = h\alpha,$$

Where R is the radius of the helix; h is the pitch, given by the vertical separation between the helix loops; and α is a point of the curve (Figure 13.4). The curvature, κ,

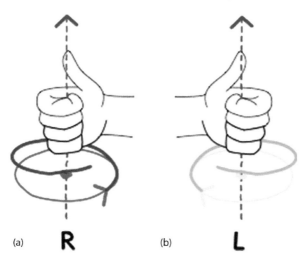

(a)   **R**   (b)   **L**

FIGURE 13.3   (a,b) Handedness of a curve. Helices are right- or left-handed when they follow the movement of a right(R) or left-(L)closing hand, respectively, with the upward thumb pointing in the evolution of the curve path.

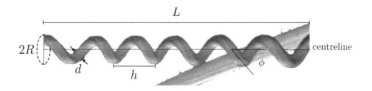

FIGURE 13.4 Right-handed helix from a tendril of the *Bryonia alba* to illustrate a helix and its properties: R is the radius of the helix; d is the diameter of the tendril; h is the pitch; and L is the helical length.

and torsion, τ, can be defined in terms of h and R by the expressions (Gray et al., 2006), $\kappa = \frac{R}{R^2+h^2}$ and $\tau = \frac{h}{R^2+h^2}$. The sign of torsion determines the handedness of the three-dimensional curve. If τ is positive, then helices are right-handed, and when τ is negative, helices are left-handed. The angle between the tangent vector and the center line (Figure 13.4), ϕ, can be defined as: $\frac{\kappa}{\tau} = \frac{R}{h} = \tan\phi$, which means that a constant curvature-to-torsion ratio is a necessary and sufficient condition for a curve to be a helix (Lancret's theorem) (Weisstein, 2011).

In nature or human-made products, helicity can arise in different contexts. A standard way to design helical compression springs involves the winding of wire around a shaft. In some materials, a heat treatment is required to accommodate the stresses caused by the deformations and, after cooling, the springs keep their shape. Another way to produce helical springs involves the application of a constant deformation to a straight wire by driving it toward a grooved head. The feed rate and the point of contact tune the helical shape (Longe, 2001). The helicity arises in both processes by intrinsically creating a constant curvature and torsion along a straight wire. By contrast, Snir and Kamien (2005) suggested a purely entropic approach to understand the helical folding of molecular chains. The researchers started by considering long molecules as solid and impenetrable tubes immersed in a solution of hard spheres. For specific ratios between the tube and spheres radii, the helical shape promotes the increase of the entropy of the spheres, as some regions of the tube surroundings overlap, decreasing the excluded volume. This model explains the occurrence of helicity on many α helices and β sheets.

Furthermore, most coiled structures do not present a regular shape. However, even if Lancret's theorem is not verified for those structures, if their shape is close enough to the helical shape, they are often considered as helices. For instance, in the case of the DNA when the molecule opens for the replication process, as a result from winding the molecule, it will generate an additional twisting called as supercoiling. This supercoiling effect makes the double twisted strands adopt more complex shapes, like superhelices and plectonemes (Bates and Maxwell, 2005). The supercoiling of the DNA can be measured by a topological quantity that defines the number of times that strands turn in the original linear molecule: the linking number, Lk. When the DNA changes in response to the supercoiling, there is a change in the number of times strands revolve one another: the twist, Tw. Moreover, there is another quantity that describes how much the molecule coils and kinks in space: the writhe, Wr. All these quantities might not be straightforward calculated, or even hard to grasp the physical meaning in a three-dimensional space. However, a critical result from the study of the DNA supercoiling is that the sum of the twist and writhe numbers is equal to the linking number, $Tw + Wr = Lk$. In a closed molecule, like DNA rings, the linking number remains constant, but twist and writhe may convert one into another. Furthermore, despite the linking number being only defined in closed curves, many types of DNA have anchorage points, which constrain the rotation of the DNA. Then, the anchorage points can be considered as if they were connected. Also, the control and formation of knots of topological defects were found crucial to the stabilization of colloidal buildings in liquid crystals (Jampani et al., 2011).

A noticeable singularity, which results from the conservation of the topological law, can be observed in plant tendrils, as reported by Darwin (1865). Similarly, the invariant linking number on DNA, where it remains constant upon winding, is also verified during the curling of tendrils and gift ribbons. However, to calculate the linking number, two closed curves are required. Since both ends are clamped, the same approximation for DNA can be used by considering the fixed points being connected. In the case of the gift ribbon, the two borders can be used for describing the linking number between the two curves (Pieranski et al., 2004).

In the case of some tendrils, like the ones of the passion flower, the cross section is approximately circular, which makes it harder to define two curves along the filament. Some other tendrils have cross sections where two curves can be easily defined, like in the case of the rectangular cross section of cucumber tendrils. Nevertheless, the conservation law implies that the number of right- and left-handed loops must be the same during the curling of tendrils, ribbons, or any winding filament constrained at two points, given there is no twist initially. The mechanism underlying the curling behavior in tendrils can be understood with returning to the example of the gift ribbon shown in Figure 13.2. In the step where the dull side of the ribbon is run over with a blade, the elastic properties of the dull side change relatively to the bright side. This modification leads to an increase of the intrinsic curvature, which is related with the final number of loops. In case of the tendrils, the number of loops increases over time, even if no significant alteration of the end-to-end distance is observed. Therefore, the elastic properties must be changing over time. Indeed, observations performed by Gerbode et al. (2012) demonstrated the differences between straight and coiled cucumber tendrils. Figure 13.5 shows the difference between straight and coiled tendrils. By inspection of the cross section of both tendrils (Figure 13.5b and e), it is possible to observe a major variation at the top left region of images. This modification is supposedly due to the strong lignification on the side of the tendril (Bowling and Vaughn, 2009; Gerbode et al., 2012; Meloche et al., 2006).

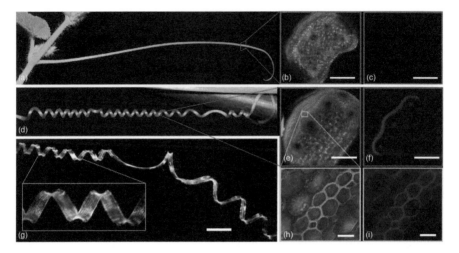

FIGURE 13.5 Coiling mechanism of tendrils: (a) early tendril present residual to none curvature and lack lignified g-fiber cells; (b) darkfield and (c) UV autofluorescence showing no lignin in a tendril cross section. After attaching to a support, the tendril curls due to the strong lignification of some cells, which can be observed in (e) and (g) darkfield and (f) and (h) UV autofluorescence; (i) extracted ribbons retain the helical morphology. Scale bars, (b) and (c) 0.5 mm, (e) and (f) 100 mm, (g) and (h) 10 mm, (i) 1 mm. (Adapted from Gerbode, S.J. et al., *Science*, 337, 1087–1091, 2012. With *Science* permission.)

13.3 TENDRIL PERVERSIONS AND THEIR FUNCTIONS

Tendrils are derived from leaves, parts of leaves (petioles, stipules), or from stems. Stem tendrils can be branched or have adhesive discs, and they also can form woody clasping hooks if they become thickened. In some cases, the true identity of tendrils is hard to interpret (Bell and Bryan, 1991). Perversions appear typically in filamentary or ribbon-like tendrils with diameters of the order of a few millimeters and a wide range of lengths that can vary, for example, from a few centimeters in *Bignonia unguis* to approximately a half meter in *Vitis vinifera* (Jaffe and Galston, 1968). These filaments present inconstant directional ellipsoidal movements, circumnutation movements, to increase the possibilities of interacting with surrounding supports. For instance, Figure 13.6 shows the attachment of a tendril on an external support. After making contact, the filament starts to revolve around the support (Figure 13.6a,b). A good circumvolution of the tendril's free end is important to guarantee that the tendril keeps attached under stimuli. Afterwards, this tendril starts coiling, which imposes the appearance of a perversion connecting right- and left-handed helices (Figure 13.6c–e). By contrast, tendrils that have not attached remained unlooped, within the observed time frame. With the increasing number of loops, there is a decrease in helical pitch and radius, which brings the plant closer to the support. Since the contact was made above the height of the plant, the tendril brought the plant to a higher position, relative to the soil, and takes advantage of better exposure to the sun.

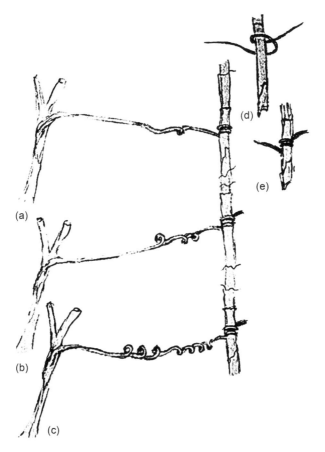

FIGURE 13.6   Sketches of the movements of tendrils after touching a support: (a) after contacting with the support; (b) the straight tendril starts to coil around it; and (c–e) tightens. Afterwards, the tendril winds and forms helices with opposite handednesses linked by perversions.

Moreover, the presence of perversions on structures can change on how the plant deals with imposing stresses. For small deformations, a hook behavior is observed for helices without and with perversions. However, for higher displacements, helices with no perversions have a stiffer response to stresses than helices with perversions (McMillen and Goriely, 2002). Therefore, it is expected that perversions allow tendrils to hold and accommodate higher strains. Furthermore, a curious phenomenon also occurs depending on the age and state of hydration of tendrils. Usually, if a relaxed tendril with perversions is stretched, it is expected that perversions unwind, revolving in the opposite direction to when they formed. By contrast, and counterintuitively, in tendrils from *Cucumis sativus* and *Echinocystis lobata*, within a certain age, perversions overwind and form more loops, upon moderate stretching.

13.4 BIOINSPIRED MATERIALS

Nature developed numerous strategies to deal with different tasks. During the many million years of evolution, organisms and other types of organic structures developed and improved a variety of ingenious mechanisms. For instance, self-cleaning surfaces (Cheng et al., 2006), the ability of animals to change their body shape and color (Kim et al., 2013) or generate light for communication purposes (Carlson and Copeland, 1985) are just a few examples of the immense diversity. Investigating the mechanisms involved in the natural processes is essential to bring forth new concepts into real applications and devices (Vullev, 2011). In the case of plant tendrils, the emergence of an intrinsic curvature happens due to an asymmetric contraction along the filament. Many works described methods for producing artificial fibers with helical shapes by modifying the structure similarly.

Liu et al. (2014) obtained ribbons with many perversions by using two elastomers of silicone with different lengths. The lengths of both ribbons were matched by stretching the ribbon with smaller length and, afterwards, both elastomers were glued together. Upon release, the resulting elastomer displayed a similar behavior with the one found on the tendrils. Clamping both ends during the release promoted the generation of perversions. The intrinsic curvature was controlled by changing the initial length of the shorter ribbon: the smaller the length, the higher the pre-strain and, therefore, the intrinsic curvature. These results show intricate three-dimensional shapes produced from linear structures by playing only with asymmetry and topological constraints. However, glueing fibers at the micro and nanoscale might be impractical to accomplish. Nevertheless, the same type of approaches can be used with techniques that allow obtaining fibers with smaller diameters.

Microfilaments with helical shapes can be produced by using the electrospinning technique. Electrospinning is a method in which strong electric fields are used to constrain and manipulate a solution, upon evaporation of the solvent, to obtain fibers with diameters ranging from micrometers to nanometers. A basic electrospinning apparatus consists of pumping a solution through a needle. Then, a strong electric field deforms the liquid to a conical shape, the Taylor cone, and forces the propulsion of a thin jet. The solvent evaporates during the flight, and a solid thread is collected in a target. Electrospinning is not only a simple and low-cost method for generating micro and nanofibers but also remarkably versatile: it can be combined with other techniques. A few examples include conventional sol-gel processing to obtain fibers with different porosities or hollow structures, different solutions can be simultaneously ejected with heterogeneous cores by using more complex containers, and jets can be manipulated by dynamically changing the electric field or by using collectors with different morphologies to obtain meshes of fibers with different arrangements and alignments.

A straightforward approach to reproduce the asymmetric contraction observed on the plant tendrils is to use an electrospinning apparatus that allows producing fibers with heterogeneous cores. In co-electrospinning, two polymers are pumped through needles with coaxial and side-by-side configurations. In a typical electrospinning experiment, jets may undergo bending instabilities due to the accumulation of electric charges and modify the jet shape. Therefore, using two polymer solutions with different physical properties should induce helicity on electrospun filaments. Wu et al. (2015) produced helical shapes using rigid and flexible polymers, Nomex and polyurethane (PU), respectively, in an off-centered core-shell co-electrospinning set (Figure 13.7a). The coiling of fibers depended on the applied voltage, conductivity, and the relative amount of used polymers. Figure 13.7b shows meshes of helical filaments displaying helical shapes, including perversions. Chen et al. (2009) produced nanosprings by also using Nomex and PU. Solutions were pumped independently through a coaxial needle. Although mechanical properties were different for both polymers, straight fibers were obtained. By adding lithium chloride (LiCl) to the PU solution, fibers displayed buckling and, in some cases, tight nanocoil shapes. The addition of LiCl reportedly enhanced conductivity and modified the shrinking thermoelastic components causing the formation of nanosprings. Fibers presented diameters ranging from 250 to 500 nm. There are some drawbacks associated with

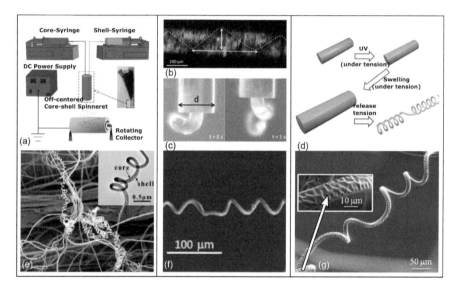

FIGURE 13.7 Micro/nano helical electrospun fibers: (a) co-electrospinning apparatus (Wu et al., 2015) used to obtain off-centered fibers, shown in (b) SEM pictures; (c) APC/DMAc solution under shear confined in a glass capillary tube, seen by polarizing optical microscopy (POM), between cross polarizers, showing a helical disclination along the tube (Canejo and Godinho, 2013); (d) same APC/DMac solution as in (c) at the end of the glass capillary tube; (e) scanning electron microscopy (SEM) pictures of APC electrospun filaments, with right- and left-handed helices connected by a perversion; (f) procedure to obtain helical fibers by selective irradiation with UV light (Trindade et al., 2013); (g) SEM pictures showing wrinkling patterns in the irradiated regions of the fibers. Image (a) (Reprinted with permission from Wu, H. et al., *Ind. Eng. Chem. Res.*, 54, 987–993, 2015. Copyright 2015 American Chemical Society.) Image (b) (Adapted from Canejo, J. and Godinho, M., *Materials*, 6, 1377–1390, 2013. With permission from the Multidisciplinary Digital Publishing Institute.) Image (c) (From Trindade, A.C. et al.: First curl, then wrinkle, *Macromolecular Rapid Communications*, 34, 1618–1622, 2013. Copyright Wiley-VCH Verlag GmbH & Co. KGaA. Reproduced with permission.)

the technique, namely matching a suitable solvent to different polymers, adjust parameters like the conductivity of solutions and feed rates, and deal with different solvent evaporations and interactions in the boundaries of the two solutions (stress and miscibility). Nevertheless, many works report the production of helical fibers using coaxial electrospinning.

In a different approach to the multicore needles, Li et al. (2012) used two spin-nerets containing two different solutions, PU, and polyacrylonitrile (PAN). Since the two spinnerets were not in contact with each other, solutions could be spun on ideal conditions and prevent gel formation or precipitation of the polymer. Polymers were under voltages with opposite polarities, so after leaving the spinnerets, jets attracted into each other and formed a bicomponent fiber. The average diameter of the fibers was around 800 nm and presented helical shapes, including the presence of perversions.

Another strategy reported in the literature uses a common component of plants: cellulose. Cellulose is widely available, and cellulose derivatives can readily form lyotropic, in common organic solvents, and thermotropic liquid crystalline phases. For instance, the ester acetoxypropylcellulose (APC), dissolved in dimethylacet-amide (DMAc), can be used to produce electrospun filaments from the liquid crystal phase. Obtained filaments presented right- and left-handed helical shapes connected by perversions (Canejo et al., 2008; Godinho et al., 2009). The intrinsic curvature of the system was attributed to the existence of a disclination line along the capillary from which the solution is extruded before generating the filament (Figure 13.7c,d). The appearance of the disclination line is due to the homeotropic boundary condi-tions imposed by the glass of the capillary. The high shear rates imposed by the elec-trospun process are responsible for the deviation of the defect line from the center of the tube. When the solvent evaporates, the filament contracts asymmetrically due to the different mechanical characteristics of the topological defect and the soft bulk liquid crystal phase of APC. Despite the pristine solutions of APC/DMAc being right-handed, filaments adopted both configurations proving that the chirality of the initial solution is not responsible for the chiral shapes of fibers (Figure 13.7e).

Another way used to promote the existence of micro/nanofilaments with intrin-sic curvature consists in the selective modification of fibers. Trindade et al. (2013) described a method where straight fibers were first electrospun and then irradiated with UV radiation. First, PU and polybutadienediol (PBDO) were dissolved in tol-uene. Using a catalyst, a chemical reaction was allowed to occur during1 h, and then the solution was inserted on a syringe. Electrospun fibers presented no helicity. However, a second crosslinking could be further created by using UV light. A thin layer with different mechanical properties was created by activating the extra reac-tive sites, as schematically shown in Figure 13.7f. Then, fibers were swollen in tolu-ene to remove the sol fraction. Finally, releasing the stretched filaments with both ends clamped generated right- and left-handed helices and perversions (Figure 13.7g). Filaments developed a half-wrinkled surface by swelling and drying, putting in evi-dence which part of the fiber was irradiated. These wrinkling patterns can also be found in old tendrils.

The modification of PU/PBDO fibers has some resemblance to what happens in tendrils. Both PU/PBDO fibers and tendrils contain roughly the same elements on their cross sections but have portions with different mechanical properties. In the PU/PBDO case, there is a second cross-linking of the fiber network, while in tendrils case, there is a lignification of gelatinous fiber cells.

Tendrils influenced many works, and the similarities between produced fibers and tendrils are self-evident. Furthermore, besides making a bridge with nature, engineering is also concerned with the the adaptation of a given effect and, if possible, improving and creating new ones. For instance, Silva et al. (2016) used the same system shown in Figure 13.7c to produce fibers containing multiple helical shapes (Silva and Godinho, 2017) and perversions with different geometries (Silva et al., 2016). In the first case, fibers were irradiated in specific portions for different times, as shown in Figure 13.8a. Molecular dynamic simulations were used to describe the dynamics of perversions in elastic filaments with intrinsic curvatures. Results displayed a good agreement between the filament shapes predicted by numerical simulations and observed experimentally, Figure 13.8b and c, respectively. Elastic fibers were modeled by considering arrangements of beads bonded by harmonic potentials. To generate the intrinsic curvature in these elastic filaments, the equilibrium bond distances of the beads on one side of the filaments were decreased. The lower the distance of the modified layer, the higher the intrinsic curvature. By releasing filaments at a constant rate, maintaining the ends fixed, regions with higher intrinsic curvature started to curl first. For enough release, regions with lower intrinsic curvature curled later. In the second case, by changing the side where fibers were irradiated and keeping

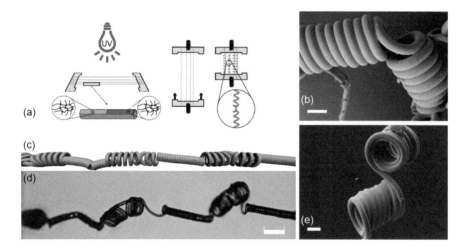

FIGURE 13.8 Method involving UV irradiation of electrospun fibers, which curl and wrinkle, after swelling followed by drying: (a) scheme of the experimental procedure; (b) and (c) simulation and experimental results of helices with different curvatures along the same filament, respectively; (d) and (e) SEM pictures of filaments showing symmetric and antisymmetric perversions, respectively. Scale bars: (c) 200 μm; (d) and (e) 10 μm. (a) to (c) (From Silva, P.E.S. and Godinho, M. H.: Helical microfilaments with alternating imprinted intrinsic curvatures, *Macromol. Rapid Commun.*, 38, 1600700, 2017. Copyright Wiley-VCH Verlag GmbH & Co. KGaA. Reproduced with permission.) (d) and (e) (Reprinted by permission from Macmillan Publishers Ltd. *Sci. Rep.*, Silva, P.E.S. et al., 2016, Copyright 2016.)

the same UV irradiation times, the shape of the tendril perversion was generalized. The most distinctive types of perversions occur when irradiation is applied on the same side and on opposite sides of the fibers: symmetric and antisymmetric perversions (see Figure 13.8d,e, respectively). Symmetric perversions are the same as those found in tendrils. Antisymmetric perversions present a different shape; upon release, the center of the perversion remains aligned with both ends and adopts a different final configuration from the symmetric perversion, as can be readily observed in Figure 13.8d,e. Using the selective method for irradiation has great potential to produce fibers with complex designs, which can be useful for specific actuation devices and smart textiles.

13.5 CONCLUSIONS

The primary advantage of the shape of tendrils seems to be their mechanical properties to facilitate support and survival of the plant. However, helical shapes can serve as an inspiration to produce right- and left-handed filaments at different scales from non-chiral objects, and perversions are a source of motivation to produce new geometric shapes connecting filaments with opposite handedness.

In fact, at the micro/nanoscale, it is crucial to tune and control the shape of filaments, and in this sense, the shapes of the tendrils and the mechanisms involved in their formation were used to produce non-woven membranes with various characteristics. Methods to do the fine-tuning of the curvature of the helices in synthetic fibers were achieved. New routes were opened to the production of intricate no-ticked tissues by precise control of the appearance of perversions along the filaments.

Depending on the structures imprinted in the helical micro/nanofibers, innovative ways on the manufacture of textiles are envisaged. The fabrics produced find application in different domains, which include energy conversion, crude oil contamination clean up, tissue engineering, and intelligent devices.

ACKNOWLEDGEMENTS

This work was funded by FEDER funds through the COMPETE 2020 Program and National Funds through FCT—Portuguese Foundation for Science and Technology under projects numbers POCI-01-0145-FEDER-007688 (Reference UID/CTM/50025), PTDC/CTM-BIO/6178/2014, M-ERA-NET2/0007/ 2016 (CellColor).

REFERENCES

Bates, A. D., and Maxwell, A. (2005). *DNA Topology* (2nd ed). New York: Oxford University Press.

Bell, A. D., and Bryan, A. (1991). *Plant Form: An Illustrated Guide to Flowering Plant Morphology*. New York: Oxford University Press.

Bowling, A. J., and Vaughn, K. C. (2009). Gelatinous fibers are widespread in coiling tendrils and twining vines. *American Journal of Botany*, 96(4), 719–727. doi:10.3732/ajb.0800373.

Canejo, J., and Godinho, M. (2013). Cellulose perversions. *Materials*, 6(4), 1377–1390. doi:10.3390/ma6041377.

Canejo, J. P., Borges, J. P., Godinho, M. H., Brogueira, P., Teixeira, P. I. C., and Terentjev, E. M. (2008). Helical twisting of electrospun liquid crystalline cellulose micro- and nanofibers. *Advanced Materials*, *20*(24), 4821–4825. doi:10.1002/adma.200801008.

Carlson, A. D., and Copeland, J. (1985). Flash communication in fireflies. *The Quarterly Review of Biology*, *60*(4), 415–436. doi:10.1086/414564.

Chen, S., Hou, H., Hu, P., Wendorff, J. H., Greiner, A., and Agarwal, S. (2009). Polymeric nanosprings by bicomponent electrospinning. *Macromolecular Materials and Engineering*, *294*(4), 265–271. doi:10.1002/mame.200800342.

Cheng, Y. T., Rodak, D. E., Wong, C. A., and Hayden, C. A. (2006). Effects of micro- and nano-structures on the self-cleaning behaviour of lotus leaves. *Nanotechnology*, *17*(5), 1359–1362. doi:10.1088/0957-4484/17/5/032.

Darwin, C. (1865). On the movements and habits of climbing plants. *Journal of the Linnean Society of London, Botany*, *9*(33–34), 1–118. doi:10.1111/j.1095-8339.1865.tb00011.x.

Domokos, G., and Healey, T. J. (2005). Multiple helical perversions of finite, intristicallycuverd rods. *International Journal of Bifurcation and Chaos*, *15*(03), 871–890. doi:10.1142/S0218127405012430.

Gerbode, S. J., Puzey, J. R., McCormick, A. G., and Mahadevan, L. (2012). How the cucumber tendril coils and overwinds. *Science*, *337*(6098), 1087–1091. doi:10.1126/science.1223304.

Godinho, M. H., Canejo, J. P., Pinto, L. F. V., Borges, J. P., and Teixeira, P. I. C. (2009). How to mimic the shapes of plant tendrils on the nano and microscale: Spirals and helices of electrospun liquid crystalline cellulose derivatives. *Soft Matter*, *5*(14), 2772–2776. doi:10.1039/B821631B.

Goriely, A., and Tabor, M. (1998). Spontaneous helix hand reversal and tendril perversion in climbing plants. *Physical Review Letters*, *80*(7), 1564–1567. doi:10.1103/PhysRevLett.80.1564.

Gray, A., Abbena, E., and Salamon, S. (2006). *Modern Differential Geometry of Curves and Surfaces with Mathematica* (3rd ed). Boca Raton, FL: Chapman & Hall.

Jaffe, M. J., and Galston, A. W. (1968). The physiology of tendrils. *Annual Review of Plant Physiology*, *19*(1), 417–434. doi:10.1146/annurev.pp.19.060168.002221.

Jampani, V. S. R., Škarabot, M., Ravnik, M., Čopar, S., Žumer, S., and Muševič, I. (2011). Colloidal entanglement in highly twisted chiral nematic colloids: Twisted loops, Hopf links, and trefoil knots. *Physical Review E*, *84*(3). doi:10.1103/PhysRevE.84.031703.

Kim, S., Laschi, C., and Trimmer, B. (2013). Soft robotics: A bioinspired evolution in robotics. *Trends in Biotechnology*, *31*(5), 287–294. doi:10.1016/j.tibtech.2013.03.002.

Li, C., Wang, J., and Zhang, B. (2012). Direct formation of "artificial wool" nanofiber via twospinneret electrospinning. *Journal of Applied Polymer Science*, *123*(5), 2992–2995. doi:10.1002/app.34944.

Liu, J., Huang, J., Su, T., Bertoldi, K., and Clarke, D. R. (2014). Structural transition from helices to hemihelices. *PLoS One*, *9*(4), e93183. doi:10.1371/journal.pone.0093183.

Liu, S., Yao, Z., Chiou, K., Stupp, S. I., and Cruz, M. O. de la. (2016). Emergent perversions in the buckling of heterogeneous elastic strips. *Proceedings of the National Academy of Sciences*, *113*(26), 7100–7105. doi:10.1073/pnas.1605621113.

Longe, J. L. (2001). *How Products are Made. Vol. 6: An Illustrated Guide to Product Manufacturing*. Detroit, MI: Gale Group.

McMillen, T. and Goriely, A. (2002). Tendril perversion in intrinsically curved rods. *Journal of Nonlinear Science*, *12*(3), 241–281. doi:10.1007/s00332-002-0493-1.

Meloche, C. G., Knox, J. P., and Vaughn, K. C. (2006). A cortical band of gelatinous fibers causes the coiling of redvine tendrils: A model based upon cytochemical and immunocytochemical studies. *Planta*, *225*(2), 485–498. doi:10.1007/s00425-006-0363-4.

Pieranski, P., Baranska, J., and Skjeltorp, A. (2004). Tendril perversion—A physical implication of the topological conservation law. *European Journal of Physics*, *25*(5), 613. doi:10.1088/0143-0807/25/5/004.

Silva, P. E. S., Trigueiros, J. L., Trindade, A. C., Simoes, R., Dias, R. G., Godinho, M. H., and de Abreu, F. V. (2016). Perversions with a twist. *Scientific Reports*, *6*, 23413. doi:10.1038/srep23413.

Silva, P. E. S., and Godinho, M. H. (2017). Helical microfilaments with alternating imprinted intrinsic curvatures. *Macromolecular Rapid Communications*, *38*(5), 1600700. doi:10.1002/marc.201600700.

Snir, Y., and Kamien, R. D. (2005). Entropically driven helix formation. *Science*, *307*(5712), 1067–1067. doi:10.1126/science.1106243.

Trindade, A. C., Canejo, J. P., Teixeira, P. I. C., Patricio, P., and Godinho, M. H. (2013). First curl, then wrinkle. *Macromolecular Rapid Communications*, *34*(20), 1618–1622. doi:10.1002/marc.201300436.

Vullev, V. I. (2011). From biomimesis to bioinspiration: What's the benefit for solar energy conversion applications? *The Journal of Physical Chemistry Letters*, *2*(5), 503–508. doi:10.1021/jz1016069.

Weisstein, E. W. (2011). *The CRC Concise Encyclopedia of Mathematics*. London, UK: Chapman & Hall.

Wu, H., Zheng, Y., and Zeng, Y. (2015). Fabrication of helical nanofibers via co-electrospinning. *Industrial & Engineering Chemistry Research*, *54*(3), 987–993. doi:10.1021/ie504305s.

14 Fibonacci Sequence
A General Account

Monoranjan Ghose and Bir Bahadur

CONTENTS

14.1 INTRODUCTION

The origins of Fibonacci sequence appear in Indian mathematics, in connection with Sanskrit prosody (Singh, 1985; Knuth, 1968). In the Sanskrit tradition of prosody, there was interest in enumerating all patterns of long (L) syllables that are two units of duration, and short (S) syllables that are one unit of duration. Counting the different patterns of L and S of a given duration results in the Fibonacci numbers: the number of patterns that are m short syllables long is the Fibonacci number F_{n+1} (Knuth, 2006). Before Fibonacci wrote his work, the sequence F_n had already been discussed by Indian scholars, who had long been interested in rhythmic patterns. Both Gopala (before AD 1135) and Hemachandra (c. 1150) mentioned the numbers 1, 2, 3, 5, 8, 13, and 21 explicitly (Singh, 1985).

Outside India, the Fibonacci sequence first appeared in the book *Liber Abaci* (1202) by Fibonacci (Pisano, 2002). Fibonacci considers the growth of an idealized (biologically unrealistic) rabbit population, assuming that: a newly born pair of rabbits, one male, one female, are put in a field; rabbits are able to mate at the age of one month so that at the end of its second month a female can produce another pair of rabbits; rabbits never die and a mating pair always produces one new pair (one male, one female) every month from the second month on. The puzzle that Fibonacci posed was: how many pairs will there be in one year?

Fibonacci, as already stated, was an Italian mathematician of the early 13th century, born in 1175 and died in 1240, most likely in the Republic of Pisa. His father's name was Guglielmo Bonacci. Fibonacci was the nickname of the mathematician. He was often known as Leonardo of Pisa, because he was from Pisa. Throughout his youth, Fibonacci used to accompany his father on trips to Africa and along Mediterranean coast. During these tours, he encountered a wide variety of foreign merchants. These businessmen taught him a numeric system other than the traditional Roman numerals used throughout Europe and other regions of the Western civilization. This system is known as Hindu-Arabic numerals, composed of a set of ten numeric symbols, along with a "positional decimal" system that allows for fine numerical adjustments (ones, tens, and hundreds places). The system was devised between the second and third centuries by Indian mathematicians, and later was adopted by the Arabs in the 800s, and used within Indian- and Arabic-controlled region ever since. Fibonacci immediately recognized the advantages of the alternative system and published his seminal work, *Liber Abaci*.

There is a sequence of numbers named after him: start with 0, 1, and then each term is the sum of the previous two. $0 + 1 = 1$, $1 + 1 = 2$, $1 + 2 = 3$, $2 + 3 = 5$, $3 + 5 = 8$, $5 + 8 = 13$, etc. So the Fibonacci sequence comes true as 1, 1, 2, 3, 5, 8, 13, 21, 34, 55, 89, 144, 233…and so on. The sequence is infinite. Written as a rule, the expression is $x_n = x_{n-1} + x_{n-2}$.

The Fibonacci spiral occurs in many places in nature, such as in snail shells or even the spiral of a hurricane. However, it is not a "true" mathematical spiral, which is composed of fragmented segments of a circle and does not scale down in size (Knott, 2010). The spiral creates a line from the middle of the spiral and increases by a factor of the golden ratio, 1.618, in every square (Norton, 1999). It appears in flora, such as in the arrangement of seeds on flowering plants, the number of petals in sunflowers, or even "phyllotaxis," which is the arrangement of leaves around the stem (Britton, 2011). According to scientists, certain flora develop in the most efficient way based on the biochemistry of plants as they develop new structures, such as leaves or flowers, which provides an evolutionary advantage in promoting the plant's survival (Rehmeyer, 2007).

Boman et al. (2017) created a model (cell division type) for tissue development based on the biology of cell division that builds upon the cell maturation concept for asymmetric cell division. Their model output on the number of cells generated over time fits specific Fibonacci p-number sequences depending on the maturation time.

14.2   FIBONACCI NUMBERS AND THE GOLDEN RATIO IN PLANTS

The Fibonacci numbers appear everywhere in Nature, from the leaf arrangement in plants, to the pattern of the florets of a flower, the bracts of a pinecone, or the scales of a pineapple. Many plants show the Fibonacci numbers in the arrangement of the leaves around the stem. Some pine cones and fir cones also show the numbers, as do daisies and sunflowers. Sunflowers can contain the number 89, or even 144. Many other plants, such as succulents, also show the numbers. The arrangement of leaves on a plant stem is known as phyllotaxis. Plant growth is governed by the Fibonacci sequence, which can be understood as a law of accumulation.

If we take the ratio of two successive numbers in the Fibonacci's series, (1, 1, 2, 3, 5, 8, 13, 21, 34, 55, 89, 144, 233…) and divide each by the number before it, we will find the following series of numbers:

$$^1/_1 = 1, \,^2/_1 = 2, \,^3/_2 = 1.5, \,^5/_3 = 1.666, \,^8/_5 = 1.6, \,^{13}/_8 = 1.625, \,^{21}/_{13} = 1.61538,$$
$$34/21 = 1.619, 55/34 = 1.617, 89/55 = 1.618…$$

The golden ratio can be approximated by a process of successively dividing each term in the Fibonacci sequence by the previous term. With each successive division, the ratio appears to be settling down to a particular value, which we call **the golden ratio** or **the golden number**. It has a value of approximately **1.61803399**.

By dividing a circle into golden proportions, where the ratio of the arc length is equal to the golden ratio, we find the angle of the arcs to be 137.5°. In fact, this is the angle at which adjacent leaves are positioned around the stem. This phenomenon is observed in many types of plants.

14.3 FIBONACCI SEQUENCE IN PLANT BRANCHES

The branching rates in plants occur in the Fibonacci pattern, where the first branch level has one branch (the trunk), the second has two branches, then 3, 5, 8, 13, and so on (Figure 14.1). This design provides the best physical accommodation for the number of branches, while maximizing sun exposure. The spacing of leaves around each branch or stalk spirals with respect to the golden ratio. The branch growth can be easily studied on Sneezewort (*Achillea ptarmica*). New shoots grow out at a point where the leaf meets the main stem of a plant.

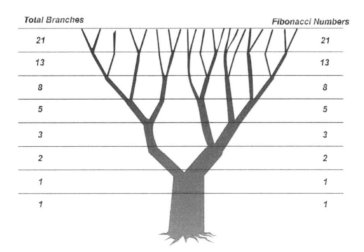

FIGURE 14.1 Branching system in *Achillea ptarmica* showing Fibonacci numbers. (Courtesy of Sandeep Ahirwar, http://ultraxart.com/author/Sandeep, 1993.)

14.4 FIBONACCI SEQUENCE IN PHYLLOTAXIS

Charles Bonnet (1754) coined the word "phyllotaxis," which means "arrangement of leaves." In the 1830s, two brothers (Bravais and Bravais, 1837) discovered that each new leaf on a plant stem was arranged at a particular angle with each other. This angle was calculated to be around 137.5°. Botanists examined few common plants again and again year after year and observed that the plant kingdom also follows the rules of the Fibonacci sequence. The leaves on this plant are staggered in a spiral pattern to permit optimum exposure to sunlight. If we apply the golden ratio to a circle, we can see how it is that this plant exhibits Fibonacci qualities.

Leaf arrangement is classified into three types (Figure 14.2).

Alternate—Stem with the alternate spiral arrangement of leaves arranged in alternating order and have only one leaf per node.

Opposite—In plants with opposite leaf arrangement, two leaves are placed opposite to each other at a node.

Whorled—In the whorled arrangement, three or more leaves are placed at a node.

In alternate phyllotaxis, plants show either left or right spirality with a ratio of about 50:50 in the population. To verify whether the spirality is genetically controlled or not, Allard (1946) studied two species of tobacco, *Nicotiana tabacum* and *Nicotiana rustica*. He determined spirality in 23,507 plants of different varieties and strains of both species. The x^2 (chi-square) values of the distribution of clockwise and counterclockwise spirality indicated that all the samples were drawn from populations having a 50:50 ratio. This appears to be an inherent condition in tobacco as in other

Leaf Arrangements

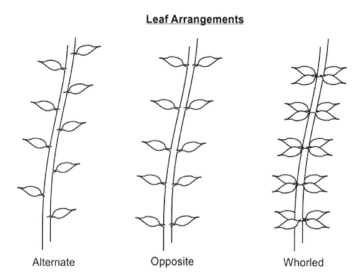

Alternate   Opposite   Whorled

FIGURE 14.2 Leaf arrangement on plants. (Courtesy of Sandeep Ahirwar, http://ultraxart.com/author/Sandeep, 1993.)

plants studied so far. The selected parents of Maryland Mammoth tobacco did not transmit any of their individual characteristics of spirality to their progenies. He postulated that the distribution of these two spiral alternatives is fundamentally one of chance.

Elhoumaizia et al. (2002) reported that in the date palm (*Phoenix dactylifera*) the most obvious contact parastichies are, in order of Fibonacci spiral, 13, 8, and 5, and offshoots show a clockwise or counterclockwise phyllotaxis independently of the mother plant.

Mitchison (1977) gave an explanation for the spiral leaf arrangement found in many plants. He concluded that Fibonacci phyllotaxis follows as a mathematical necessity from the combination of an expanding apex and a suitable spacing mechanism for positioning new leaves. He considered an inhibitory spacing mechanism at some length; the same treatment would apply to depletion of, or competition for, a compound by developing leaves. The mathematical principles involved are clear when it is assumed that only two leaves (the contacts) position a new leaf.

Holly leaves are arranged on the stem in a regular spiral, which may be right- or left-handed, and each has two spiny edges that can also be distinguished as right and left (Dormer, 1955). Many attempts have been made to explain regular occurrence of a standard type of spiral arrangement. The start of a spiral is not a matter of chance, because although right- and left-handed spirals are about equally frequent, statistical investigations show that the direction of the spiral on a lateral branch is correlated with that of the spiral on the parent shoot. The extent of this correlation is controlled partly by genetic factors but can also be modified by experimental treatment (Dormer, 1955).

The basic phyllotactic pattern in *Citrus* is 3/8. The direction of spirality of the first shoot growth is a random phenomenon. Within a given growth flush, the direction of spirality is generally uniform and constant. Passing from one growth flush to another, the direction of spirality is reversed. This reversal is constant and regular in the successive growth flushes as the phyllotaxy are traced up a branch or throughout the tree. The direction of spirality and placement of thorn in *Citrus* are regularly alternate (Schroeder, 1953).

A good deal of literature (Wright, 1873; Henslow, 1876; de Candolle, 1881; Weisse, 1894; Strasburger, 1930; Douady and Couder, 1996a, 1996b, 1996c) on the causes of phyllotaxy occur, which can be grouped into four categories: (1) general or non-specific, (2) unknown influences that circulate through the plant or its stem, (3) leaf and leaf-primordial positions, and (4) changes resulting from the application of X-rays, apical incision, or hetero-auxine. But no conclusive decision could be made from these assumptions.

Hofmeister (1868) formulated a theory that each new leaf-primordium forms in the largest gap between the previously formed primordium. He assumed that the primordial positions were controlled by a tension produced by the developing primordia and that a new primordial position is not established until a position is reached in which the tension is below the level of inhibition.

Davies (1937) reported that normal leaf arrangement is not always maintained in *Ailanthus altissima*. Abnormal leaf patterns were frequent. The spiral arrangement of leaves on the stem occurs either in a clockwise or counter-clockwise direction.

He showed that 51.6% of stems had leaves arranged spirally in a clockwise direction and 48.4% in a counter-clockwise direction. The normal leaf arrangement in *A. altissima* is eight-ranked. The angle of divergence is approximately 135° or three-eighths of a circumference.

Kundu and Sarma (1965) made selections for three years for the right and left arrangement of leaf spiral in a cultivated inbred of *Corchorus capsularis*. The progenies raised from the seeds of a single-selfed plant produced approximately equal proportions of right- and left-spiralled plants each year, which indicate that genetic control of the character difference may either be extremely loose or even absent.

An interesting observation by Sarma (1968) is found on the direction of leaf spiral in *C. capsularis* related to environmental conditions. Generally, both the species of jute, *C. capsularis* and *Corchorus olitorius* possess a right- and left-handed arrangement of leaves with 2/5 and 3/8 phyllotaxy. An experiment was done on *C. capsularis* at three different plots: (1) centrally located plot—where plants received moderate to bright light on all exposed parts throughout the day; (2) east-exposed plot—plants received bright light in the forenoon and bright to moderate light in the afternoon; (3) west-exposed plot—plants received light from the midday and afternoon sun but received little sun in the morning. The results show that the centrally located plot in which the plants were exposed to moderate to bright light throughout the day showed an excess of right spirals (56.04%), while the other two plots contained an excess of plants with left spirals (56.96% and 58.61%, respectively). The chi-square test for independence of phyllotaxy and the direction of spiral is showing that the association between left spiral and 2/5 phyllotaxy, and right spiral and 3/8 phyllotaxy is highly significant, which is largely determined by environmental conditions, of which the direction of sunlight may be considered as one.

Bahadur et al. (1978) reported that both right- and left-spiralled plants occur in *Cycas revolute* and *Cycas circinnalis*. There are two kinds of leaves in each species—small scale leaves and large foliage leaves. The scale leaves exhibit clear spirality, which may be right or left spiralled. The foliage leaves are produced in five distinct whorls of alternating spirals, which also may be in the clockwise or counter-clockwise direction. The left-spiralled:right-spiralled ratio is 1.14 and 1.22 for *C. revolute* and *C. circinnalis*, respectively, which indicate a slight excess of left-handed plants. Davis (personal communication) commented on the paper that the male cone of *Cycas* species shows a Fibonacci series of 8, 13, and 21.

The phyllotaxis of palms is always alternate, that is, a single leaf is produced at each node. In *Areca catechu*, two consecutive leaves are placed at an angular deflection of about 143°. The leaves in this case are seen to be spirally arranged. To determine the direction of the spiral, any two consecutive leaves on the trunk are examined. If the younger leaf lies nearer the older one along the right-hand side of the observer looking from the mid-position of the older leaf, then the spirality of the palm is regarded as right-handed. If nearer by the left-hand, it is considered as having a left-handed foliar spiral (Davis and Kundu, 1966). If an imaginary line is drawn from the base of any leaf, and this line is passed through the bases of the successive leaves, it is seen that the spiral line moves either clockwise (left-spiralled) or counterclockwise (right-spiralled) around the stem. In a coconut tree, *Cocos nucifera*, the leaves are arranged along five distinct spirals, which are right-handed or left-handed.

FIGURE 14.3   A left-spiralled coconut palm showing a bunch of fruits hanging on the right side of the subtending leaves.

In a left-handed-bearing palm, the bunch of fruits hangs on the right side of the sub-tending leaves (Figure 14.3). On the other hand, in a right-spiralled palm, the fruit bunch hangs on the left side of the leaves, if viewed from below the subtending leaf and towards its base. From the base of any specific leaf, the sixth younger leaf will fall above the first leaf in the spiral. Although this asymmetry is not genetically deter-mined (Compton, 1912; Davis, 1962; Kundu and Sharma, 1965), it has been reported that fruit yield is associated with the foliar arrangement in some palms (Davis, 1972).

An interesting experiment was done by Ghose et al. (1996) on the effect of foliar arrangement of leaf epidermal structures in Areca palm (*A. catechu* Linn. Figure 14.4) to determine: (1) whether the result in the left-spiralled (FSL) and right-spiralled (FSR) plants are significantly different; (2) whether the observations obtained from the right side of the leaf and those from the left side are significantly different; and (3) whether the spirality of the plants and side of the leaves are correlated. The result showed the number of stomata per unit area of 1 mm^2 of FSR plants, on an average, possessed 130.87 and 130.38 stomata, and FSL plants 117.13 and 116.37 stomata per unit area on the left and right sides of lamina, respectively. It revealed that the overall difference between FSR and FSL plants was 11.88%, which was highly

FIGURE 14.4 Six-month-old left-spiralled (FSL) and right-spiralled (FSR) seedlings of *Areca catechu*. (From Ghose, M. et al., *Acta Bot. Neerl*. 45, 303–308, 1996.)

statistically significant ($P \leq 0.01$). The difference in width of guard cells was significant ($P \leq 0.01$) between FSR (12.8 µm) and FSL (13.71 µm) plants. It indicated that FSL plants possessed the widest but minimum number of stomata, and FSR plants possessed the narrowest but maximum number of stomata.

Davis (1972) reported that fruit yield is associated with the foliar arrangement in *A. catechu* and *C. nucifera*. Although foliar spirality does not follow Mendelian inheritance (Compton, 1912; Davis, 1962, 1972; Rao, 1980), it has a significant role in determining stomatal frequency in leaves, which might have correlation with fruit yield. The right-spiralled plants of *Cajanus cajan* show higher stomata and higher yield of pods (Bahadur et al., 1980). The phyllotaxy of each leaf spiral in coconut is nearly two-fifths, that is, any two successive leaves make an angle of deflection of a little less than 144° (Davis, 1962). However, Sampson (1923) described it as 142°, and Patel (1938) mentioned that the angle varying from 137° to 141° left or right of the previous leaf. Patel also stated that this angle gets reduced as the tree grows older.

FIGURE 14.5   A Rattan palm *Daemonorhops jenkinsianus*, showing single left spiral leaf arrangement.

The number of foliar spirals varies among different palm species. In *A. catechu* (Figure 14.6a) *Calamus* sp., *Daemonorhops jenkinsianus* (Figure 14.5), *Chamaedorea costaricana*, and *Ptychosperma macarthurii*, the leaves are arranged in a single spiral. These palms possess a few green leaves, and the internodes are much longer in comparison to the thickness of the stem. *Arenga pinnata* has two spirals. The palmyra palm (*Borassus flabellifer*) and *Corypha eleta* have three distinct spirals. *C. nucifera* and a few other palms possess five distinct spirals. The African oil palm (*Elaeis guineensis*) and wild date palm (*Phoenix sylvestris*) have eight foliar spirals each. Palms with higher foliar spirals, for example, *E. guineensis* (Figure 14.6e) and *Copernicia prunifera*, normally have eight spirals, but also show five spirals moving opposite to eight spirals. On the trunk of *C. prunifera*, foliar spirals numbering 8, 5, and 3 are also found. The direction of foliar spirals in palms is not genetically controlled. A study was made in coconut in which "right-handed" seed parents and "right-handed pollen parents were bred under control conditions; this study showed that, about 50% of the progeny produced right-spiralled and the rest left-spiralled. Similarly, the progenies of the parental combinations, Left X Left, Left X Right, and Right X Left, all showed a 1:1 ratio indicating that they are not influenced by their parents for their foliar spirality (Davis, 1962). In any locality the right-handed and left-handed palms are distributed equally (Davis, 1963).

Palms having fewer leaves show smaller number of foliar spirals, for example, *A. catechu* (Figure 14.6a) having one spiral, and those bearing larger numbers of leaves show greater number of spirals, for example, *C. nucifera* (Figure 14.6d)

FIGURE 14.6 (a) "Left–handed" *Areca catechu*; (b) two-spiralled trunk of *Arenga pinnata*; (c) stem of *Borassus flabellifer* with three spirals; (d) crown of *Cocos nucifera* showing "left-handed" foliar spiral; and (e) trunk of *Elaeis guineensis* displaying eight and five spirals. (From Davis, T.A., *Fibonacci Quart.* 9, 237–244, 1971.)

possesses five spirals (Davis, 1970). Similarly, *P. macarthurai* has a single spiral, *A. pinnata* or *Arenga saccarifera* (Figure 14.6b) has two, *B. flabellifer* (Figure 14.6c) and *C. eleta* have three spirals, *E. guineensis* (Figure 14.6e) bears eight spirals, and *Phoenix canariensis* (Figure 14.7) has thirteen spirals. All these foliar spirals are Fibonacci numbers. It is interesting to note that foliar spirals 4, 6, 7, 9, 11, or 12 are not known in any palms.

FIGURE 14.7   The leaf scars on the trunk of *Phoenix canariensis* showing 3, 5, 8, and 13 foliar spirals. (From Davis, T.A., *Fibonacci Quart.*, 9, 237–244, 1971.)

14.4.1 GOLDEN PROPORTION

Davis (1970) suggested a model for the arrangement of leaves in any palm crown in which two consecutive leaves had shown to subtend 137.5° between them, and this angle made a golden proportion of 0.618 with the remaining 222.5° angle to complete one full revolution (see Figure 14.8). The angular deflection of 137.5° has been chosen arbitrarily. With this angular proportion, no two leaves in the diagram (Figure 14.8) would exactly superimpose each other till the 145th leaf. No palm is likely to have 100 functioning leaves at a time, and this gives the leaves scope for maximum exposure to sunlight. However, there are exceptional palm species where the leaves are arranged in vertical rows. The number of rows in a species shows a Fibonacci number.

The central most point in the drawing (Figure 14.8) represents the aerial view of a palm trunk, and the radial lines, its leaves. The outermost leaf, which is the oldest, is numbered 1. Leaf number 2 is drawn at an arbitrary angular deflection of 137.5 to the left of leaf No. 1. Since leaf No. 2 is nearer to leaf No. 1 by the left-hand side looking from the tip of leaf No. 1, this crown may be regarded as representing a left-spiralled palm. Similarly, leaf No. 3 will be nearer to leaf No. 2 by the left, and subsequent leaves are also similarly located. In another palm, leaf No. 2 can as well be nearer to leaf No. 1 by the right, in which case the diagram will represent a right-spiralled crown. The younger leaves are represented progressively by shorter

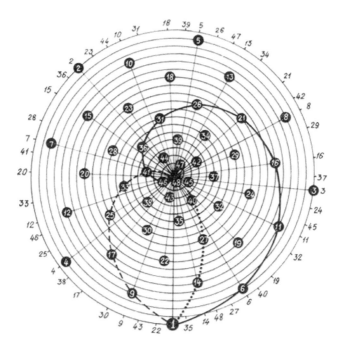

FIGURE 14.8 Schematic representation of a palm crown. (From Davis, T.A., *Acta Bot. Neerl.*, 19, 49–256, 1970.)

radial lines, and leaf No. 48 is the youngest visible leaf in this crown. The tips of all the leaves are connected by a line that forms a clockwise (left-handed) coil, and this will represent the only visible spiral in some palms such as *A. catechu*. In a palm with two foliar spirals, one spiral will comprise leaves 1, 3, 5, 7, 9, and so on, while the second spiral will comprise all the even-numbered leaves. Here, both the spirals move counter clockwise. In a palm bearing three spirals, the first spiral will connect leaves 1, 4, 7, 10, 13, and so on. The second spiral will have leaves 2, 5, 8, 11, 14, and so on, while the third spiral will comprise leaves 3, 6, 9, 12, 15, and so on. All the three spirals run clockwise as opposed to the direction of the two spirals. It is to be noted that no palm shows four clear spirals. In a five-spiralled crown, leaves 1, 6, 11, 16, 21, and so on, constitute one of the spirals, the other four starting with leaves 2, 3, 4, and 5, respectively. All the five spirals move counter clockwise, which is oppo-site to the direction of the three spirals. In a palm with eight spirals, leaves 1, 9, 17, 25, 33, and so on, will form one of the spirals and the remaining spirals commence from leaves 2, 3, 4, 5, 6, 7, and 8. The eight spirals move opposite to the five spirals. Similarly, if the diagram is to represent a 13-spiralled palm, it will comprise leaves 1, 14, 27, 40, etc., and the spirals will move opposite to the eight spirals, and so, the 21 spirals move at a slant opposite to the 13 spirals. Hence, in this diagram, the more obvious numbers of foliar spirals synchronize with the Fibonacci Numbers. Foliar spirals representing the Fibonacci numbers 1, 3, 8, 21, move clockwise.

In a palm crown having four or five leaves, only the single spiral is discern-ible, and two spirals may be clear if the leaf number goes up to seven or eight.

Three spirals may be possible in a crown having about 10 to 12 leaves, and 5 spirals in a crown having about 20 leaves. Therefore, as the number of green leaves in a crown increases; higher orders of foliar spirals are displayed.

It is reported that the Fibonacci phyllotaxis gives optimum illumination to the photosynthetic surface of plants (Church, 1904; Thompson, 1917; Richards, 1951; Leppik, 1961).

In some exceptional palms, the arrangement of leaves does not show any spiral mechanism. Here the leaves are arranged vertically one above another along two or more rows. In *Wallichia disticha* (Figure 14.9a), there are two vertical rows of leaves and the angle between any two consecutive leaves is 180°. In *Neodypsis decaryi* (Figure 14.9b), the leaves fall along three vertical rows, and the angle between two successive leaves is 120°. In *Syagrus treubiana* (Figure 14.9c), there are five vertical rows of leaves, and the angular deflection between any two consecutive leaves is 144°. In some palms of Canary Island, 13 vertical rows of leaves occur with an angular deflection between 2 consecutive leaves at 138.5°. The number of vertical rows of leaves in these palms is all Fibonacci numbers (Davis, 1971). If thousands of trunks of different palm species are examined, it is possible to get examples of 13 and 21 vertical rows of leaves, and each leaf in them will be disposed at 138.5° and 137.14°, respectively from its immediate younger one (Davis, 1970). If the figures of angular deflections of different palms (180°, 120°, 144°, 135°, 138.5°, 137.14°...) are examined, one finds the alternate numbers turning out to be more than 137.5° and the others less, and the difference between two numbers progressively gets narrower,

FIGURE 14.9 (a) *Wallichia disticha* with two vertical rows of leaves; (b) *Neodypsis decaryi* with three vertical rows of leaves; and (c) trunk of *Syagrus treubiana* with five vertical rows of leaf bases. (From Davis, T.A., *Fibonacci Quart.*, 9, 237–244, 1971.)

FIGURE 14.10 Palm *Nypa fruticans* leaves coming out from horizontally spreading rhizomes.

and reaching 137.5°. This narrow angle makes a proportion of 0.618 with the remaining wider angle 222.5° to complete one full revolution. This is known as "golden proportion."

The palm *Nypa fruticans* (Figure 14.10) does not form an aerial trunk; instead, horizontally spreading rhizomes bear the upward slanting alternate leaves that bear more or less a golden proportion to the remaining wider angle.

In a bearing palm, almost every leaf subtends a spadix, which may be female, male, or androgynous. The tendency of the spadix is to lean towards the direction of the leaf just younger to its own subtending leaf. Thus, in a right-spiralled coconut, the spadix falls on the left side of the subtending leaf, and the spadix hangs to the right of the leaf in a left-spiralled palm. On the other hand, the situation in *E. guineensis* (oil palm) is reversed; the bunch and the foliar spirals will seem to move along the same direction (Davis, 1962).

14.5 FIBONACCI SEQUENCE IN REPRODUCTIVE PARTS

14.5.1 INFLORESCENCE

14.5.1.1 Sunflower Spirals

The leaves of sunflower plant are arranged in alternate phyllotaxis in a single spiral, running clockwise or counter clockwise. Any two consecutive leaves make an angular deflection of about 137.5°. At its maturity, the plant possesses greatly reduced leaves when the main stem flattens out into the flower-bearing disc known as capitulum. At the base of the disc, the reduced leaves are arranged in spiral patterns and form involucral bracts. Each of the petal-like structures at the periphery of the disc represents a female or sterile flower known as the ray floret. Bordered by the ray florets on the disc are the numerous regular bisexual flowers known as the disc florets. On the developing capitulum, the flowers start blooming one by one at a place approximately 137.5° away from its immediate older one in relation to the central point. As the flowers get differentiated, the tip of the meristematic axis rotates in such a way that the older florets move away from the growing point in logarithmic spirals that approximate to an Archimedes' spiral (Mathai and Davis, 1974). The ratio of any two consecutive Fibonacci Numbers (higher ones) turns out to be about 0.618, which is known as golden ratio. It is due to this peculiar situation that shoots with closely set alternate leaves on the head of a sunflower, or other similar organs, show a spiral mechanism, and the number of spirals match a Fibonacci number.

Mathai and Davis (1974) explained nicely the configuration of ray and disc florets on the sunflower head with drawings. Figure 14.11 gives the false appearance that the basic pattern of configuration is formed by the several arcs. But these arcs are the resultants of a single basic arrangement of the individual flowers. Most of the earlier workers (Bonnet, 1754; Church, 1904, 1920; Coxeter, 1953; Gardner, 1969) overlooked that the individual flowers/seeds are more or less of the same size and shape whether they are formed near the periphery or near the centre of the disc. Considering all these factors, a small head of a sunflower with only 234 flowers/seeds has been constructed and shown in Figure 14.11. In this figure, a line AB has been drawn as the reference axis, and a point 0 is drawn on it to represent the very tip (growing point) of the shoot that enlarges into the capitulum. A circle with a fixed diameter is drawn touching AB at 0. A second circle of the same diameter is drawn at an angular deflection of 137.507° from the line connecting the centre of the first circle and 0. The second circle, which just touches the first one, can be placed either to the left of

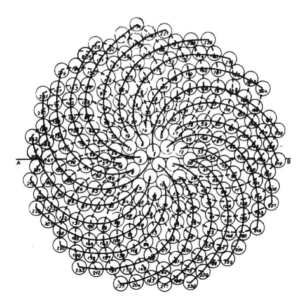

FIGURE 14.11   A schematic head of sunflower showing 34 and 21 spirals. (From Mathai, A.M., and Davis, T.A., *Math. Biosci.,* 20, 117–133, 1974.)

the first one or to its right. This decides whether the capitulum becomes right-handed or left-handed. In the present case, the second circle is shown on the left of the first. Successive circles of the same diameter have been drawn one after another on the left of the preceding ones at the same angular deflection of 137.507°. Although the subsequently formed circles in the picture seem to move away from the centre than the early formed ones, a reverse course takes place on the head. That is, flower No. 234 is the oldest, and No. 1 the youngest. At the time when the second flower (circle) is differentiated at the tip of the stem, the first one rotates away from its original position by about 137.5°. Thus, when more and more florets are formed, the already-formed ones are pushed towards the periphery with minimum space between them. Some of the earliest formed flowers turn into the ray florets, and the latter ones into disc florets. It is evident that the configuration in the capitulum does not depend on the size of the circle as long as every circle has the same diameter. Figure 14.11 exhibits the following property. By starting with any circle at the periphery and joining the centres of adjacent circles whose numbers differ by 21, the continuous-line redial arcs are obtained. Similarly, starting from any circle at the periphery, the circles whose numbers differ by 34 are connected. The dotted-line arcs, which move opposite to the 21 arcs, are obtained. The two kinds of arcs pass through the entire circle.

In Figure 14.12, the slightly modified pattern of Figure 14.11 is shown with individual seeds on the capitulum boldly without wasting of space. As the radial arcs are drawn just bordering the circles, the space available for a seed is represented by each circle. Thirty-four clockwise and 21 counterclockwise arcs are seen at the periphery. The 34 arcs, each after passing through 4 or 5 seeds, merge with the set of 13 arcs moving along the same direction but showing a greater degree of curvature. From where the 13 arcs take over from the 34, the original 21 arcs are seen moving

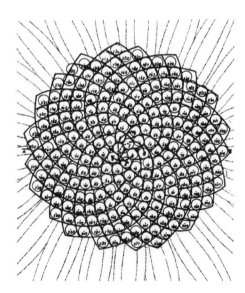

FIGURE 14.12 A reconstructed head of a sunflower. (From Mathai, A.M., and Davis, T.A., *Math. Biosci.*, 20, 117–133, 1974.)

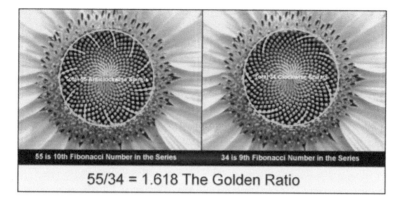

55/34 = 1.618 The Golden Ratio

FIGURE 14.13 Sunflower head showing the golden ratio and ray and disc florets. (Courtesy of Sandeep Ahirwar, http://ultraxart.com/author/Sandeep, 1993.)

opposite to the 13 arcs. Thus, the 34 and 21 arcs of the periphery get reduced to 21 and 13. At some level still closer to the centre, arcs 13 and 8 are seen, and this combination progressively gets reduced to 8 and 5, 5 and 3, and finally to 3 and 2.

In Figure 14.13, we can see the spirals arranged in two directions (clockwise and anti-clockwise). On summing the numbers of spirals in each direction, we can see the two consecutive Fibonacci numbers. If we divide the clockwise spirals and anti-clockwise spirals, we will get the golden ratio. These spirals are arranged according to the golden angle. This provides a biological advantage because it maximizes the number of seeds that can be packed into a seed head.

Most of the earlier workers (Bonnet, 1754; Church, 1904, 1920; Coxeter, 1953; Gardner, 1969) thought that all the arcs start from the periphery and reach the centre of the capitulum. However, Richards (1948, 1950) seems to be the only worker whose illustrations on the phyllotaxis could throw considerable light on this peculiarity.

It can be interpreted from Figure 14.12 that in capitulums bearing equal-sized seeds, the numbers of visible arcs will vary with the diameter of the disc, a smaller one showing lower numbers and larger ones higher numbers, all matching with terms in the Fibonacci sequence. The oldest seed of the capitulum in Figure 14.12 is just older to 233 members. A head having 233 numbers (as in Figure 14.12) is a Fibonacci number. Hence, a head having 233 to 144 (next Fibonacci number) seeds will show 34 and 21 arcs at its periphery. Another having 89 seeds will show 21 and 13 spirals, one with 34 seeds will show 13 and 8 spirals, and so on to smaller numbers of arcs as the numbers of individual seeds per head decrease. The periphery of the capitulum in Figure 14.12 has 21 projections. The flower at each projection develops into a ray floret. In a smaller head, the number of projections is reduced to 13, 8, and 5 and may show an equal number of ray florets. The spiral numbers synchronizing the alternate terms of the Fibonacci sequence, that is, 21, 8, 3, and 1, commencing from the periphery move counter clockwise. This peculiarity has a striking similarity with a basic property of Fibonacci numbers. The ratio between consecutive Fibonacci numbers swings steadily between the plus and minus sides of the golden ratio (0.618).

The ratio of 1/1 is 1.0, which is on the plus side; 1/2 = 0.5 is on the minus side; 2/3 = 0.667 is on the plus side; 3/5 = 0.60 is on the minus side, and so on.

The Fibonacci sequence allows for the maximum number of seeds in an arrangement of seeds in the seed heads of sunflowers (Grob et al., 2007). There is no crowding in the centre and no spaces at the edges, meaning that the sunflower has the perfect space organization for its seeds (Grob et al., 2007). This is due to a growth characteristic of sunflowers (Segerman, 2010). Individual seeds grow at the centre of the flower and continue adding seeds, pushing those on the edge outwards (Heimbuch, 2012). As it develops in a Fibonacci arrangement, seeds are always grouped uniformly and they stay compact (Heimbuch, 2012).

To optimally fill the sunflower head with the highest amount of seeds, the most irrational number should be used, which is phi, otherwise known as the golden ratio (Winer, 2013).

Using this irrational number to spread out the seeds makes it harder to see the number of lines in the spiral of a sunflower; however, this means that the distribution is maximally efficient (Winer, 2013). If the spaces between different spirals were easy to see, it would signify that this space is "wasted" and therefore not being effectively utilized (Winer, 2013).

By manipulating the stress on Ag core/SiO_2 shell microstructures, Li et al., (2007) obtained a series of Fibonacci spirals (3×5 to 13×21) of definite chirality as a least-elastic energy configuration. They (Li et al, 2007) demonstrated that the Fibonacci spiral patterns of definite chirality can be reproduced through stress manipulation on the AG core/SiO_2 shell microstructures. The Fibonacci spirals occur uniquely on conical supports-spherical receptacles result in triangular tessellations, and slanted receptacles introduce irregularities. Swinton et al. (2016) worked on 657 sunflowers and evaluated the occurrence of Fibonacci structure in the spirals of sunflower seed heads (capitulum). They observed 768 clockwise or anti-clockwise parastichy numbers of which 565 were Fibonacci numbers, and 67 had Fibonacci structure of a predefined type. Some of these seed heads were approximately Fibonacci, some without Fibonacci structure, some parastichy numbers equal to one less than a Fibonacci number, and some one more than a Fibonacci number.

14.5.2 FIBONACCI NUMBER IN PETALS

Plants where the number of flower petals follows Fibonacci numbers are very common in nature. One can easily find flowers with 0, 1, 2, 3, 5, 8 or more petals. Flowers with five petals are most common in nature (Sandeep, 1993). The followings are the examples of some plants that follow Fibonacci numbers in their number of petals: cord-grass (*Spartina alterniflora*) and cotton weed (*Achillea maritima*) have 0 petals. White calla lily (*Zantedeschia aethiopica*) has one petal. Lilies (*Lilium*) and irises (*Iris*) have three petals. Buttercups (*Ranunculus*) and wild roses (*Rosa acicularis*) have five. Larkspur (*Delphiniums*) and bloodroot (*Sanguinaria canadensis*) have eight. Black-eyed Susan (*Rudbeckia hirta*), mayweed (*Anthemis cotula*), and *Cinerarias* have 13. Chicory (*Cichorium*), Shasta daisy (*Leucanthemum*), and aster (*Asteraceae*) have 21 petals. Daisy (*Bellis perennis*) and pyrethrum (*Chrysanthemum* or *Tanacetum*) have 34 petals. Helenium (*Helenium autumnale*) has 55 petals, and other types can have 89 petals.

Some of these flowers are very specific and always have a Fibonacci number of petals while others are not as precise with the number, but still average a Fibonacci number (Peterson, 2006). In such cases, it is more likely that underdevelopment occurred and the flower has one or two petals less than it is meant to (Peterson, 2006). As with most flora, the reason for these amounts of petals has to do with maximizing the efficiency of light during the growth process of plants (Knott, 2010).

As per Darwinian processes, each petal is placed at 0.618034 per turn (out of a 360° circle). This angular arrangement exposes the plants completely to receive sunlight and rain in highly efficient manner.

Davis and Bose (1971) reported that in most Aroids, clear spirals occur on the arrangement of the flowers. The numbers of these spirals generally synchronize with Fibonacci numbers, which may run clockwise or counter clockwise. But in some species they do not. According to the size of the inflorescence, the numbers of spirals vary, the thinner ones having smaller numbers. If the numbers of spirals in a species are, say, 5 and 8, some individuals have the five spirals moving clockwise and the eight spirals, counter clockwise. In other individuals of the same species, the reverse is the situation.

In *Ptychosperma macarthurii*, a mirror-image pattern occurs in some of its organs. The male and female flowers are distributed on the flowering spikes in triads, each comprising two males bordering a female. One of these males shows clockwise aestivation of the calyx and the other counter clockwise imbrication. The six perianth members are distinguished into the calyx and corolla. In female flowers, both the whorls imbricate, and in male flowers, the calyx imbricates while the corolla remains valvate.

Seed arrangements can be found in all different types of flowers like sunflowers, which were discussed in a previous section; the majority of flowers have two sets of spirals that go in opposite directions (Knott, 2010). The number of spirals going in both ways is almost always adjacent Fibonacci numbers (Peterson, 2006). Very often it happens that there are 55 and 34 spirals.

14.5.3 CONES AND FRUITS

Beal (1873) examined a large number of cones (155) of several species of Coniferae to see if there was any variation in their scales or leaf arrangement. He found 74 cones of a Norway spruce that showed 5 parallel spirals to the right and 8 to the left; while 74 showed eight spirals to the right and 5 to the left. Five cones possessed seven spirals to the right and four spirals to the left. One cone had four spirals to the right and six to the left, and one cone had six spirals to the right and four to the left. He also noticed that there were spirals with three rows, 8 and 21 one way, and 5 and 13 the other way. Other cones showed 3 and 7 one way and 11 the other.

In pine cones or pineapples, we see a double set of spirals—one going in a clockwise direction and one in the opposite direction. When these spirals are counted, the two sets are found to be adjacent Fibonacci numbers (Figure 14.14).

All pine cones develop in a spiral shape that starts from the base where it was connected to the tree (Carson, 1978). There are usually two sets of spirals which circulate in different directions up to the top of the pine cone (Dunlap, 1997).

FIGURE 14.14   Pine cones with double set of spirals.

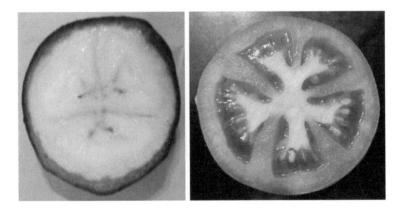

FIGURE 14.15   Cross sections of banana and tomato fruits showing three and five sections.

In some cases, a pine cone is found not to incorporate the sequence, which is usually due to deformities caused by pests (Carson, 1978). The spiralled bracts going in opposing directions are adjacent Fibonacci numbers (Simmons, 2011). Dunlap (1997) reported that the same pine cone that has 8 spirals in one direction yet 13 in the other. The sequence is an approximation to an irrational number, meaning that the bracts of pine cones should not line up. If they did, the pinecone would be weakened and susceptible to breakage (Collins, 2011).

Inside the fruit of many plants we can observe the presence of Fibonacci order. The banana has three sections; the tomato has five sections (Figure 14.15).

14.5.4   VENATION

The golden ratio can be easily seen in the venation of plants leaf. This has been drawn over this figure, which is showing the hidden divine ratio in the veins of a leaf (Figure 14.16).

FIGURE 14.16 Golden ratio on leaf vein. (Courtesy of Sandeep Ahirwar, http://ultraxart.com/author/Sandeep, 1993.)

14.6 FIBONACCI RECTANGLES AND SHELL SPIRALS

Figure 14.17 shows the Fibonacci numbers 1, 1, 2, 3, 5, 8, 13, 21, if we start with two small squares of size 1 next to each other. On top of both of these we draw a square of size 2 (=1 +1). We can now draw a new square—touching both a unit square and the latest square of side 2—so having sides 3 units long; and then another touching both the 2-square and the 3-square (which has sides of 5 units). We can continue adding squares around the picture; each new square having a side that is as long as the sum of the latest two squares' sides. This set of rectangles whose sides are two successive

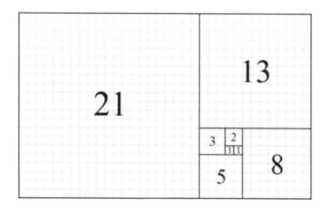

FIGURE 14.17 Fibonacci rectangles. A tiling with squares whose side lengths are successive Fibonacci numbers. (Courtesy of https://commons.wikimedia.org/wiki/File:34 * 21-Fibonacci Blocks.png#mw-head.)

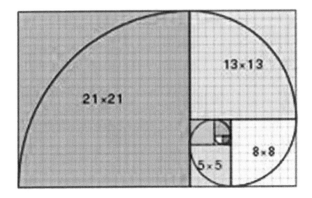

FIGURE 14.18 Fibonacci spiral. (Courtesy of https://commons.wikimedia.org/wiki/File:FibonacciSpiral.svg#mw-head.)

Fibonacci numbers in length and that are composed of squares with sides that are Fibonacci numbers, we can call the **Fibonacci rectangles**.

The Fibonacci spiral (Figure 14.18) is an approximation of the golden spiral created by drawing circular arcs connecting the opposite corners of squares in the Fibonacci tiling; this one uses squares of sizes 1, 1, 2, 3, 5, 8, 13 and 21 (Tiner, 2004). The spiral is not a *true* mathematical spiral (since it is made up of fragments that are parts of circles and does not go on getting smaller and smaller), but it is a good approximation to a kind of spiral that does appear often in nature. Such spirals are seen in the shape of shells of snails and sea shells (Figure 14.18) and in the arrangement of seeds on flowering plants too. The spiral-in-the-squares makes a line from the centre of the spiral increase by a factor of the golden number in each square. The points on the spiral are 1.618 times as far from the centre after a quarter turn. In a whole turn, the points on a radius out from the centre are $1.618^4 = 6.854$ times farther out than when the curve last crossed the same radial line.

Cundy and Rollett (1961) say that this spiral occurs in snail shells and flower heads referring to D'Arcy Thompson's *On Growth and Form* "The Equiangular Spiral." Here, Thompson is talking about a class of spiral with a constant expansion factor along a central line and not just shells with a phi expansion factor.

In a cross section of a nautilus sea shell, we can see the spiral curve of the shell and the internal chambers that the animal using it adds on as it grows. The chambers provide buoyancy in the water. If we draw a line from the centre out in any direction and find two places where the shell crosses it so that the shell spiral has gone around just once between them, the outer crossing point will be about 1.6 times as far from the centre as the next inner point on the line where the shell crosses it. This shows that the shell has grown by a factor of the golden ratio in one turn.

This factor varies from 1.6 to 1.9 and may be due to the shell not being cut exactly along a central plane to produce the cross section.

It is *incorrect* to say this is a phi spiral. First, the "spiral" is only an approximation as it is made up of separate and distinct quarter circles; second, the (true) spiral increases by a factor phi every *quarter turn* so it is more correct to call it a phi^4 spiral.

The curve of this shell is called equiangular or logarithmic spirals and are common in nature, though the "growth factor" may not always be the golden ratio.

14.7 CONCLUSION

The Fibonacci numbers coined by Leonardo Bonacci, whose nickname was Fibonacci, occur in every sphere of natural world. In 1202, Fibonacci completed his book, *Liber Abaci*, in which he demonstrated how Hindu–Arabic numerals made highly important tasks faster and easier. He showed how the new system improved critical functions, such as commercial bookkeeping, computation of interest, monetary exchange, and the conversion and calculation of weights and measures. All of this became more efficient and effective in a series of 10 digits: 0 through 9 are used. Fibonacci put these numbers to good use, developing the Fibonacci sequence (e.g., 0, 1, 1, 2, 3, 5, 8, 13, 21, 34, 55, 89, 144, etc.). Mathematicians now apply this sequence as a key element in such practices as "planning poker" software often used to estimate developmental goals, even the basis of a geometric spiral employed by carpenters to build large tree houses. However, the greatest contribution of Fibonacci to mankind is the use of Hindu–Arabic numerals. Because of his vision, later mathematicians like Isaac Newton were able to create fields such as calculus, which led to everything from advanced physics to flight, telephones, binary code and computers, the internet, and medical advances. Plant growth is governed by the Fibonacci sequence, which can be understood as a law of accumulation. The branching rates, phyllotaxis, number of petals, arrangement of seeds on a sunflower seed head, and spirality of leaves on plant stems follow Fibonacci numbers. The Fibonacci spiral in the nautilus sea shell approximates the golden spiral, which is an example of a logarithmic spiral, very common in nature.

REFERENCES

Allard, H. A. (1946). Clockwise and counter clockwise spirality in the phyllotaxy of tobacco. *Journal of Agricultural Research* **73**: 237–242.

Bahadur, B., Reddy, N. P., and Kumar, P. V. (1978). Bioisomerism in cycads. *Current Science* **47**: 404–406.

Bahadur, B., Rao, K. L. and Rao, M. M. (1980). Asymmetry and yield in pigeonpea. *Cajanus cajan* (L.) Millsp. International Workshop on Pigeonpea, Icrisat, Patancheru, India, **2**: 15–19.

Beal, W. J. (1873). Phyllotaxis of cones. *The American Naturalist* **7**: 449–453.

Boman, B. M., Dinh, T.N., Decker, K. Emerick, B., Raymond, C. and Schleiniger, G. (2017). Why do Fibonacci numbers appear in patterns of growth in nature? A model for tissue renewal based on asymmetric cell division. *The Fibonacci Quarterly* **55**: 30–45.

Bonnet, C. (1754). Recherches sur l'Usage des Feuilles dans les Plants, Gottingen and Leiden.

Bravais, L. and Bravais, A. 1837. Essai sur la disposition des feuilles curvisériées. Annales des Sciences Naturelles Botanique 7: 42–110; 193–221; 291–348; 8: 11–42.

Britton, J. (2011). Fibonacci Numbers in Nature. Available at: http://britton.disted.camosun.bc.ca/fibslide/jbfibslide.htm (Accessed: January 19, 2014).

Candolle, M. C. de. (1881). Considerations l'etude de la phyllotaxie. *Archives des Sciences Physiques et Naturelles* **5**: 358–399.

Carson, J. (1978). Fibonacci numbers and pineapple phylloaxy. *The Two-Year College Mathematics Journal* **9**(3): 132–136.

Church, A. H. (1904). On the Relation of Phyllotaxis to Mechanical Laws. Williams & Norgate, London, UK.

Church, A. H. (1920). On the Interpretation of Phenomena of Phyllotaxis. New York.

Collins, D. (2011). Mona Lisa and Fibonacci Pinecones. Available at: http://www.warren-wilson.edu/~physics/PhysPhot OfWeek/2011PPOW/20110225 Fibonacci Pinecone/ (Accessed: January 19, 2014).

Compton, R. H. (1912). A further contribution to the study of right- and left-handedness. *Journal of Genetics* **2**: 53–70.

Coxeter, H. S. M. (1953). The golden section, Phyllotaxis and Wythoff's game. *Scripta Mathematics* **19**: 135–143.

Cundy, H. M. and Rollett, A. P. (1961). *Mathematical Models*, 2nd ed. Clarendon Press, Oxford, UK, p. 70.

Davies, P. A. (1937). Leaf arrangements in *Ailanthus altissima. American Journal of Botany* **24**: 401–407.

Davis, T. A. (1962). The non-inheritance of asymmetry in *Cocos nucifera. Journal of Genetics* **58**: 42–50.

Davis. T. A. (1963). The dependence of yield on asymmetry in coconut palms. *Journal of Genetics* **58**: 186–215.

Davis, T. A. (1970). Fibonacci numbers for palm foliar spirals. *Acta Botanica Neerlandica* **19**: 49–256.

Davis, T. A. (1971). Why Fibonacci sequence for palm leaf spirals? *The Fibonacci Quarterly* **9**: 237–244.

Davis, T. A. (1972). Effect of foliar arrangement on fruit production in some tropical crop plants. Tropical ecology with an emphasis on organic production, Athens, GA, pp. 147–164.

Davis, T. A. and Bose, T. K. (1971). Fibonacci system in aroids. *Fibonacci Quarterly* **9**: 253–263.

Davis, T. A. and Kundu, A. (1966). Aestivation of perianth of *Areca catechu* L. *Journal of Bombay Natural History Society* **63**: 270–282.

Dormer, K. J. (1955). Mathematical aspects of plant development. *Discovery* **16**.

Douady, S. and Couder, Y. (1996a). Phyllotaxis as a dynamical self-organizing process Part I: The spiral modes resulting from time-periodic iterations. *Journal of Theoretical Biology* **178**: 255–273.

Douady, S. and Couder, Y. (1996b). Phyllotaxis as a dynamical self-organizing process Part II. The spontaneous formation of a periodicity and the coexistence of spiral and whorled patterns. *Journal of Theoretical Biology* **178**: 275–294.

Douady, S. and Couder, Y. (1996c). Phyllotaxis as a dynamical self-organizing process Part III. The simulation of the transient regimes of ontogeny. *Journal of Theoretical Biology* **178**: 295–312.

Dunlap, R. A. (1997). The Golden Ratio and Fibonacci Numbers. Google eBooks [Online]. Available at: http://books.google.de/ books?id=Pq 2AekTsF6oC&pg=PA130&dq= fibonacci+pinecone &hl=en&sa=X&ei=M4frUruJMYKVtQa ayIHQDA&ved=0C C4Q6AEwAA#v=onepage&q=fibonacci%20pinecone&f=false (Accessed: January 5, 2014).

Elhoumaizia, M. A., Lecoustreb, R. and Oihabic, A. (2002). Phyllotaxis and handedness in date palm (*Phœnix dactylifera* L.). *Fruits* **57**: 297–303.

Gardner, M. (1969). The multiple fascination of the Fibonacci sequence. *Scientific American* **220**: 116–120.

Ghose, M., Bhattacharya, K., Ghosh, S. S., Roychoudhury, K., and Datey, A. (1996). Effect of foliar arrangement on the leaf epidermal structures in Areca palm (*Areca catechu* Linn.). *Acta Botanica Neerlandica* **45**: 303–308.

Grob, V., Pfeifer, E. and Rutishauser, R. (2007). Sympodial construction of Fibonacci type Leaf Rosettes in *Pinguicula Moranesis. Annals of Botany* **100**(4): 857–863.

Heimbuch, J. (2012). Nature Blows My Mind! The Hypnotic Patters of Sunflowers. Available at: http://www.treehugger.com/slideshows/natural-sciences/nature-blows-my-mind-hypnotic-patterns-sunflowers/ (Accessed: January 24, 2014).

Henslow, G. (1876). On the origin of the prevailing systems of phyllotaxis. *Transactions of the Linnean Society of London.* 2nd ser., Botany, pt. 2, **1**: 37–45.

Hofmeister, W. (1868). Handbuch der Physiologischen Botanik I-2., Leipzig 1868.

Knott, R. (2010). Fibonacci numbers and nature. [Online]. Available at: http://www.maths.surrey.ac.uk/hosted-sites/R. Knott/Fibonacci/fibnat.html (Accessed: January 5, 2014).

Knuth, D. E. (1968). *The Art of Computer Programming*, **1**, Addison Wesley, Reading, MA.

Knuth, D. E. (2006). *The Art of Computer Programming, 4. Generating All Trees – History of Combinatorial Generation*, Addison-Wesley, Reading, MA, p. 50.

Kundu, B. C. and Sharma, M. S. (1965). Direction of leaf spiral in *Corchorus capsularis* L. *Transactions of the Bose Research Institute* **28**: 107–112.

Leppik, E. E. (1961). Phyllotaxis, anthotaxis and semataxis. *Acta Biotheoretica* **14**: 1.

Li, C., Ji, A. and Cao, Z. (2007). Stressed Fibonacci spiral patterns of definite chirality. *Applied Physics Letters* **90**: 164102.

Mathai, A. M. and Davis, T. A. (1974). Constructing the sunflower head. *Mathematical Biosciences* **20**: 117–133.

Mitchison, G. J. (1977). Phyllotaxis and the Fibonacci series. *Science* **196**: 270–275.

Norton, A. (1999). Fibonacci and the golden ratio. *Mathematics Magazine* **75**(3): 163–172.

Patel, J. S. (1938). *The Coconut, A Monograph.* Government Press, Madras, India.

Peterson, I. (2006). Fibonacci's Missing Flowers. Available at: https://www.sciencenews.org/article/fibonaccis-missing-flowers (Accessed: January 28, 2014).

Pisano, L. (2002). *Fibonacci's Liber Abaci: A Translation into Modern English of the Book of Calculation, Sources and Studies in the History of Mathematics and Physical Sciences*, Sigler, L. E, trans, Springer.

Rao, M. M. (1980). Studies on seedling and corolla handedness in Papilionaceae with special reference to *Vigna* and *Phaseolus* spp. and its possible influence on yield. PhD thesis, Kakatiya University Warangal.

Rehmeyer, J. (2007) The mathematical lives of plants. Available at: http://www.sciencenews.org/article/mathematicallives-plants (Accessed: January 19, 2014).

Richards, F. J. (1948). The geometry of phyllotaxis and its organ. *Society for Experimental Biology Symposium* **2**: 217–245.

Richards, F. J. (1950). Phyllotaxis: Its quantitative expression and relation to growth in the apex. *Philosophical Transactions of the Royal Society A (London) B* **235**: 509–564.

Sampson, H. C. (1923). *The Coconut Palm.* John Bale, Sons & Danielsson, London, UK.

Sandeep, A. (1993). http://ultraxart.com/fibonacci-numbers-in-plants-design-of-leaf-petals-branches-and-flowers/

Segerman, H. (2010). The sunflower spiral and the Fibonacci metric. [Online]. Available at: http://ms.unimelb.edu.au/~segerman/papers/sunflower_spiral_fibonacci_metric.pdf (Accessed: January 24, 2014).

Schroeder, C. A. (1953). Spirality in citrus. *Botanical Gazette* **114**: 350–352.

Sharma, M. S. (1968). A preliminary study on the direction of leaf spiral and phyllotaxy in jute. *Journal of Genetics* **58**: 49–52.

Simmons, J. R. (2011). Fibonacci numbers and nature. Available at: http://jwilson.coe.uga.edu/EMAT6680/Simmons/.Essay1/6690ProjectFibonacciF.htm (Accessed: January 19, 2014).

Singh, P. (1985). The so-called Fibonacci numbers in ancient and medieval India. *Historia Mathematica* **12** (3): 229–244. doi:10.1016/0315-0860(85)90021-7

Strasburger, E. (1930). *Textbook of Botany.* 6th English ed., Revised by W. H. Lang. London, UK.

Swinton, J., Ochu, E., The MSI Turing's Sunflower Consortium. (2016). Novel Fibonacci and non-Fibonacci structure in the sunflower: Results of a citizen science experiment. *Royal Society Open Science* **3**: 160091. doi:10.1098/rsos.160091.

Thompson, D. W. (1917). *On Growth and Form.* Cambridge University Press, London, UK.

Tiner, J. H. (2004). Exploring the world of mathematics: From ancient record keeping to the latest advances in computers. New Leaf Publishing Group.

Weisse, A. 1894. Neue Beitrage zur mechanischen Blattstellungslehre. *Jahrbücher für Wissenschaftliche Botanik* **26**: 236–294.

Winer, M. C. (2013) Fibonacci Numbers in sunflowers. Available at: http://www.martinc-winer.com/fibonacci-numbersin-sunflowers/ (Accessed: January 24, 2014).

Wright, C. (1873). The basis and origin of the arrangement of leaves in plants. *Memoirs American Academy of Arts and Sciences* **9**: 379–415.

15 Sunflower Spirals

Oleh Bodnar

CONTENTS

15.1 INTRODUCTION

The theme of the following research is the biological phenomenon called phyllotaxis. The whole range of clear regularities is realized in phyllotaxis—spiral symmetry, golden section, Fibonacci sequence. The phenomenon in question has often been within the focus of researchers' attention. Among them there are such distinguished scientists as J. Kepler, G. Weil, and many other researchers who represented not only the sciences but also the artistic field. The fact that phyllotaxis has mysterious "divine proportion" principle of spiral organization in the phenomenon of animate and inanimate nature, in particular, in flora, in space structures, and in different earth elements, this is what made phyllotaxis an attractive field of research and made it possible to see in this phenomenon the fundamental law of nature. Nowadays phyllotaxis research has become a separate scientific area. It is mainly of mathematical character and is developing around the task of studying geometric regularities and building adequate mathematical models of this phenomenon.

However, one can state that history of the problem under research has two different stages that correspond to the general sequence of changes in understanding of space. Until a certain period, the problem of phyllotaxis was about studying "static geometry" of the phenomenon. Even assumptions about the mechanism of biological growth are based on the basis of "static geometry" data. Thus, from external properties, symmetry of resemblance in particular, the conclusion is made that phyllotaxis growth is based on the law of logarithmic spiral. That means that classic approach, which, by the way, is also developed by the theoreticians of proportions, formally exhausts the geometric side of the phyllotaxis problem.

The past decades developed new understanding of the phyllotaxis problem and new approach to its solution, particularly, in the field of theoretical (mathematical) biology, where phyllotaxis research was most active, and geometric phenomena of nature in general. One understands the necessity to study this issue from the viewpoint of non-classic geometric understanding. At the same time, modern scientists ground their ideas on the forecast of V. Vernadski about the non-Euclid character of animate nature geometry; the issue of phyllotaxis deciphering is viewed within the context of the idea of dynamic (spatial-temporal) geometry. It turns out that the growth mystery is not solved directly with the help of static geometry features. Research of the growth algorithm is conducted taking into account these or those dynamic (biological) properties of the phenomenon. However, despite numerous attempts, none of the developed phyllotaxis models is now recognized in biology as a satisfactory one. And the matter is not only in the evaluation of the biologists. So far, the study of phyllotaxis has failed to explain the issues regarding appearance in this phenomenon of the golden section regularities and the Fibonacci sequence. The essence of geometry of the phenomenon is not clear either. That means that there are no answers to those principal questions that draw attention of the scientists and directly are related to the problem of understanding of space that is discussed in the monograph. All this brought about the necessity to carry out special research of phyllotaxis.

15.2 STRUCTURAL AND NUMERICAL PROPERTIES AND DYNAMIC SYMMETRY OF PHYLLOTAXIS

The main task of the research is formulated on the basis of minimum initial information, which is sufficient to show the essence of the main issue of the problem and build the research "from scratch," making it independent of the research experience accumulated in the field. Though, as it has been mentioned already, this experience is quite substantial. In particular, in the second half of the 20th century, the research of phyllotaxis was in the center of attention of Coxeter (1953, 1966), Adler (1974, 1977), Jean (1976), Petukhov (1988), etc. who in their works created the modern "mathematical and biological" theory of phyllotaxis.

It is necessary to mention that the findings of this research were obtained by the author and published for the first time in 1989 (Bodnar, 1989). At that time they were new. At any rate, none of the researchers describing phyllotaxis applied Minkowski's geometry and the system of hyperbolic trigonometry. We shall briefly present this research.

It is known from biology that mutual arrangement of various primordia that appear on the cones of spindles is characterized by the spiral symmetry. This principle of the situation that was named phyllotaxis is also clearly observed in dense inflorescences and infructescences, for instance, on sunflower discs, pine cones, and many other types of bioforms (Figure 15.1).

On the surfaces of phyllotaxis forms, dense inflorescences and infructescences in particular, one can clearly observe left- and right-wound spiral-like rows of structural elements (primordia, seeds, leaves). The symmetry order of phyllotaxis forms is usually shown as the ratio of numbers that corresponds to the quantity of left and right adjacent numbers of recurrent rows, which accept the rule: $U_n = U_{n-2} + U_{n-1}$. Widely spread are Fibonacci series: ..., 0, 1, 1, 3, 5, 8, 13, 21, 34, ... Numbers of Lucas series ..., 1, 3, 4, 7, 11, 18, 29, 34, ... are also quite frequently realized, seldom— numbers that belong to the series ..., 4, 5, 9, 14, 23, ... The order of symmetry in case of Fibonacci phyllotaxis (F-phyllotaxis) is expressed by the ratios: $\frac{1}{2}, \frac{2}{3}, \frac{3}{5}, \frac{5}{8}, \frac{8}{13}, \dots$.

As a rule, the off-shoots of plants and trees are characterized by the low order of symmetry while the inflorescences and infructescences by high symmetry order. For instance, the order of symmetry with the sunflower can be up to $\frac{55}{89}, \frac{89}{144}$ and even $\frac{144}{233}$.

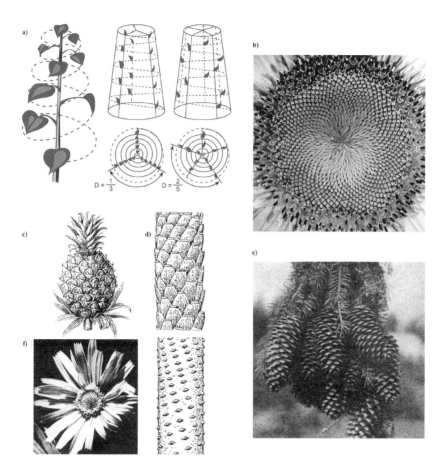

FIGURE 15.1 Examples of phyllotaxis structures: (a) leaves arrangement schemes; (b) sunflower disc; (c) pineapple fruit; (d) palm trunk; (e) camomile flower; and (f) pine cones.

An interesting indicator of phyllotaxis structures is the so-called D divergence, which is the angle of splitting of two subsequent primordia. Divergence measured in circle fractions in case of F-phyllotaxis is always expressed by the same number that is the pattern order of symmetry, that is, it can be equal $\frac{1}{2}, \frac{2}{3}, \frac{3}{5}, \frac{5}{8}, \frac{8}{13}, \dots$ This series of fractions tends to the limit ≈ 0.618 of the circle, with which the full plane angle appears to be divided as to the golden section Φ.

Some types of phyllotaxis patterns subsequently change (expand) the order of their symmetry in the process of growth. This peculiarity of phyllotaxis is called dynamic symmetry. As an example, one can take the sunflower. The sunflower heads arranged at the various levels of one and the same stem have different symmetry: the higher the level, that is, the older the disk, the higher its order of symmetry is. In symmetry dynamics, the following sequence is realized:

$$\dots \to \frac{5}{8} \to \frac{8}{13} \to \frac{13}{21} \to \frac{21}{34} \to \dots$$

With the change of symmetry, the divergence angle changes, respectively. Still on all the disks regardless of the number of spirals the so-called conforming (angular) characteristics of spiral patterns appear to be the same, that is, the spirals cross at a straight angle.

This data is enough to identify the objective of the research, which is geometrical deciphering of the phyllotaxis pattern formation process, and, consequently, the key issue is how the symmetry is changing in the process of phyllotaxis pattern growth.

At the beginning it is necessary to do geometrical stylization of phyllotaxis surface: let us represent it as a regular flat grid (Figure 15.2). This grid is numbered in such a way that the numbers of vertices characterize their distance from the straight line 00′; at the same time per unit is accepted the distance of Point 1 from the straight line 00′. Due to this numbering rule, the order of symmetry of cylindrical grid in the system of numerical denomination is expressed in the following way: points adjacent to 0 are numbered 5, 8, and 3 (also −5, −8, and −3), that is, such numbers that characterize numerical structure of helical spiral of cylindrical grid.

It is obvious that for the grids with different symmetry the numbers of neighboring vertices of O point will be different.

We shall do the comparative analysis of the series of grids illustrating the sequential stages of symmetry change in pyllotaxis (Figure 15.3). According to their

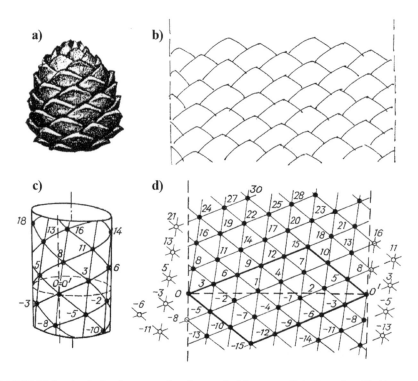

FIGURE 15.2 Analysis of structural and numerical features of phyllotaxis grid: (a) general look of cedar cone; (b) scheme of unrolling; (c) cylindrical grid—idealized form of cedar cone; and (d) unrolling of cylindrical grid.

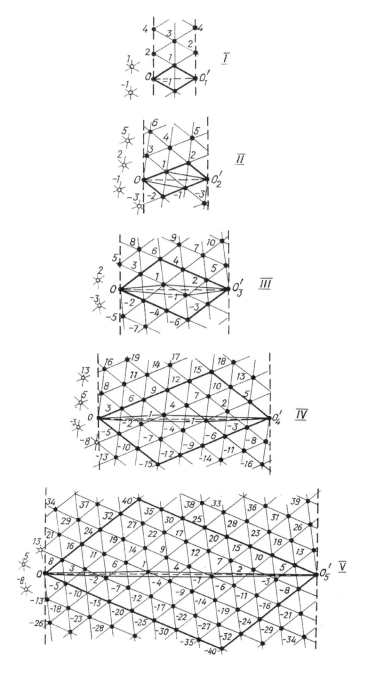

FIGURE 15.3   The series of unrollings illustrating sequential stages of symmetrical transformation of cylindrical grids. Each unrolling depicts parallelogram 010′1 (use bar kine above 1 as per in next).

metrical features, all the grids are the same. Thus, the peculiarity of natural pyllotaxis grids to periodically renew their conforming characteristics is taken into account.

Let us trace the regularity of transformation of parallelogram $010'\bar{1}$. We shall start from the scheme III. One can see that the sides 01 and $0'\bar{1}$ of this parallelogram are equal to the sections $00'_1$ and $00'_2$, which are the diagonals of this parallelogram at the two previous stages of its transformation, that is, at stages I and II. The sides of parallelogram $010'_4\,\bar{1}$ at the fourth stage appear to be equal the sections $00'_2$ and $00'_3$. This regularity can be traced also at the next stages. Let us provide separate illustration of this regularity (Figure 15.4).

The building rule is that the sides of every parallelogram (starting with the third one) are diagonals of two foregoing parallelograms. As any two successive parallelograms have three common vertices, it is obvious that all parallelograms have equal squares. Preserving the square is the first interesting feature of parallelogram dynamic transformation (Figure 15.5). The second feature is that straight lines are parallel: at any transformation stage parallelogram still remains a parallelogram. Here one can draw the key supposition of the research – preservation of square and parallelism of straight lines are the features of hyperbolic rotation. It means we deal with hyperbolic transformation. It is necessary to specify this idea.

Figure 15.5 illustrates "binding" of the grid to the scheme of hyperbolic transformation. Thus, we can state that hyperbolic rotation is the basis of symmetry

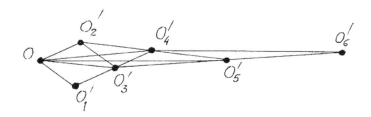

FIGURE 15.4   Research of elementary parallelogram transformation regularities.

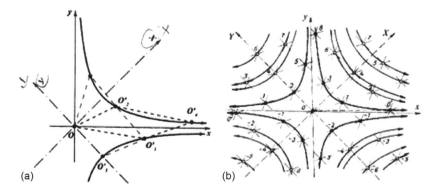

FIGURE 15.5   (a) Transformation of elementary parallelogram by means of hyperbolic motion and (b) scheme of grid transformation by hyperbolic rotation.

transformation illustrated on Figure 15.3. This is the key finding of the research that allows us to develop an entirely new look at the issue of phyllotaxis geometry.

15.3   ANALYTICAL AND GEOMETRIC RESEARCH

First of all, within the framework of the idea of hyperbolic rotation it is necessary to conduct a specific analytical and geometric research of a regular grid. Immediately an interesting fact is revealed: in metrical characteristics of the grid organically present is the value of the golden section.

Let us have a look at Figure 15.6. Here the arrangement of vertices corresponds to Figure 15.5 and is characterized by the following conditions:

$$x_a = x_{N1}, \; y_A = 1, \; 0A = 0N_1 = \sqrt{2}$$

Points M_1 and M_2 are symmetrical as to $0A$; $OM_1M_2N_1$; $OM_2M_2N_1$; $OM_2M_3N_2$ are parallelograms, that means that $ON_1 = M_1M_2 = 0A = \sqrt{2}$.

Let us determine the abscissa of the point M_2, marking $x_{M_2} = x$. From the condition of symmetry of the points M_1 and M_2, it goes that $x_{M_1} = y_{M_2} = x^{-1}$ and that the section M_1M_2 is tilted at an angle of 45° to the axis $0x$. It is evident that the difference of abscissas of the points M_1 and M_2 is equal to 1. We shall write down and solve the equation:

$$x - x^{-1} = 1,$$

$$x^2 - x - 1 = 0,$$

$$x = \frac{1 + \sqrt{5}}{2}.$$

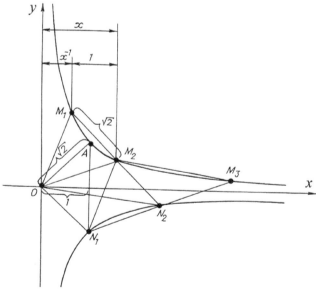

FIGURE 15.6   Analysis of metrical features of the grid.

As it follows from the statement of the problem $x_{M_2} = \frac{1+\sqrt{5}}{2} = \Phi$. Now one can easily make conclusions. The coordinates x and y of the arbitrary grid can be expressed in formula:

$$x = \frac{a}{\sqrt{2}} * \Phi^n, \; y = \frac{a}{\sqrt{2}} * \Phi^{-n} \tag{15.1}$$

where a is the semi-axis of hyperbole that the point under consideration belongs to. Moreover, if one takes the angle dividing two neighboring transformation stages (for instance, stages I and II, II and III, III and IV, etc.—Figure 15.3), in other words, the smallest angle, the rotation to which will result in the self-alignment of the grid as the angular unit (module) of hyperbolic rotation, then the exponent n will coincide with the value of hyperbolic angle (in modules), characterizing the position of the point under consideration (Figure 15.7).

In the coordinate system XOY (Figure 15.5b), the formulae of the coordinates of the arbitrary point look like hyperbolic functions:

$$X = a * \frac{\Phi^n + \Phi^{-n}}{2}, \; Y = a * \frac{\Phi^n - \Phi^{-n}}{2} \tag{15.2}$$

As one can see, these are unusual hyperbolic functions as both the base Φ and the value of hyperbolic angle unit here differ from the accepted ones.[1]

Therefore, these functions have independent signs and names:

$$\text{for } \frac{\Phi^n + \Phi^{-n}}{2} = Gch, \; n = \text{the golden cosine,}$$

$$\text{for } \frac{\Phi^n - \Phi^{-n}}{2} = Gch, \; n = \text{the golden sine,}$$

$$\text{for } \frac{\Phi^n - \Phi^{-n}}{\Phi^n + \Phi^{-n}} = Gch, \; n = \text{the golden tangent, etc.}$$

Formulae (15.2) of the coordinates of the arbitrary point in the system XOY now can be written like that:

$$X = a \cdot Gch \; n, \; Y = a \cdot Gsh \; n. \tag{15.3}$$

---

[1] *The classic variant hyperbolic angle unit is the square of the so-called coordinate rectangular of the hyperbole point M that is formed by the coordinate axes and the straight lines that were drawn through the point M parallel to the coordinate axes. If the hyperbole equation is $xy = 1$, then the square of the coordinate rectangular is numerically equal to 1. If unit square pertains to the hyperbolic sector 0AM (A—hyperbola vertex), then $x_M = e$, and $y_M = e^{-1}$.*

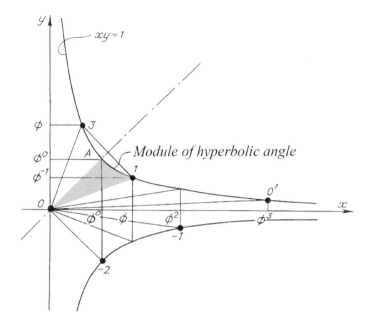

FIGURE 15.7   Determining the module of hyperbolic angle.

Changes of arbitrary point coordinates caused by the hyperbolic rotation are expressed by the formulae:

$$x' = X \cdot Gch\ n + Y \cdot Gsh\ n,$$
$$y' = X \cdot Gsh\ n + Y \cdot Gch\ n. \qquad (15.4)$$

In the process of hyperbolic rotation, the grid undergoes deformation, but periodically, through rotation module, the same metric states are repeated.

Figure 15.8 illustrates the subsequent deformation stages of the elementary triangle of the grid, which are fixed through half-module of the rotation. Metric peculiarities of the triangle are obvious. Figure 15.9 shows full depictions of the grid in two extreme states. In one of, them the elementary triangle gets the sizes $\sqrt{3}, \sqrt{3}, \sqrt{2}$. In the other state, when the grid becomes a square one, the triangle becomes a right isosceles triangle. These two states set the general grid triangle deformation range.

As one can see, transformation of the grid by means of hyperbolic rotation is characterized by periodicity. Two rotation modules are considered to make a full period (cycle). It is necessary to note that the grid state is repeated through one module. However, two such states do not coincide; they are in mirror position as to the hyperbole symmetry axes.

It is important to state: hyperbolic rotation is symmetry transformation of a regular grid. This transformation is not considered in the classical theory of symmetry.

It is also necessary to focus on the following issues: (1) unlike the circular rotation when the grid knots preserve strict stable arrangement, the process of hyperbolic rotation leads to the changing of the mutual arrangement of the knots; (2) for the

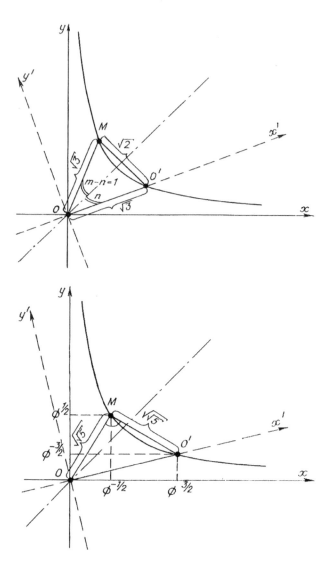

FIGURE 15.8 Characteristic conditions of grid elementary triangle.

symmetric transformation to happen, the grid should be aligned with the hyper-plane in a special way. In the general case, the asymptotes of hyperboles–trajectories of the grid knots should not go through the grid knots with the exception of point 0—the beginning of coordinates. Figure 15.10 (as compared to Figure 15.9) illustrates the example of such a "non-phyllotaxis" transformation.

Following the principle of grid numeration, the number of any of its points in the coordinate system where the abscissa axis coincides with the direction $00'$ (Figures 15.2 and 15.5), is numerically equal to its ordinate. We have developed formulae to describe the coordinates of arbitrary vertex in the system of movable coordinates $x'0y'$ (Figure 15.11). With any position of the coordinates, the ordinate of

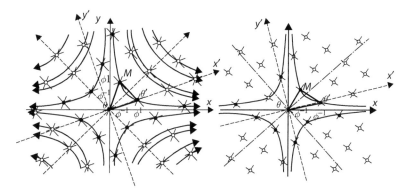

FIGURE 15.9   Two extreme conditions of the grid.

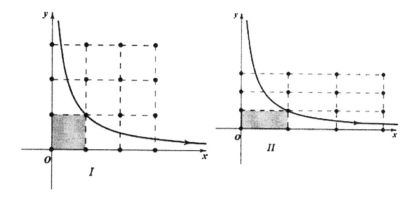

FIGURE 15.10   Example of non-phyllotaxis transformation when asymptotes coincide with the basic direction of the grid.

Point 1 nearest to $0x'$ is taken as a unit of measurement. Thus, for vertices that belong to the hyperbole $xy = 1$, we have:

$$x'_M = \frac{2}{\sqrt{5}} * Gch(m-n)$$

$$y'_M = \frac{2}{\sqrt{5}} * Gcsh(m+n) \tag{15.5}$$

where $m =$ hyperbolic angle $X0M$; $n =$ hyperbolic angle $X00'$. For vertices that are situated on the hyperbole $xy = -1$, the coordinate formulae will be the following:

$$x'_M = \frac{2}{\sqrt{5}} * Gch(m-n),$$

$$y'_M = \frac{2}{\sqrt{5}} * Gsh(m+n), \tag{15.6}$$

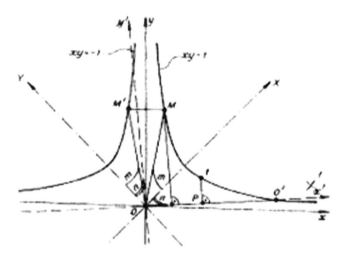

FIGURE 15.11 Determining coordinates of the arbitrary point in the system of movable coordinates **X′oy′**.

Thus, in Formulae (15.5) and (15.6), we have an interpretation of the numbers of those vertices of the grid that slide on the hyperboles nearest to the asymptotes. But the analysis of Figure 15.5 shows that the numbers of these vertices are Fibonacci numbers. Taking for the sake of simplicity $m + n = k$ and keeping in mind that the successive vertices are situated on the neighboring branches through rotation module, one can come to the following correspondence:

$$F_1 = \frac{2}{\sqrt{5}} * Gch1 = 1, F_2 = \frac{2}{\sqrt{5}} * Gch2 = 1,$$

$$F_3 = \frac{2}{\sqrt{5}} * Gch3 = 2, F_4 = \frac{2}{\sqrt{5}} * Gch4 = 3,$$

$$F_5 = \frac{2}{\sqrt{5}} * Gch5 = 5, F_6 = \frac{2}{\sqrt{5}} * Gch6 = 8,$$

$$F_7 = \frac{2}{\sqrt{5}} * Gch7 = 13, F_8 = \frac{2}{\sqrt{5}} * Gch8 = 21,$$

$$F_k = \frac{2}{\sqrt{5}} * Gchk, F_{k+1} = \frac{2}{\sqrt{5}} * Gch(k+1) \qquad (15.7)$$

In the grid system, different recurrent sequences of numbers are realized on different hyperboles. Introducing the hyperbole scale ratio (g) one gets the generalized variant of Formulae (15.7):

$$U_k = \frac{2}{\sqrt{5}} * Gchk, U_{k+1} = \frac{2}{\sqrt{5}} * Gch(k+1) \qquad (15.8)$$

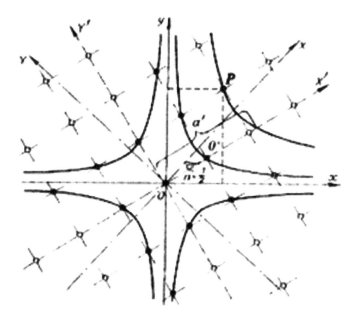

FIGURE 15.12   Analysis of the grid in the system of coordinates **X'OY'**.

There is another way of representation of the general member U_k of recurrent sequence:

$$U_k = A * F_k + B * F_{k+1} \qquad (15.9)$$

Here A and B are the so-called initial sequence elements; if the recurrent sequence is viewed as an infinite sequence, for instance, ..., 12, –7, 5, –2, 3, 1, 4, 5, 9, 14, ..., then initial will be two first numbers of sequence part of fixed signs. In this case, $A = 3$, $B = 1$. In the Fibonacci sequence, it will be $A = 1, B = 0$.

There is another way of obtaining trigonometric interpretation of Fibonacci numbers. There is one interesting arrangement of coordinate axes—$X'OX'$—that is given by the directions of square grid (Figure 15.12). Here hyperbolic angle XOX' is equal to half of the module. If the unit of measurement is the side of cell square, then the coordinates of X' and Y' grid knots will be integers. It is evident. For the arbitrary vertex P we have:

$$X'_p = a' * \frac{2}{\sqrt{5}} * Gch \left(m - \frac{1}{2}\right),$$

$$Y'_p = a' * \frac{2}{\sqrt{5}} * Gch \left(m + \frac{1}{2}\right). \qquad (15.10)$$

Here a' hyperbole radius (the orbit of point P) that coincides with OX' axis. If the counting of angle is done with respect to OX' axis, then formulae will look like (15.10):

$$X'_p = a' * \frac{2}{\sqrt{5}} * Gch \, (\psi - 1),$$

$$Y'_p = a' * \frac{2}{\sqrt{5}} * Gch\psi. \tag{15.11}$$

where $\psi = m + \frac{1}{2}$.

This will result in Formulae (15.7) and (15.8) that correspond to the special case, in particular, when the vertices belonging to the hyperbole that is the nearest to the asymptotes, for which $a' = 1$. In other words, the vertices coordinates belonging to the single-scale hyperbole are expressed in Fibonacci numbers.

Between the integral-valued coordinates X' and Y' there is a dependence:

$$X'^2 + X' * Y' - Y'^2 = a'^2 \tag{15.12}$$

In fact, it expresses the hyperbole equation in reference to the coordinates $X'OY'$. At the same time, as the variables X' and Y' successively take on the values of the adjacent numbers of a certain recurrent series, the Formula (15.12) illustrates an important feature of recurrent series, which, taking into account the specificity of Formulae (15.10) and (15.11), is recorded in the following way:

$$\left| u_k^2 + u_k u_{k+1} u_{k+1}^2 \right| = const. \tag{15.13}$$

It is obvious that every series has its own corresponding constant.

Thus, we have got new mathematical interpretation of numerical properties of phyllotaxis dynamic symmetry. As we see, symmetry indicators in phyllotaxis are the integral-valued expressions of golden hyperbolic functions, and dynamics of these indicators is presupposed by the regularity of integral-valued growth of golden functions. In this case, Formula (15.13) expresses the numerical constant of dynamic symmetry. In particular, for F-phyllotaxis, this constant will be 1:

$$\left| F_k^2 + F_k F_{k+1} F_{k+1}^2 \right| = 1. \tag{15.14}$$

15.4 INTERPRETATION OF PHYLLOTAXIS SPATIAL MODEL

The elementary spatial geometric model of phyllotaxis form is a cylinder (Figure 15.14). Let us consider Figure 15.13. We shall focus on the deformation character of the elementary parallelogram $011'0'$ on the cylinder involute. Here there are two important peculiarities. First, in the process of transformation, the parallelogram square is preserved: $0P \bullet 00' = \text{const}$; second, the deformation brings about the parallel shift of the bases $00'$ and $11'$.

On the cylinder surface, the elementary parallelogram is transformed into the so-called elementary belt; therefore, it is clear that in the process of transformation

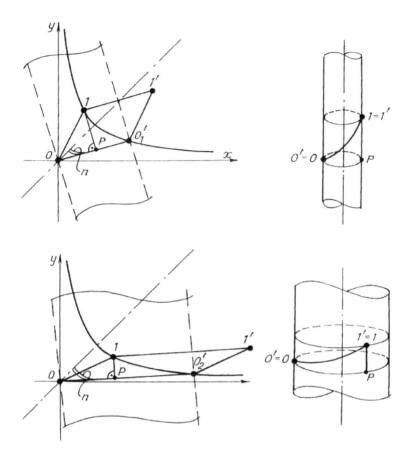

FIGURE 15.13   Analysis of hyperbolic transformation of cylindrical surface.

such an elementary belt will concentrically increase and decrease as to its height (the square will be left unchanged), and simultaneously there will be relative angular displacement of its bases that corresponds to the parallel shift on the involute. Such angular displacement will take place in every elementary belt, and this will stipulate the torsion effect on the cylinder surface in general.

The formula of divergence angle change comes from the ratio $\dfrac{OP}{OO'}$:

$$\frac{OP}{OO'} = \frac{\Phi^{2n-1} - \Phi^{-2n+1}}{\Phi^{2n} + \Phi^{-2n}} = \frac{Gsn(2n-1)}{Gch2n}.$$

The transformations result in:

$$D = \frac{1 - \sqrt{5} * Gth2n}{2}. \tag{15.15}$$

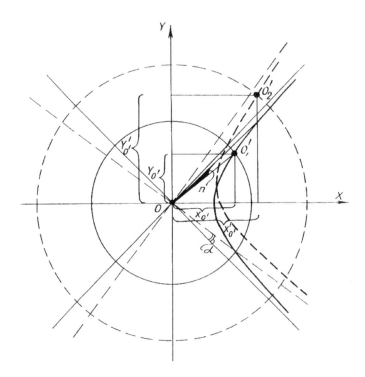

FIGURE 15.14   Analysis of transformation of cylinder transverse circumference.

Also

$$\lim_{n \to \infty} \frac{1-\sqrt{5} * Gth2n}{2} = \frac{1-\sqrt{5}}{2} = -\Phi^{-1} \tag{15.16}$$

As we see, the D angle limit corresponds to the known value of the so-called ideal angle of F-phyllotaxis. It is worth noting that in various "non-Fibonacci" cases, the divergence angle limit has various values, but in all the cases it changes according to the hyperbolic tangent law.

Analysis of Figure 15.14 allows us to understand the transformation character of the cylinder surface in the process of symmetrical transformation of the cylinder grid. It is possible to state that this transformation results in three interrelated motions: concentric widening of the cylinder, compression along the axis, and torsion.

We shall specify geometric regularity of cylinder transverse circumference transformation. The radius $r = 00'$ of the circumference is increased according to the hyperbolic cosine law:

$$r = 00' = \sqrt{2Gch2n}.$$

It is obvious that any point of the circumference can be simultaneously presented as the end of the hyperbole movable radius, as well as the end of the circumference movable radius.

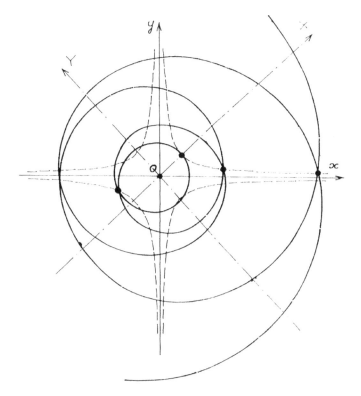

FIGURE 15.15   Composite spiral for which $\dfrac{\omega}{\omega_h} = \pi$.

The trajectory of such a motion of the point is a self-intersecting spiral, which can be called a composite one because it illustrates the composition of two rotations (Figure 15.15). Thus, the very composite spiral is a graphic expression of geometrical law of cylinder transverse circumference transformation. It finds its laconic representation in the formula:

$$\frac{\omega_h}{\omega} = const, \qquad\qquad (15.17)$$

which shows the main peculiarity of composite transformation, namely, consistency of angle speeds of hyperbolic rotation $\omega_h = n/t$ and circular rotation $\omega = a/t$.

Thus, we have presented the main findings of the phyllotaxis research. It is necessary to stress they are about cylindrical phyllotaxis. But this is only a preliminary stage of theoretical idealization of phyllotaxis. Cone is considered to be more adequate generalization of phyllotaxis pattern. Cylinder as well as disk is considered to be a special case of cone determined by the extreme values -0 and $\pi/2-$ of angle of tilt of cone generator to its axis.

We shall remind the essence of the existing ideas about geometrical features of phyllotaxis cone grids and dynamic mechanism of their pattern formation. It is generally considered that the structure of phyllotaxis cone grids (arrangement of vertices) is subject to the law of logarithmic spiral. Thus, to the logarithmic spiral also belong parastichy that form grids on cones and disk-like patterns and the so-called basic (or genetic) spiral that successively runs through all the vertices of the grid. Such grids will be called logarithmic.

Structural and semantic character of the logarithmic grid is determined by two indicators: q_π – ratio of local similitude that can be found from the ratio $q_\pi = \rho_k/\rho_{k+1}$, where ρ_k and ρ_{k+1} – the distance of two successive primordia to the cone vertex and D divergence angle. Here there is dependence:

$$q_\pi = q^D, \qquad (15.18)$$

where q= ratio of the basic spiral similitude.

Correspondingly, the algorithm of logarithmic grid pattern (Figure 15.16) formation goes to indicating the primordia motion trajectory, that is, basic logarithmic spiral and the so-called primordial growth interval Δt, or to the intensity of their reproduction N (it is meant that $N = 1/\Delta t$). It is supposed that in nature such a process is stipulated due to equal, in terms of time, primordia growth, and thus, the determination conditions for grid-formation process are as follows: first, invariability of local similitude ratio ($q_\pi = \rho_k/\rho_{k+1} = const$) and second, constant speed of primordia circular rotation ω_3:

$$\omega_3 = \frac{D}{\Delta t} = const. \, \omega_3 = \frac{D}{\Delta t}.$$

So, this is the classical model of phyllotaxis pattern formation.

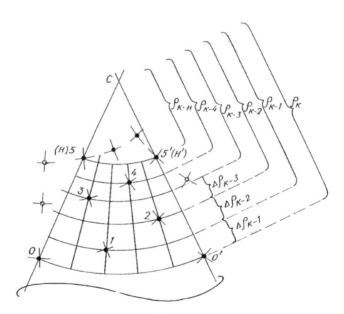

FIGURE 15.16   Analysis of the general case of logarithmic grid.

The findings of cylindrical phyllotaxis research result in a principally different idea of cone grid modeling. The idea is that the fundamental regularity of the structural organization of cone phyllotaxis is composite spiral. Parastichy in reality are composite and not logarithmic spirals. Composite is also genetic spiral.

Grids that are formed according to the composite spiral law are called natural (Figure 15.17). They resemble logarithmic ones but, in fact, they are incompatible with them. The reasons and character of this incompatibility can be explained on the comparative drawing of logarithmic and composite spiral (Figure 15.18). As we see, the logarithmic spiral is a curvilinear asymptote for the composite one. Different features of logarithmic and natural grids are explained by different mathematical nature of these two curves. The natural grid is characterized by the scale, which is determined by the value of the smallest radius of genetic spiral. Evidently, the notion of scale does not have any sense in reference to the logarithmic grid.

Of principal importance for the natural grids is the fact that indicators q_π and D in their structure are not observed in the ideal situations. They only come close to the nominal values as they withdraw from the center. In the center, zone divergences from logarithmic regularity are evident. In fact, this is violation of similitude. We can specify the regularity of divergence. For instance, for q_π this regularity is determined by the change of ratio:

$$\frac{\rho_k}{\rho_{k+1}} = \frac{\sqrt{2Gch2n}}{\sqrt{2Gch2(n-\Delta)}},$$

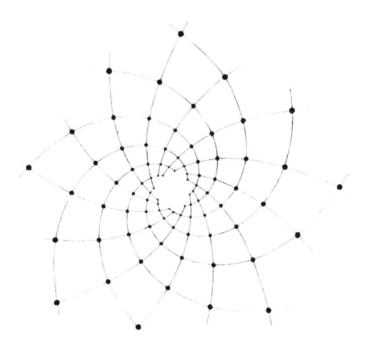

FIGURE 15.17   Example of natural grid with 8:13 symmetry.

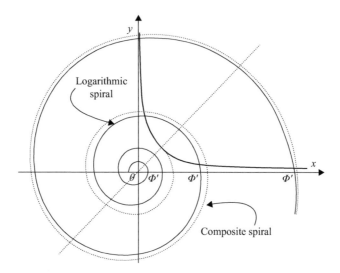

FIGURE 15.18   Comparative drawing of logarithmic and composite spiral.

where Δ = the difference of angular hyperbolic coordinates of two successive vertices that belong to the base of arbitrary parallel belt.

We shall once again stress that the process of natural grid pattern formation is
determined by the condition $\omega/\omega_h = const$. The process parameters are the speeds
ω, ω_h, and primordium's growth interval Δt or the indicator of their reproduction intensiveness $N = 1/\Delta t$. In every specific case, the values of ω, ω_h, and Δt
are constant; from them, consequently, one gets nominal characteristics q_π and
$\Delta\alpha = D$, that determine the symmetry of the grid restored. So, $\Delta\alpha = \omega * \Delta t$ and
$q_\pi = \Phi * \Delta^n$, where $\Delta^n = \omega_h * \Delta t$ is the angular hyperbolic interval between the
successive primordia.

15.5   CONCLUSION

Mathematical research of the botanic notion phyllotaxis revealed that growth
mechanism of spiral and symmetric plant structures is characterized by Minkowski
geometry. As a result, spiral lines, which are, in particular, formed by the blocking
elements of the dense florets are not logarithmic spirals (as it was considered before)
but the so-called composite spirals—total trajectory of point motion along the circumference and hyperbole at the same time.

REFERENCES

Adler, I., 1974. A model of contact pressure in phyllotaxis. *Journal of Theoretical Biology*,
     45(1), pp. 1–79. doi:10.1016/0022-5193(74)90043-5.
Adler, I., 1977. The consequences of contact pressure in phyllotaxis. *Journal of Theoretical
     Biology*, 65(1), pp. 29–77. doi:10.1016/0022-5193(77)90077-7.

Bodnar, O. Ya. 1989. Geometric Model of the Uniform Growth (laws of harmony in natural formation). – (dep. 19.06.89, # 54-Т3 89), М., (Боднар О. Я. Геометрическая модель однообразного роста (законы гармонии в природном формообразовании). – М., депонюВНИИТЗ, № 54-Т3 89, 1989. С. 60).

Bodnar, O. Ya. 1992a. Golden Section in Nature and Arts. – Technical Aesthetics. – # 1. (Боднар О. Я. Золотое сечение в природе и искусстве. – Техническая эстетика. – 1992. – № 1).

Bodnar, O. Ya. 1992b. Dynamic Symmetry. – Kiev, (Боднар О. Я. Динамическая симметрия. – Киев, 1992).

Bodnar, O. Ya. 1992c. Geometry of Phyllotaxis. – Proceeding of Academy of Science of Ukraine. –# 9. (Боднар О. Я. Геометрия филлотаксиса. – Доклады Академии наук Украины. – 1992. – № 9).

Coxeter, G. 1953. The Golden Section, Phyllotaxis and Wythoff's Game // Scr. Math. 19.

Coxeter, G. 1966. Introduction to Geometry. – M. (КокстерГ. С. Введениевгеометрию. – М., 1966).

Jean, R. V. 1976. La G-entropie en phyllotaxie // Rev. bio-math. – # 5.

Jean, R. V. 1986. A Basic Theorem on and a Fundamental Approach to Pattern Formation on Plants // Mathematical Biosciences. 79. – # 2.

Petukhov, S. V. 1981. Biomechanics, Bionics and Symmetry. – M. (Петухов С. В. Биомеханика, бионика и симметрия. – М.,1981).

Petukhov, S. V. 1988. Geometries of living nature and algorithms of self-organization. – M. (Петухов С. В. Геометрии живой природы и алгоритмы самоорганизации. – М., 1988.)

16 Determination of Parastichy Numbers and Its Applications

Riichirou Negishi

CONTENTS

16.1 INTRODUCTION

Sunflower florets or seeds are arranged in spirals on the head inflorescence. Spiral arrangements are characterized by the number of curves winding clockwise (CW) and counterclockwise (CCW), and the number of curves forming spirals is called parastichy number. The parastichy numbers often demonstrates the Fibonacci number. Also, the ratio of CW/CCW or CCW/CW in the spiral arrangement is approximately

241

FIGURE 16.1 (a) The pattern of seeds on a sunflower head and (b) ramenta on a pineapple, displaying as Fibonacci numbers. Each number is counted along black lines.

equal. In the 18th century, Johannes Kepler observed that Fibonacci numbers were common in plants (Adler et al., 1997). When we visually count the spirals toward the outer region on a sunflower head, we find 21/34, 34/55, and 55/89 (CW or CCW) as the parastichy numbers (Figure 16.1a). Furthermore, when we visually count the helical numbers along the black line on pineapple skin, we find 8 and 13 corresponding to Fibonacci numbers (1, 1, 2, 3, 5, 8, 13, 21, 34, 55, 89,...) (Figure 16.1b). These parastichy numbers have attracted much attention from researchers for centuries (Vogel, 1979).

Sunflower seed spirals have been studied by a number of scientists following Hofmeister's (1868) systematic description of the phyllotaxis mechanism including spiral formation (Adler et al., 1997; Mathai and Davis, 1974). Alan Turing sketched seed patterns and studied Fibonacci phyllotaxis (Turing, 1952, 1956). Van der Linden (1990) obtained a sunflower spiral-like formation via the dislodgement model without using divergence angles. Dunlap (1997) emphasized the fundamental properties of Fibonacci numbers and their application to diverse fields including mathematics, computer science, physics, and biology. Negishi and Sekiguchi (2007) has proposed an application that offered the possibility of transmitting information at high speed or prohibiting the moiré fringe in image processings.

Such spirals and parastichy numbers have also drawn attention in various model systems including laboratory experiments. For example, Douady and Couder (1992) successfully obtained Fibonacci spirals with drops of ferrofluid under the influence of a magnetic field. Spiral structures are emerging as powerful nanophotonic platforms with distinctive optical properties for multiple engineering applications (Agrawal et al., 2008; Trevino et al., 2008; Liew et al., 2011; Negro et al., 2012).

Adler (1974) proposed a theorem to determine the number sequences for various divergence angles. Jean (2009) summarized the relationship between the divergence angles and the number sequences using Adler's theorem. Table 16.1 shows the calculated number sequences for various divergence angles using Adler's theorem. In Table 16.1, the Fibonacci sequence is denoted by F, the Lucas sequence is denoted

TABLE 16.1

Number Sequences for Various Divergence Angles

Divergence Angle	Sequences
137.51°	$F,G(1,2)$: 1, 2, 3, 5, 8, 13, 21, 34, 55, 89, 144, 233, 377, 610, 987, 1597,...
99.50°	$L,G(1,3)$: 1, 3, 4, 7, 11, 18, 29, 47, 76, 123, 199, 322, 521, 843, 1364,...
77.96°	$G(1,4)$: 1, 4, 5, 9, 14, 23, 37, 60, 97, 157, 254, 411, 665, 1076, 1741,...
64.08°	$G(1,5)$: 1, 5, 6, 11, 17, 28, 45, 73, 118, 191, 309, 500, 809, 1309, 2118, ...
54.40°	$G(1,6)$: 1, 6, 7, 13, 20, 33, 53, 86, 139, 225, 364, 589, 953, 1542, 2495, ...
47.25°	$G(1,7)$: 1, 7, 8, 15, 23, 38, 61, 99, 160, 259, 419, 678, 1097, 1775, 2872, ...
151.14°	$G(2,5)$: 2, 5, 7, 12, 19, 31, 50, 81, 131, 212, 343, 555, 898, 1453, 2351, ...
158.15°	$G(2,7)$: 2, 7, 9, 16, 25, 41, 66, 107, 173, 280, 453, 733, 1186, 1919, 3105, ...
162.42°	$G(2,9)$: 2, 9, 11, 20, 31, 51, 82, 133, 215, 348, 563, 911, 1478, 2385, 3863, ...
68.75°	$2\ G(1,2)$: 2, 4, 6, 10, 16, 26, 42, 68, 110, 178, 288, 466, 754, 1220, 1974, ...

by L, and the generalized Fibonacci sequences are denoted by G (Koshy, 2001). Two successive numbers can be expressed as follows:

$$\lim_{n\to\infty} \frac{G_{n+1}}{G_n} \to \tau = 1.61803 \qquad (16.1)$$

For a divergence angle of 137.51°, which is approximately equal to the golden angle (137.507764...°), the parastichy numbers demonstrate the Fibonacci sequence (F). When the divergence angle is 99.50°, the parastichy numbers give the Lucas sequence (L). Generalized Fibonacci sequences (G) appear for specific divergence angles, such as 77.96°, 64.08°, and so forth. The point distribution in this simulation fills the plane mostly uniformly.

Even spirals with significantly different structures can be obtained by making a slight difference in divergence angles. For example, in the sunflower model simulation, when a divergence angle is slightly smaller (137.4°) than the golden angle (Figure 16.2a in Section 16.2.1), the parastichy numbers at the outer region consist of the Fibonacci number (F) 55 and the Lucas number (L) 76. In addition, when the divergence angle is 137.45°, the parastichy number at the outer region only consists of F 55 (Figure 16.2b), and when the divergence angle is 137.8°, the parastichy numbers at the outer region are L 47 and 81 in G (2, 5), consisting all of F, L, and G sequences (Figure 16.2c). In these cases, the parastichy numbers cannot be calculated using Adler's theorem. Since novel Fibonacci and non-Fibonacci structures in sunflowers have recently been reported (Swinton et al., 2016), we need a practical method to systematically compute parastichy numbers of any spiral patterns.

Vogel (1979) was one of the first researchers to develop a mathematical spiral model to approximate the complex arrangements of florets on a sunflower head. However, Vogel's spirals lack both translational and orientational symmetry in reality. Accordingly, the Fourier space of Vogel's spirals do not exhibit well-defined Fourier peaks, though they show diffuse circular rings, similar to the electron diffraction patterns observed in amorphous solids and liquids (Trevino et al., 2008).

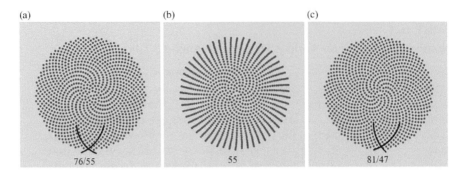

FIGURE 16.2 Simulated point pattern of a sunflower model with a divergence angle of: (a) 137.4°; (b) 137.45°; and (c) 137.8° at $n = 1000$.

This suggests that point distances in a short range are required to analyze spirals. Liew et al. (2011) applied the Fourier–Bessel transform to understand the structural complexity of the golden angle spiral. Negishi et al. (2017) has showed that the parastichy number can be directly measured using the discrete Fourier transform. Fourier transforms are widely used to grasp the characteristics of periodic and aperiodic patterns in natural phenomena and to analyze crystal structures via X-ray diffraction (Authier, 2001; Kikuta, 2011).

Through this chapter, applicability of the proposed method is tested with sunflower and pineapple models. The discrete Fourier transform is applied to the simulated point patterns, and their parastichy numbers with various divergence angles are carefully analyzed.

16.2 SIMULATION MODELS

In this study, two simulation models are used. One is called a sunflower model when the filling points form a circle; the other model is called a pineapple model when the filling points form a rectangle (Negishi et al., 2017).

16.2.1 Sunflower Model (S-Model)

For a sunflower model, the point positions can be determined by the following equation in polar coordinates (Figure 16.3):

$$(r, \theta) = \left(n^p, n\phi \right) \tag{16.2}$$

Here, n is an integer, p is a constant scaling factor, and ϕ is the divergence angle. When $n = 1000$, $p = $ constant, and $\phi = \phi_c$, the points form the spiral shape (Figure 16.4). When $p = 0.2$, dominant parastichy numbers can be visually counted as 55/89 in the outer region (Figure 16.4a). When $p = 0.5$, dominant parastichy numbers can be counted as 21/34, 34/55, and 55/89 in the inner, middle, and outer regions, respectively (Figure 16.4b). And when $p = 1.0$, the numbers in the outer region is 34/55

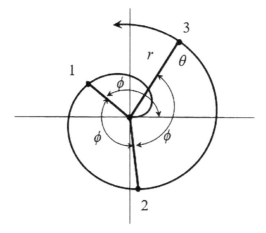

FIGURE 16.3   Spiral trajectory of a sunflower model.

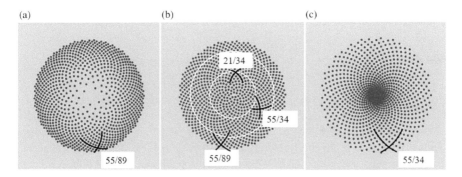

FIGURE 16.4   Simulated point pattern of a sunflower model with $n = 1000$, $p = 0.5$, and $\phi = \phi_\tau$. Each number is counted along black lines: (a) $p = 0.2$; (b) 0.5; and (c) 1.0.

(Figure 16.4c). These numbers correspond to Fibonacci numbers. The parastichy numbers vary depending on the values of n, p, and ϕ, as described later. In the following discussion, $p = 0.5$ is selected because the points can be uniform in the plane. In order to know parastichy numbers by varying p, we will discuss in Section 16.5.

16.2.2   PINEAPPLE MODEL (P-MODEL)

A pineapple model is expressed by points on the surface of a cylinder (Figure 16.5). The height l and the argument angle θ are determined by n, p, and ϕ. Point positions on the cylinder for a pineapple model can be generated as follows: $(l,\theta) = (n^p, n\phi)$ Figure 16.5 shows point positions with $n = 20$, $p = 1$, and $\phi = \phi_\tau$. The points are patterned after the cylinder being vertically sliced open when $n = 1000$ and $\phi = \phi_\tau$. Figure 16.6a–c correspond to $p = 0.5$, 1, and 2, respectively. When $p = 1$, the points fill a rectangle having a horizontal side of 2π and a vertical side of l. The points

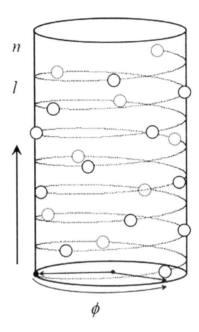

FIGURE 16.5   A pineapple model expressed by points on the surface of the cylinder.

form a series of straight lines (Figure 16.6b). The parastichy numbers around the top of the cylinder are visually counted as 13, 21, 34, and 55, which correspond to Fibonacci numbers (F). When $p = 0.5$ and $p = 2.0$, the points form a series of curves (Figure 16.6a and c). In the following discussion, $p = 1.0$ is selected because of a straight line that the points form. We will discuss parastichy numbers by varying p in Section 16.5.

    The parastichy numbers of pineapple models also vary depending on the values of n, p, and ϕ.

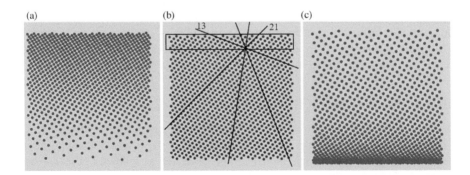

FIGURE 16.6   Simulated point pattern of a pineapple model with $n = 1000$, $p = 1$, and $\phi = \phi_\tau$. The black rectangle shows a measured area to find the parastichy numbers: (a) is $p = 0.5$; (b) 1.0; and (c) 2.0.

16.3 ANALYSIS PROCEDURE USING A DISCRETE FOURIER TRANSFORM

16.3.1 THE SUNFLOWER MODEL

In this section, we propose a method in which discrete Fourier transform is used to compute parastichy numbers in sunflower models. We focus on a short-range arrangement of the seeds (points) at a given radius since the overall arrangement of seeds on the sunflower head (the parastichy number) is a variable spiral structure. Figure 16.7a shows the simulated point pattern in the outer region of the pattern in (Figure 16.4b) when $n = 1000$, $p = 0.5$, and $\phi = \phi_\tau$. Our method is based on a discrete Fourier transform analyzing a set of distances from each point to its closest point in the spiral pattern. A Fourier transform peak position (spatial frequency) corresponds to a parastichy number. The number of sample points for the discrete Fourier transform is selected to be a power of 2, that is, 256 points in Figure 16.7a. First, an azimuth angle θ for each point is calculated, and the sample data is arranged in ascending order of their angles. Then, the distance from each point to its closest point with a larger angle is measured, and a dataset is created. Figure 16.7b shows the state from the first to the fifth closest point measured from their reference points. Using the Wolfram mathematical software package, a one-dimensional discrete Fourier transform is applied to the dataset. Figure 16.8 shows the absolute value of the Fourier transform spectra. Due to conjugate symmetry for real sequences, the half of the data range is shown. The same procedure is carried out to the second, third, fourth, and fifth closest points. Finally, the Fourier data is summed from the first closest to the fifth closest points to obtain a more accurate power spectrum. As a result, we found that the sum up to the fourth closest point reflects the accurate results. For that reason, hereafter we will analyze parastichies using the sum from the first to the fourth (Figure 16.9).

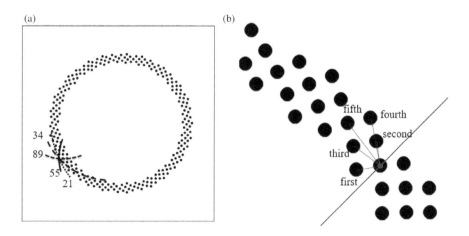

FIGURE 16.7 (a) Simulated point pattern in the outer region of a sunflower model with $n = 1000$, $p = 0.5$, and $\phi = \phi_\tau$, showing parastichy numbers of 21/34, 34/55, and 55/89; and (b) closest points for target points.

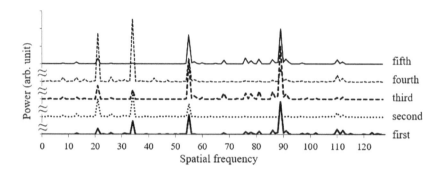

FIGURE 16.8 One-dimensional discrete Fourier transform results for the first, second, third, fourth, and fifth closest points in a sunflower model with $n = 1000$, $p = 0.5$, $\phi = \phi_\tau$, and 256 sample points.

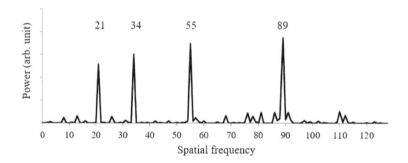

FIGURE 16.9 The Fourier transform result (sum of the data from the first closest points to the fourth closest points in Figure 16.8).

In Figure 16.9, four large peaks can be observed with spatial frequencies of 21, 34, 55, and 89. These values agree with the parastichy numbers visually counted in Figure 16.7. This indicates that parastichy numbers can be determined using the Fourier transform method presented.

16.3.2 THE PINEAPPLE MODEL

We used the same method to determine parastichy numbers in a pineapple model. The number of sample points is 256 points in the black rectangle (Figure 16.6).

16.4 ANALYTICAL RESULTS

16.4.1 PARASTICHY NUMBERS NEAR THE GOLDEN ANGLE IN SUNFLOWER MODELS

This section compares the parastichy numbers determined at the golden angle ϕ_τ (the Sunflower model in Section 16.3.1) with those determined at near the golden angle, ϕ_τ for the sunflower model. The numbers are both computed using the Fourier transform method. When a point pattern is simulated with $\phi = 137.45°$, which is

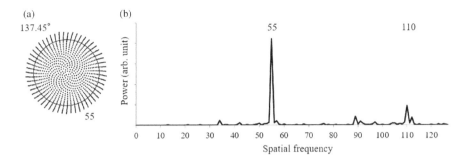

FIGURE 16.10 (a) Simulated point pattern for a sunflower model with $\phi = 137.45°$ and $n = 1000$ and (b) Fourier transform result for the area outside the black circle in panel (a).

slightly smaller than ϕ_τ interestingly, the points radiated outward in all directions (Figure 16.10a). The Fourier transform results from the outer region showing two large peaks at 55 and 110 (Figure 16.10b). The number 55 is a Fibonacci number, and the number 110 should be the second harmonic wave of 55. Harmonic waves seem to appear as a characteristic feature of the Fourier transform. Figure 16.11a is the simulated point pattern for the case of $\phi = 137.55°$, which is slightly greater than ϕ_τ. The Fourier transform result from the outer region show the large peaks at 34 and 89 (Figure 16.11b). These numbers agree with those counted visually in the point pattern of Figure 16.11a and are Fibonacci numbers.

When the angles are changed from 137.25° to 137.75° in increments of 0.05° when $n = 1000$ and $p = 0.5$, the Fourier transform results show Fourier peaks at 21, 55, and 76 for a divergence angle of 137.35°. The parastichy numbers 21 and 55 are Fibonacci numbers, while the number 76 is a Lucas number. In addition, the spatial frequency of 97 is observed at a divergence angle of 137.30°. The number 97 belongs to the generalized Fibonacci sequence $G(1, 4)$ according to Table 16.1. Even with a slight change in the angles, like 137.25° to 137.75°, the parastichy numbers can be accurately counted and classified via the presented Fourier transform method (Figure 16.12).

How about when the divergence angle is near 99.50°? Figure 16.13 shows the Fourier results when the angles are changed from 99.25° to 99.75° in increments of 0.05° with $n = 1000$ and $p = 0.5$. When the divergence angles are 99.45° and 99.50°,

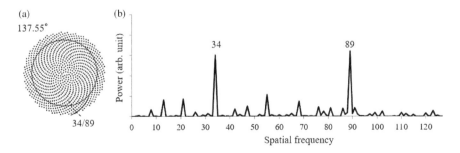

FIGURE 16.11 (a) Simulated point pattern for a sunflower model with $\phi = 137.55°$ and $n = 1000$ and (b) Fourier transform result for the area outside the black circle in panel (a).

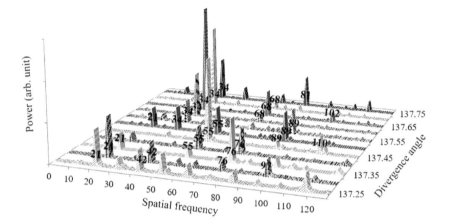

FIGURE 16.12 Fourier transform results near the golden angle ϕ_τ in sunflower models with $n = 1000$ showing various parastichy numbers.

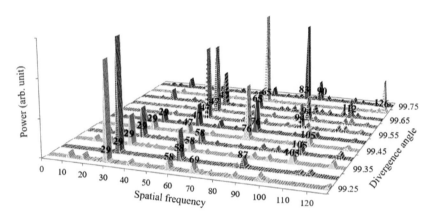

FIGURE 16.13 Fourier transform results near the divergence angle $\phi = 99.50°$ in sunflower models with $n = 1000$ showing various parastichy numbers.

Fourier peaks 29, 47, and 76 are noted as Lucas numbers. At 99.55° and 99.60°, the maximum peak shifts to 47, and the harmonic wave, 94, also increases. When the divergence angle reaches 99.65°, the peak 65 (= 13 × 5) appears, and when the angle is 99.70°, the peak becomes extremely strong. And a large peak 83 appears at 99.75°. On the other hand, the frequency 29 becomes strong at 99.35° and 99.40°, and the second harmonic wave 58 also appears. Furthermore, when divergence angles are 99.25° and 99.30°, the peak 29 is extremely strengthened and the harmonic wave 58 is enlarged. The parastichy numbers resulted from angle changes around 99.50° are centered on the Lucas sequence.

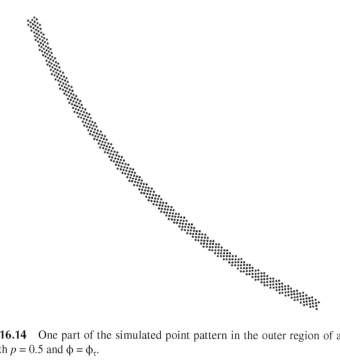

FIGURE 16.14 One part of the simulated point pattern in the outer region of a sunflower model with $p = 0.5$ and $\phi = \phi_\tau$.

When parastichy numbers are visually counted from point patterns, points are highly likely to be miscounted due to extremely dense points (Figure 16.14). The Fourier transform method is used to assess its effectiveness to determine the parastichy numbers with dense points. The number of sampled points for the Fourier transform is 4096 in the outer region. Large Fourier peaks are observed at 610, 987, and 1597, corresponding to Fibonacci numbers (Figure 16.15). This result shows that the parastichy numbers with even a large point pattern can be accurately computed using the proposed method. Lucas numbers, 843 and 1364, though they were relatively small, are also included.

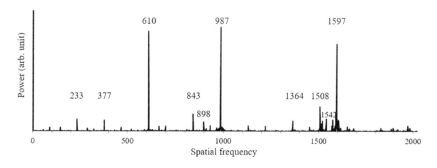

FIGURE 16.15 Fourier transform result for $n = 100000$, $p = 0.5$, and $\phi = \phi_\tau$.

16.4.2 Parastichy Numbers for Wide Divergence Angles in Sunflower Models

This section discusses the effectiveness of the proposed determining method for various divergence angles. The results are summarized in Table 16.1. The point patterns are simulated using the sunflower model with divergence angles of 47.25°, 54.40°, 64.08°, 77.96°, 99.50°, 151.14°, 158.15°, and 162.42°. The parastichy numbers are computed using the discrete Fourier transform (Figure 16.16). All the results agree with those in Table 16.1. When the divergence angle is 47.25°, large Fourier peaks appear at 38 and 61. These numbers belong to G (1,7). For 54.40°, Fourier peaks appear at 20, 33, and 53. These numbers are included in G (1,6). For 64.08°, Fourier peaks 17, 28, and 45 are included in G (1,5); peaks 37 and 60 for 77.96° are included in G (1,4); and peaks 11, 18, 29, and 47 for 99.50° belong to Lucas sequences, that is, G (1,3). Furthermore, peaks (31, 50) for 151.14° belong to G (2,5); (25, 41) for 158.15° belong to G (2,7); and (31, 51) for 162.42° belong to G (2,9). All peaks numbers agree with the results in Table 16.1.

In addition, we examine the parastichy numbers for a divergence angle, 99.65°. It is slightly larger than a divergence angle, 99.50° as we have shown in Figure 16.13. Even though the angle difference between them is only 0.15°, the parastichy numbers cannot be inferred from Adler's theorem. The simulated point pattern and the results by the Fourier transform are shown in Figure 16.17a. The visually counted parastichy numbers are 47, 65, and 112. The Fourier transform results demonstrate the same numbers (Figure 16.17b). This indicates that the proposed Fourier method is highly effective to elucidate a complex structure of generalized Fibonacci sequence with arbitrary divergence angle.

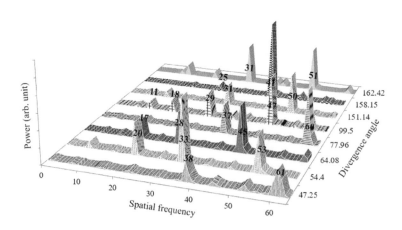

FIGURE 16.16 Fourier transform results showing various spatial frequency (parastichy numbers) when the divergence angles are 47.25°, 54.40°, 64.08°, 77.96°, 99.50°, 151.14°, 158.15°, and 162.42°.

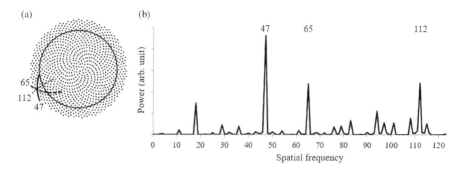

FIGURE 16.17   (a) Simulated point pattern and (b) Fourier transform result for a divergence angle, 99.65° for the region outside the black circle in panel (a).

16.4.3   PARASTICHY NUMBERS IN PINEAPPLE MODELS

In this section, the parastichy number of the pineapple model (see pineapple model in Section 16.2.2) is investigated using the presented Fourier transform method. Point patterns of a pineapple model when $n = 1000$ and $p = 1$ is simulated by making changes in the divergence angles. Figure 16.18a and b show the point pattern with $\phi = \phi_\tau$ and its Fourier transform result, respectively. Sample numbers are 128 points. Large Fourier peaks at 13, 21, 34, and 55 are observed, corresponding to the Fibonacci sequence (see also Figure 16.6). Thus, the proposed method is also effective in finding parastichy numbers on the plane as seen in the pineapple model. When $\phi = 137.8°$, some nonuniformity for the distribution is observed (Figure 16.19a). The Fourier peaks of 13, 34, and 47 are observed (Figure 16.19b), and they agree with the parastichy numbers visually counted (panel a). Because the number 47 is a Lucas number, the parastichy numbers are a combination of Fibonacci and Lucas numbers in this case.

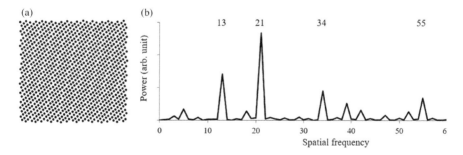

FIGURE 16.18   (a) Simulated point pattern of a pineapple model with $n = 1000$, $p = 1$, and $\phi = \phi_\tau$; and (b) Fourier transform result for panel (a).

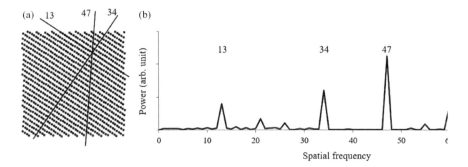

FIGURE 16.19 (a) Simulated point pattern of a pineapple model with $n = 1000$, $p = 1$, and $\phi = 137.8°$; and (b) Fourier transform result for panel (a).

When the divergence angles are changed from 137.1° to 137.9° by increments of 0.1°, we recognize some transitions in the parastichy numbers. The transition from $F21$ to $F34$ occur near $\phi = 137.6°$, and $L47$ is observed at divergence angles greater than 137.8°. Here, we recognize some shifts in the parastichy number (Figure 16.20).

The transformation results at 4096 points for $n = 100,000$, $p = 1$, and $\phi = \phi_\tau$ (Figure 16.21). Fourier peaks as the spatial frequencies appear at Fibonacci numbers 610, 987, and 1597. Although the magnitude of these powers is different from those in Figure 16.15, they all peak at the same spatial frequency.

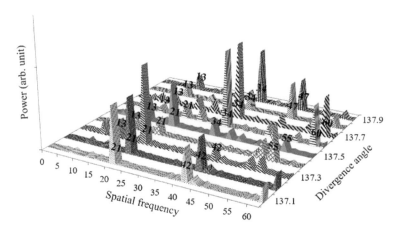

FIGURE 16.20 Fourier transform results for pineapple models with several divergence angles ranging from 137.1° to 137.9°.

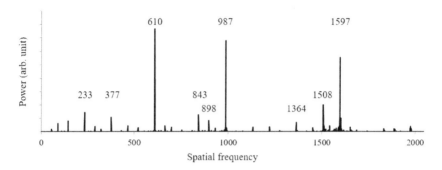

FIGURE 16.21 Fourier transform result for $n = 100000$, $p = 1$, and $\phi = \phi_\tau$.

16.5 DISCUSSION

16.5.1 DIFFERENCE IN THE NUMBERS OF SAMPLE POINTS

This section presents how a different number of sample points affect the Fourier transform results. Figure 16.22a shows the simulated point pattern for a sunflower model with $n = 1000$, $p = 0.5$, $\phi = \phi_\tau$, and 499 sample points. Figure 16.22b shows the Fourier transform result. Fourier peaks at 21, 34, 55, 89, and 144 and are compared with the result with 256 sample points (Figure 16.9); we found that there was a remarkable match between them, except for the power. Note that because the sampling window in the radial direction results in more sample points, more peaks appear.

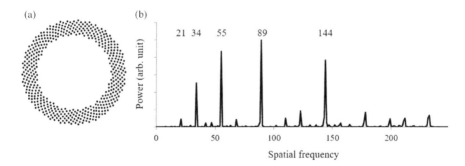

FIGURE 16.22 (a) Simulated point pattern for a sunflower model with $n = 1000$, $p = 0.5$, $\phi = \phi_\tau$, and 499 sample points and (b) Fourier transform result for panel (a).

16.5.2 DIFFERENCE IN INDEX P

This section reports how the parastichy numbers are differed by changing index p using the simulation models. As shown in Figures 16.4 and 16.6, the behavior of the point distribution largely depends on the index p. Figure 16.23a shows the point distribution when $n = 1000$, $p = 1$, and $\phi = \phi_\tau$ in the sunflower model, and Figure 16.23b illustrates the transformation result for 128 points at the outer region. According to the transformation result, the largest Fourier peak is counted as 34, which correspond to the visually counted value.

As for the pineapple model, Figure 16.24a shows the point distribution when $n = 1000$, $p = 2$, and the simulation result shows ununiformity. One hundred twenty-eight points of the upper part of the panel (a) is analyzed by Fourier transform. The parastichy numbers, 13 and 21, are visually confirmed from the distribution image, and they correspond to the transform result.

The results in Figures 16.23 and 16.24 suggest that even with changes in the p values, using the method we propose, we are still able to find parastichy numbers that can correspond to those visually counted.

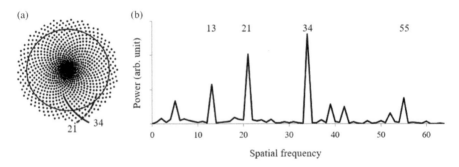

FIGURE 16.23 (a) Simulated point pattern for a sunflower model with $n = 1000$, $p = 1$, $\phi = \phi_\tau$, and 128 sample points; and (b) Fourier transform result for panel (a).

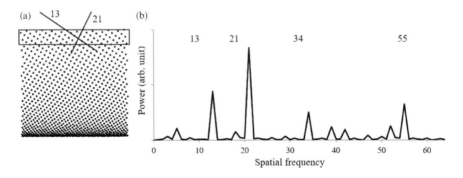

FIGURE 16.24 (a) Simulated point pattern for a pineapple model with $n = 1000$, $p = 2$, $\phi = \phi_\tau$, and 128 points in the rectangle; and (b) Fourier transform result for panel (a).

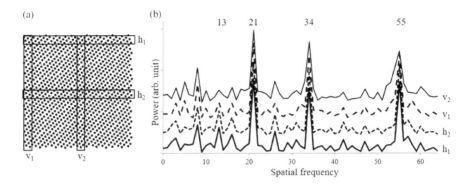

FIGURE 16.25 (a) Fourier transform result by differential sampling area for $n = 1000$, $p = 1$, and $\phi = \phi_\tau$; and (b) Fourier transform result for each area in panel (a).

16.5.3 Sampling Position

This section discusses how the sampling position affects the Fourier transform results. Black rectangles in Figure 16.25a show the sampling area with $n = 1000$, $p = 1$, and $\phi = \phi_\tau$, where h_1 and h_2 are horizontal areas and v_1 and v_2 are vertical areas. Figure 16.25b shows the transform results obtained by sampling in each area. The Fourier peaks, 21, 34, 55, and 89 agree with those visually counted. The Fourier peak 13 for the vertical area (v) is smaller than that of the horizontal area (h). This is because the cycles of the distance interval at the specified place is different between vertical and horizontal regions.

16.6 APPLICATIONS

16.6.1 A Part Sampling

This section presents the effectiveness of a part sampling of an image. In general, a sampling area is assigned as a part of the target image. When 512 points, which is 1/8 of 4096 points in Figure 16.14, are Fourier transformed, the peaks appear at 76, 123, and 199 (Figure 16.26) corresponding to approximately 1/8 of 610, 987, and 1597 in Figure 16.15b, respectively.

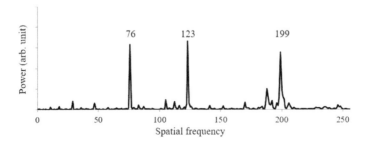

FIGURE 16.26 Fourier transform result for Figure 16.14.

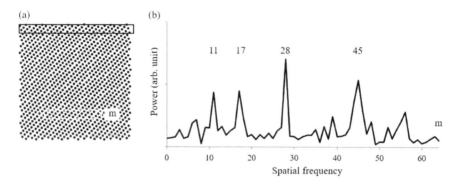

FIGURE 16.27 (a) m is a sampling area at 128 points in $n = 1000$, $p = 1.0$, and $\phi = \phi_r$; and (b) Fourier transform result for the area m in panel (a).

From these results, $199/123 = 1.618$ calculated from Figure 16.26 corresponds to $1597/687 = 1.618$ in Figure 16.15b. The number value matches up to the third decimal place. This indicates that the proposed method can accurately estimate feature quantities of images even with only a part of target sampling region.

How about planar images? A small rectangle area (m) has been sampled from the image in Figure 16.25a as a target area for this study (Figure 16.27). Though the height for the sampled rectangle is the same as the upper rectangle in Figure 16.27, the width is on-half of the upper one. The Fourier peaks appear at 11, 17, 28, and 45 in Figure 16.27b are approximately one-half of the peak values of 21, 34, 55, and 89 in Figure 16.25b. Even with a plane image being analyzed as the sample region, peak values can still be accurately found according to a size of the target region, and that can be treated as the feature quantities of an image.

16.6.2 EVALUATION OF DISTURBANCE

This section shows the effectiveness of the proposed method in evaluating disturbance. Figure 16.28 shows distribution examples when the divergence angle is nearly the golden angle with $n = 1000$ of the sunflower model. Two hundred fifty-six points are extracted from each figure, and they are analyzed by Fourier transformation using the proposed method. The results are shown in Figure 16.29. The two numbers represent the largest peak and the second largest peak of each image. Because of distribution nonuniformity in (a), disturbance can be easily identified compared to (c) with the golden angle (Figure 16.28). The ratio of the two numbers is also large for (a). On the other hand, even though (b) and (d) appear to be inhomogeneous compared with (c), it is difficult to distinct quantitative differences between (b) and (d). However, Fourier peak patterns of (b) and (d) exhibit a clear difference between them. Therefore, the proposed method is effective to accurately evaluate a disturbance of an image.

A distribution disturbance in the pineapple model is also investigated. Figure 16.30 shows rectangular distributions with the same divergence angle as those in Figure 16.28 when $n = 1000$. Two hundred fifty-six points of the upper part

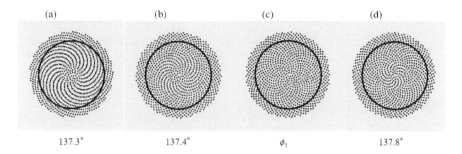

FIGURE 16.28 Differential distribution due to small change in divergence angles under each panel. Sampling areas are outer region of each black circle when the divergence angles are (a) 137.3°, (b) 137.4°, (c) ϕ_τ, and (d) 137.8°.

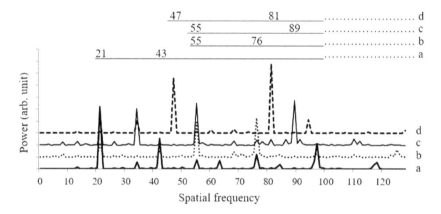

FIGURE 16.29 Fourier transformation results of Figure 16.28.

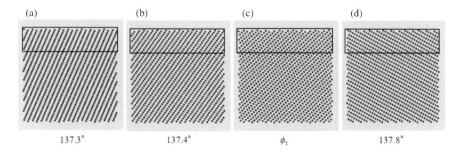

FIGURE 16.30 Differential distribution due to small change in divergence angles under each panel. Sampling areas are upper region of each black rectangle when the divergence angles are (a) 137.3°, (b) 137.4°, (c) ϕ_τ, and (d) 137.8°.

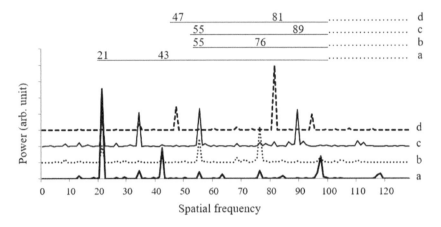

FIGURE 16.31   Fourier transformation results of Figure 16.30.

of each distribution are the sample areas. The transformation results are shown in Figure 16.31. The two numbers are the largest peak and the second largest peak of spatial frequencies of each sample. The two large Fourier peaks are somewhat different magnitude in Figure 16.29, but each position is the same. For (b) and (d), it is, again, difficult to visually identify both figures, yet the numerical values of each spatial frequency clearly differ. Thus, as conclusion, the proposed method is also highly effective to accurately assess a distribution disturbance of a planar image.

16.6.3   Applying to a Real Sunflower

Figure 16.32a is a Russian sunflower with a diameter of 0.3 m. Figure 16.32b shows seed positions extracted from the sunflower. Five hundred twelve points near the

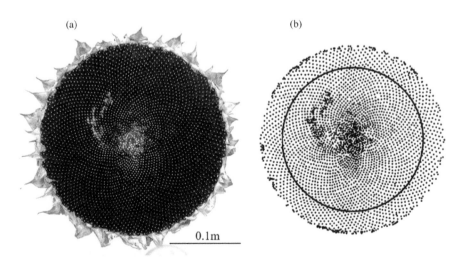

FIGURE 16.32   (a) Real Russian sunflower and (b) reading result of seeds position.

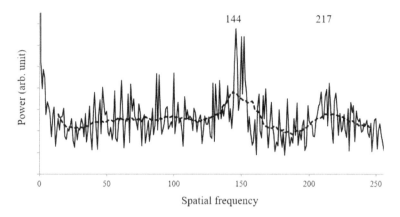

FIGURE 16.33 Transformation result of 512 points near the black circle in Figure 16.32b. Broken line indicates moving averages from the result.

black circle are sampled (Figure 16.32b) and Fourier transformed (Figure 16.33). Many peaks resulted from the transformation indicate many interval cycles of individual living seeds exist (Figure 16.33). The broken line represents the moving averages (Figure 16.33). The ratio, 217/144 = 1.507, of the two peaks is close to the golden ratio. Because the ratio suggests that the seeds are sparsely distributed throughout the head inflorescence, the results demonstrate that the proposed Fourier transform method is useful to evaluate a disturbance of individual seed positions and the distribution tendency of whole seeds.

16.7   SUMMARY

This chapter clearly shows that parastichy numbers for point patterns can be directly measured with the one-dimensional discrete Fourier transforms. The points are patterned from the simulations using the sunflower model and the pineapple model. The detailed pattern analysis using the Fourier method reveals that the parastichy numbers in general cases are a combination of Fibonacci, Lucas, and generalized Fibonacci numbers. In addition, even when the number of measurement points increases to 100,000 points, the parastichy number is still accurately determined by this method.

In the application, even when only a part of the target image is analyzed, the Fourier peaks precisely reflect the image feature quantity. It reveals that the pattern of the adjacent Fourier peaks obtained from the point distribution image is a clear indicator of image disturbance. Moreover, as applying this method to a real sunflower, the feature quantity can also be precisely evaluated.

REFERENCES

Adler, I. 1974. A model of contact pressure in phyllotaxis. *J. Theor. Biol.* 45:1–79.
Adler, I., Barabe, D., and Jean, R. V. 1997. A history of the study of phyllotaxis. *Ann. Bot.* 80:231–244.

Agrawal, A., Kejalakshmy, N., Chen, J., Rahman, B. M. A., and Grattan, K. T. V. 2008. Golden spiral photonic crystal fiber: Polarization and dispersion properties. *Opt. Lett.* 33:2716–2718.

Authier, A. 2001. *Dynamical Theory of X-Ray Diffraction*, 73–78. Oxford University Press.

Douady, S. and Couder, Y. 1992. Phyllotaxis as a physical self-organized growth process. *Phys. Rev. Lett.* 68:2098–2100.

Dunlap, R. A. 1997. *The Golden Ratio and Fibonacci Numbers*, Word Scientific Publishing.

Hofmeister, W. 1868. *Handbuch der PhysiologischenBotanik*, 437–46. Engelmann.

Jean, R. V. 2009. *Phyllotaxis*, Cambridge University Press.

Kikuta, S. 2011. *X-Ray Scattering and Synchrotron Radiation Science-Fundamentals* (*in Japanese*), 144–148, University of Tokyo Press.

Koshy, T. 2001. *Fibonacci and Lucas Numbers with Applications*, 109–115, John Wiley & Sons.

Liew, S. F., Noh, H., Trevino, J., Negro, L. D., and Cao, H. 2011. Localized photonic band edge modes and orbital angular momenta of light in a golden-angle spiral. *Opt. Express.* 19:23631–23642. doi:10.1364/OE.19.023631.

Mathai, A. M. and Davis, T. A. 1974. Constructing the sunflower head. *Math. Biosci.* 20:117–133.

Negishi, R. and Sekiguchi, K. 2007. Pixel-filling by using Fibonacci spiral. *Forma.* 22:207–215.

Negishi, R., Sekiguchi, K., Totsuka, Y., and Uchida, M. 2017. Determining parastichy numbers using discrete Fourier transforms. *Forma.* 32:19–27.

Negro, L. D., Lawrence, N., and Trevino, J. 2012. Analytical light scattering and orbital angular momentum spectra of arbitrary Vogel spirals. *Opt. Express.* 20:18209–18223.

Pennybacker, M. F., Shipman, P. D. and Newell, A. C. 2015. Phyllotaxis: Some progress, but a story far from over. *Physica. D.* 306:48–81.

Swinton, J., Ochu, E., and The MSI Turing's Sunflower Consortium. 2016. Novel Fibonacci and non-Fibonacci structure in the sunflower: Results of a citizen science experiment. *Roy. Soc. Open Sci.* 3:160091.

Trevino, J., Cao, H., and Negro, L. D. 2008. Circularly symmetric light scattering from nanoplasmonic spirals. *Nano Lett.* 11:2008–2016.

Turing, A. M. 1952. Morphogen theory of phyllotaxis, in Saunders, P. T. (Ed.), 1992, *Collected Works of A. M. Turing*: Morphogenesis, North-Holland.

Turing, A. M. 1956. *The Turing Digital Archive*.http://www.turingarchive.org.

Van der Linden, F. M. J. 1990. Creating phyllotaxis, the dislodgement model. *Math. Biosci.* 100:161–199.

Vogel, H. 1979. A better way to construct the sunflower head. *Math. Biosci.* 44:179–189.

17 Floral Symmetry

T. Pullaiah and Bir Bahadur

CONTENTS

17.1 INTRODUCTION

The term symmetry is derived from two Greek words: σύν (su'n, meaning "with") and μέτρον (métron, meaning "measure") and meant to show a relationship of commensurability. But subsequently took the general meaning of equilibrium of proportions, qualifying the harmony of the different elements of a unitary whole, thus becoming closely related to the idea of beauty and regularity (Damerval et al., 2017). Symmetry is a defining feature of floral diversity in angiosperms. Floral symmetry contributes to the attraction and beauty of flowers to the human eye. It also provides plants with an efficient means to cope with physical constraints and explore three-dimensional space, hence an integral architectural trait that accounts for the astonishing diversity of the flower form. Floral symmetry describes whether, and how, a flower, and in particular its perianth, can be divided into two or more identical or mirror-image parts. Uncommonly, flowers may have no perceivable axis of symmetry at all, typically because their floral parts are helically arranged. Sprengel (1793) was the first to point out the existence of floral symmetry, and he pointed out the regular and irregular nature of flowers. Much later, Braun (1835, 1843) introduced the word "zygomorph" for Sprengel's "irregular" flowers. In fact, symmetry is an important component of floral descriptions and has been recognized and used since Theophrastus (371–287 BC).

Variation in flower symmetry has attracted the attention of botanists for more than a century (Delpino, 1887; Church, 1908; Leppik, 1972; Stebbins, 1974). Research has centred on understanding the developmental mechanisms that establish patterns of symmetry, the ecological contexts in which alternative patterns of symmetry are favoured, and the evolutionary history of transitions between different forms. This research has provided key insights into how, when and why transitions in floral symmetry evolve. Several authors have reviewed the subject which include Endress (1995), Coen (1996), Running (1997), Lyndon (1998), Jabbour et al. (2009), Citerne et al. (2010), Reyes et al. (2016), and recently by Damerval et al. (2017).

17.2   TYPES OF FLORAL SYMMETRY IN ANGIOSPERM FLOWERS

Floral symmetry refers to if, and how, the flowers can be divided into mirror-image parts. When deciding on the symmetry of a flower, the position and shape of the more conspicuous features are considered, that is, the petals or perianth and or androecium. Angiosperm flowers usually have one or two forms of symmetry. Most flowers have radial symmetry, that is, multiple planes of symmetry, meaning they can be divided into three or more identical sectors that are related to each other by rotation about the center of the flower (polysymmetry). Such flowers are also called actinomorphic or regular flowers, for example, *Catharanthus roseus*, *Lilium*. Flowers may have bilateral symmetry. Such flowers are also called zygomorphic, for example, Orchidaceae, Lamiaceae, Scrophulariaceae, and Gesneriaceae. In this case, flowers can be divided by only a single plane into two mirror-images like a person's face (Monosymmetry) (Figure 17.1).

A few plant species have flowers with no plane of symmetry and therefore have a "handedness" (asymmetrical), for example, *Canna indica*, *Valeriana officinalis*.

The simplest kind of distinction is between polysymmetric and monosymmetric flowers, for which the terms "symmetrical" (or "regular") versus "asymmetrical" (or "irregular") have generally been used. These terms have been used since the time of Linnaeus (1751), and they are also used today by molecular developmental geneticists who work on *Antirrhinum* flowers (Luo et al., 1996). Most flowers are actinomorphic, meaning they can be divided into three or more identical radial longitudinal sectors that are related to each other by rotation about the center of the flower. Typically, each sector might contain one tepal or one petal and one sepal and so on. It may or may not be possible to divide the flower into symmetrical halves by the same number of longitudinal planes passing through the axis. Zygomorphic (from the Greek word; *zygon*, yoke, and *morphe*, shape) flowers can be divided by only a single longitudinal plane into two mirror-image halves, much like a yoke or a person's face.

Another, and highly understudied, floral symmetry pattern is present in flowers with contort aestivation and results from the mutual covering of petal flanks in the flower bud (aestivation—pattern). In contort aestivation, each petal

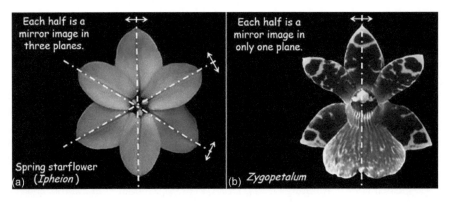

FIGURE 17.1   (a) Actinomorphic and (b) zygomorphic flowers.

overlaps only one of its neighbor petals (Schoute, 1935; Scotland et al., 1994). According to how the petals overlap, the corollas rotate clockwise or counter-clockwise (also known as left- and right-handedness, respectively) and are chiral to each other (Schoute, 1935; Scotland et al., 1994; Endress, 2001). Corolla chirality is visibly distinguishable when the aestivation pattern is still present after anthesis and mainly when individual petals are asymmetric conveying the corolla in a pinwheel appearance (Endress, 1999, 2001). Endress (2001) sum-marizes the phylogenetic distribution of contort flowers across the angiosperm clade and distinguishes unfixed species (both left and right flowers are found on the same individual) from fixed species (all individuals of that species exhibit only one floral form). No species have been observed where individuals are fixed for either left or right flowers; for example, in no species, are some individuals left and the remaining individuals right in the same population. Endress (2012) also refers to unfixed contort floral morphology as enantiomorphic, but Diller and Fenster (2016) refer to this condition as unfixed corolla chirality. Most fixed species are asterids, and most unfixed species, such as *Hypericum*, are rosids (Endress, 1999, 2001). Little is known of the adaptive biology underlying unfixed corolla chirality, although it is present within 8 taxonomic orders and 15 families within the rosids (Endress, 1999; Reyes et al., 2016).

Asymmetry in flowers can be due to perianth, stamens, or styles. Information given above deals mostly with asymmetry due to perianth while asymmetry due to style is given below.

17.2.1 ENANTIOSTYLY

Enantiostyly is the asymmetrical deflection of the style, either to the left (left styled) or to the right (right styled) side of the floral axis and has evolved in at least ten angiospermous families. Two types of enantiostyly occur: monomorphic enantiostyly, in which individuals exhibit both stylar orientations, and dimor-phic enantiostyly, in which the two stylar orientations occur on separate plants. The systematic distribution and abundance of the two forms of enantiostyly are strikingly different. Monomorphic enantiostyly is widely distributed occurring in both dicotyledons (e.g., Fabaceae, Solanaceae, and Gesneriaceae) and mono-cotyledonous families (e.g., Commelinaceae, Haemodoraceae, Philydraceae, Pontederiaceae, and Tecophilaceae). In contrast, dimorphic enantiostyly is only reliably reported from *Wachendorfia* and *Barbaretta* of the Haemodoraceae, *Heteranthera multiflora*, and possibly *Monochorea cyanea* of the Pontederiaceae and *Cyanella alba* of the Tecophilaceae (Barrett et al., 2000). Most workers have interpreted enantiostyly as a floral design that promotes outcrossing through pollinator-mediated intermorph pollination. Pollen removed by a pollinator from a flower of one type is more likely to be deposited on the opposite type (Barrett et al., 2000).

Similar to monomorphic enantiostyly (Todd, 1882; Gao et al., 2006), most unfixed chiral species show a 1:1 ratio of flower morphs within individuals (Davis, 1964; Davis and Ramanujacharyulu, 1971; Diller and Fenster, 2014). In addition, Diller and Fenster (2014) found that the corolla chirality of a flower in two neotropical

Hypericum species, *H. irazuense* and *H. costaricense*, is independent of the chirality of its closest neighbor flower, indicating a random distribution of corolla types within an individual. Some monomorphic enantiostylous species, such as *Heteranthera mexicana*, also present a random distribution of morph types as in *H. irazuense* and *H. costaricense* (Jesson et al., 2003). Despite the similarities to monomorphic enantiostyly, we do not know whether chirality variation has a parallel influence on pollen flow or movement.

Hutchinson (1964) has employed the twisting of the keel petal and styles for distinguishing different tribes of Fabaceae. Rao et al. (1979) have noted that the presence or absence of corolla handedness can be used as taxonomic pointer in delimiting taxa. On the basis of the overlapping of the vexillum or the wing petals, or both, corolla handedness in Fabaceae, Bahadur and Rao (1981) classified into four distinct types: (1) *Butea* type. In this type, none of the petals including the vexillum and the wings exhibit contortion, hence corolla handedness is absent, for example, *Butea*, *Medicago*; (2) *Cajanus* type. In *Cajanus cajan* and in several other genera, only the vexillum petal shows contortion either to the left or to the right; (3) *Phaseolus* type. In *Phaseolus* and related genera, the wing petals show contortion and the vexillum is open. Hence, handedness can be described in wing petals but not in the vexillum; (4) In the fourth type, both the vexillum and the wing petals show handedness, within this, two subtypes have been recognized: (a) *Crotalaria* subtype. In this subtype, both the vexillum and the wing petals show contortion but in the opposite direction, for example, *Crotalaria*, *Erythrina*; (b) *Gliricidia* subtype. Both the vexillum and the wing petals show contortion in the same direction (left or right), for example, *Gliricidia* and some other genera (Figures 17.2 and 17.3).

During the diversification of flowering plants (angiosperms), there have been numerous evolutionary transitions between radial flower symmetry (polysymmetry, actinomorphy) and bilateral flower symmetry (monosymmetry, zygomorphy), or in more extreme cases, flower asymmetry (Endress, 1999, 2012). Bilateral symmetry is predominant in a number of species-rich lineages—for example, Lamiales (mints and allies) and Fabaceae (legumes) in eudicots, and Orchidaceae in the monocots. Bilateral symmetry in these lineages is not only common but also highly elaborate. However, a survey of flowering plant lineages demonstrates that both elaborate and subtle forms of bilateral flower symmetry have evolved from radially symmetrical ancestors many times, and that reversals from bilateral to radial, or approximately radial symmetry, are not uncommon (Endress, 2012).

In phyllotaxis, spiral, whorled, and irregular patterns can be distinguished. Flowers with a spiral phyllotaxis may attain approximately the same three symmetry patterns as whorled flowers: (1) almost polysymmetric, if the flowers have a larger number of organs—this is the predominant case in spiral flowers (e.g., *Adonis*, *Nigella*, Ranunculaceae); (2) almost monosymmetric, if the flowers become dorsiventrally differentiated (e.g., *Aconitum*, *Consolida*, *Delphinium*, Ranunculaceae); and (3) asymmetric, if there are only few organs (e.g., *Hypserpa decumbens*, Menispermaceae). Flowers with an irregular phyllotaxis tend to be asymmetric if they have a low number of

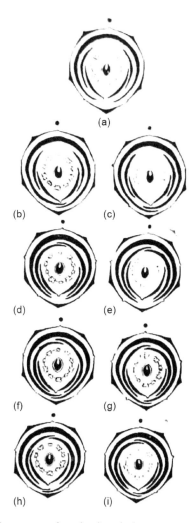

FIGURE 17.2   (a-i) Various types of aestivations in legumes.

organs (e.g., some species of *Zygogynum*, Winteraceae). They may also approach polysymmetry if they have many organs (e.g., some Annonaceae, Monimiaceae) (Endress, 1995, 1999).

The buzz-pollinated genus *Senna* (Leguminosae) is outstanding for including species with monosymmetric flowers and species with diverse asymmetric, enantio-morphic (enantiostylous) flowers. To recognize patterns of homology, Marazzi and Endress (2008) dissected the floral symmetry character complex, explored corolla morphology in 60 *Senna* species, and studied floral development of four enantiomor-phic species. The asymmetry morph of a flower is correlated with the direction of

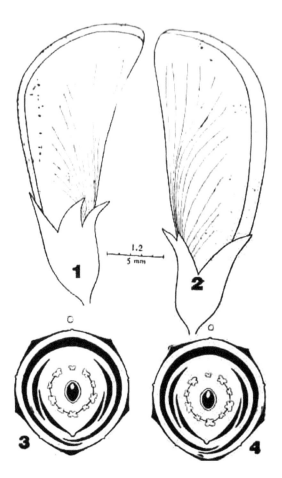

FIGURE 17.3   Left and right flowers of *Cajanus cajan* in respect of aestivation on vexillary petals.

spiral calyx aestivation. They recognized five patterns of floral asymmetry, resulting from different combinations of six structural elements: (1) deflection of the carpel, (2) deflection of the median abaxial stamen, (3) deflection or modification in size of one lateral abaxial stamen, and (4) modification in shape and size of one or (5) both lower petals. Prominent corolla asymmetry begins in the early-stage bud (unequal development of lower petals). Androecium asymmetry begins either in the midstage bud (unequal development of thecae in median abaxial stamen; twisting of androecium) or at anthesis (stamen deflection). Gynoecium asymmetry begins in early bud (primordium off the median plane, ventral slit laterally oriented) or midstage to late bud (carpel deflection). In enantiostylous flowers, pronouncedly concave and robust petals of both monosymmetric and asymmetric corollas likely function to ricochet and direct pollen flow during buzz pollination. Occurrence of particular combinations of structural elements of floral symmetry in the subclades was also shown (Marazzi and Endress, 2008).

Organ morphology, however, differs greatly between species and can even vary within a single flower. In *Antirrhinum*, for instance, the two dorsal (upper or adaxial) petals have larger lobes than the two lateral (side) petals, which in turn have larger lobes than the ventral (abaxial) petal. In addition, the lateral and ventral petals, but not the dorsal petals, contain yellow areas and hairs near the base. Because of these morphological differences along the dorsoventral axis, the flower has a single axis of symmetry and is classified as irregular (Luo et al., 1996). Regular flowers, such as *Arabidopsis*, have two or more axes of symmetry.

In its current acceptance symmetry mainly characterizes the perianth aspect, even though the most elaborate forms also include modifications at the androecium level. Angiosperm flowers are predominantly symmetric or very rarely asymmetric (Endress, 1999). It was shown that some pollinators (bees, for instance) are only attracted by symmetric flowers (Giurfa et al., 1996, 1999; West and Laverty, 1998). A symmetric flower can be polysymmetric, dissymmetric (with two perpendicular symmetry planes), or monosymmetric. Zygomorphic and dissymmetric flowers are commonly found in clades characterized by a closed ground plan (i.e., where floral organ number and arrangement are fixed within and between individuals) as in core eudicots (Lamiales, Fabaceae, Asterales) and monocots (Orchidaceae and Zingiberales). In contrast, actinomorphic flowers are widespread in open ground-plan clades, like basal angiosperms and early diverging eudicots (Endress, 1999; Ronse DeCraene et al., 2003). Zygomorphy involves more or less pronounced shape differentiation between organs within a single whorl (or along spirally inserted organs in rare cases). A strong morphological differentiation at the perianth level is often associated with alterations at the androecium level, including stamen reduction (staminodes) or even abortion (Rudall and Bateman, 2004). It must be noted that floral symmetry is never perfect, and small deviations always exist. The minimum rate of deviation is observed in flowers with a high degree of synorganization (e.g., Orchidaceae and Apocynaceae, which are, respectively, zygomorphic and generally actinomorphic) (Endress, 1999).

Asymmetric flowers are rare in angiosperms and are known to occur mostly within large families or orders with predominantly monosymmetric (zygomorphic) flowers (e.g., Leguminosae, Lamiales, Orchidaceae, Zingiberales) and only exception ally in basal angiosperms (e.g., Winteraceae; Endress, 1999). Enantiomorphy is a special kind of floral asymmetry in which flowers have two mirror-image morphs. Commonly in enantiomorphic flowers, the style is deflected to the left or to the right of the median plane, a condition known as enantiostyly, which occurs in at least 10 angiosperm families of both monocots and dicots (Jesson, 2002) and seems to have evolved from monosymmetry multiple times (Jesson and Barrett, 2003). Asymmetry can be generated by enantiostyly (Endress, 2001) or reduction, such as in Cannaceae or Valerianaceae. In these cases, asymmetry is not an irregularity but rather a highly complicated and ordered phenotype. Unordered simple asymmetric flowers occur in a few basal angiosperms, with chaotic organization where the innermost perianth organs and the stamens are irregularly arranged (Endress, 1999). Left- versus right-styled flowers may occur on different individuals (i.e., dimorphic enantiostyly) or on the same plant (i.e., monomorphic enantiostyly (Jesson and Barrett, 2003).

17.3   EVOLUTION OF FLORAL SYMMETRY

Based on the fossil record, Crepet and Niklas (2009) suggested that zygomorphy evolved from actinomorphy ca. 50 million years after the emergence of angiosperms, concomitantly with the diversification and coevolution with specialized insect pollinators towards a better arrangement of their reproductive organs with respect to pollinator body shape, maximizing pollen transfer resulting in efficient cross-pollination.

Citerne et al. (2010) reviewed the evolutionary and ecological context of floral symmetry. Zygomorphic flowers appear relatively late in the fossil record (late Cretaceous, ca. 70 mya) compared to the accepted period for angiosperm origin (early Cretaceous, ca. 150 mya) (Crepet, 1996; Endress, 1999; Crepet et al., 2004). Actinomorphy is considered as the ancestral state for angiosperms, and basal angiosperms mainly have actinomorphic flowers (Ronse De Craene et al., 2003). Zygomorphy evolved several times independently from actinomorphy throughout the angiosperms with multiple reversals towards actinomorphy (Stebbins, 1974; Donoghue et al., 1998).

Phylogenetic analyses of asymmetry variation, inheritance, and molecular mechanisms by Palmer (2004) revealed unexpected insights into how development evolves. Directional asymmetry, an evolutionary novelty, arose from nonheritable origins almost as often as from mutations, implying that genetic assimilation ("phenotype precedes genotype") is a common mode of evolution.

Sauquet et al. (2017) reported model-based reconstructions for ancestral flowers at the deepest nodes in the phylogeny of angiosperms, using the largest data set of floral traits ever assembled. They reconstructed the ancestral angiosperm flower as bisexual and radially symmetric, with more than two whorls of three separate perianth organs each (undifferentiated tepals), more than two whorls of three separate stamens each, and more than five spirally arranged separate carpels. They suggest two different evolutionary pathways for the reduction in number of whorls in early angiosperm evolution: reduction by loss of entire whorls (Magnoliidae, Monocotyledoneae) or reduction by merging of whorls concomitant with an increase in the number of organs per whorl (Pentapetalae) (Figure 17.4).

Sokoloff et al. (2018) have discussed about the spiral and cyclic nature of the angiosperm flower. Figure 17.5 shows the complex nature of distribution of spiral and whorled patterns among major clades of angiosperms.

With respect to flower symmetry development and evolution, the products of molecular phylogenetic studies allow researchers to determine how often and in which lineages transition from radial flower symmetry to bilateral symmetry (and back to radial flower symmetry) have occurred, thus providing the framework for informed choice of species when addressing comparative developmental questions (Figure 17.6; Hileman, 2014).

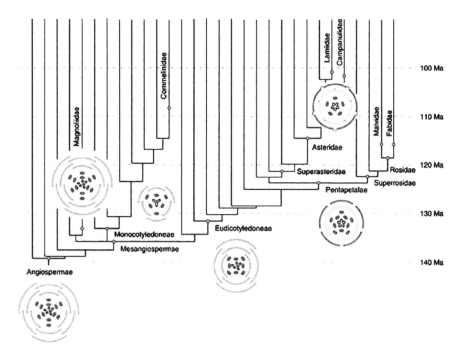

FIGURE 17.4 Simplified scenario for the earliest phase of floral diversification. (From Sauquet, H. et al., *Nat. Commun.*, 8, 16047, 2017.)

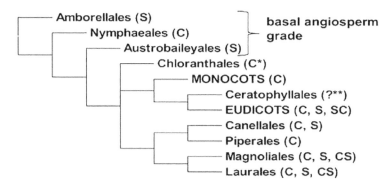

FIGURE 17.5 Simplified angiosperm phylogeny (based on APG IV, 2016) with types of floral phyllotaxis indicated for each group as following: C, cyclic; S, spiral; CS, cyclospiral (with a transition from cyclic to spiral within a flower); and SC, spirocyclic (with a transition from spiral to cyclic within a flower). Data on floral phyllotaxis are unrefined, as all kinds of uncertain or irregular phyllotaxis are not considered here * Based on perianth in *Hedyosmum*. ** Following interpretation of flowers as naked unicarpellate or unistaminate. (From Endress, P.K., and Doyle, J.A., *Taxon*, 64, 1093–1116, 2015; Sokoloff, D.D. et al., *Am. J. Bot.*, 105, 5–15, 2018.)

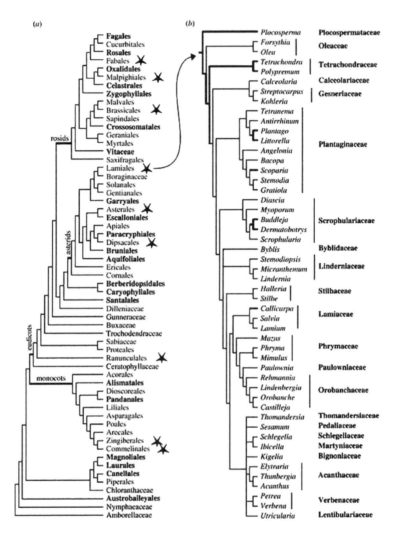

FIGURE 17.6 Evolutionary transitions in floral symmetry in a phylogenetic framework. (a) phylogeny of major angiosperm lineages (From Soltis, D.E. et al., *Am. J. Bot.*, 98, 704–730, 2011). Lineages in which elaborate bilateral flower symmetry can be found (from Endress, 2012) are in red text and (b) lamiales enlarged. (Stars indicate lineages containing species for which CYCLOIDEA homologues have been implicated in transitions to bilateral flower symmetry.) (From Hileman, L.C., *Phil. Trans. R. Soc. Lond. B Biol. Sci.*, 369, 20130348, 2014.)

17.4 POLLINATOR-MEDIATED EVOLUTION OF FLORAL SYMMETRY

Floral symmetry design plays a prominent role in plant mating system patterns (Barrett, 2002), pollination systems (Faegri and van der Pijl, 1979; Johnson and Steiner, 2000; Fenster et al., 2004), pollen transfer efficiency (Gomez et al., 2006), and angiosperm diversification rates (Sargent, 2004; van der Niet and Johnson, 2012).

Actinomorphic flowers are accessible from all sides (in front view) and are typically pollinated by a wide range of pollinators. By contrast, zygomorphic flowers are more specialized, even though they can be visited by a range of species depending on the environmental context. Symmetry patterns are often associated with directed pollinator movement and consequently pollen movement. For example, bilateral symmetry is associated with predictable placement of pollen on a pollinators body while asymmetric mirror-image flowers (enantiostyly) found on the same plant not only place pollen on specific parts of a pollinator's body but also direct pollen on to opposite sides of a bee's body, greatly reducing the opportunity for selfing through geitonogamy (Sprengel 1793; Barrett, 2002; Jesson and Barrett 2002a, 2005; Fenster et al. 2009).

Enantiostyly is a form of reciprocal herkogamy, in which floral morphs present reciprocal differences in the position of sexual elements, and occurs in monomorphic and dimorphic forms. This polymorphism maximizes cross-pollination and reduces self-pollination, being very common within the subtribe Cassiinae (Fabaceae). In addition to heterostyly, enantiostyly is considered to be a second type of reciprocal herkogamy (Webb and Lloyd, 1986) that promotes pollinations between floral forms and consequently outcrossing (Wilson, 1887). There are two components to cross-pollination and outcrossing; the donation of pollen to and the reception of pollen from other individuals. Reciprocal herkogamy will not affect the level of autogamous self-pollination more than regular herkogamy (Ganders, 1979). Therefore, recent models of the evolution of heterostyly specifically and reciprocal herkogamy in general focus on the role of these breeding systems in promoting the male component of outcrossing by increasing the efficiency of pollen donation through intermorph pollination (Webb and Lloyd, 1986).

Enantiostyly is similar to heterostyly in the reciprocal placement of anthers and stigmas in different flowers but differs in two major ways. First, with the exception of *Wachendorfia* and *Barberetta* (Ornduff and Dulberger, 1978), both floral forms (left- and right-styled flowers) occur on all individuals (Solanaceae, Bowers, 1975; *Dilatris*, Haemodoraceae, Ornduff and Dulberger, 1978; Cynastraceae, Dulberger and Ornduff, 1980; Dulberger, 1981; Leguminosae, Todd, 1882; Nyctaginaceae, Webb and Lloyd, 1986). In addition, somatic enantiostyly is associated with self-incompatibility except for two species of *Cyanella* that exhibit weak self-incompatibility (Cynastraceae, Dulberger and Ornduff, 1980), whereas heterostyly is most often associated with self- or cryptic self-incompatibility (Barrett, 1988). The consequence of these differences is that somatic enantiostyly might actually promote geitonogamous pollinations over nonreciprocal herkogamy with the subsequent increase of the selfing rate. If pollinators indiscriminately visit different floral types on a plant, then pollen dusted on one side of the pollinator will likely be deposited on the next flower with its stigma in the reciprocal position. This may impose a cost to somatic enantiomorphy if geitonogamy makes a significant contribution to self-fertilization, and if inbreeding depression exists. The adaptive significance of somatic enantiostyly in promoting cross-pollination is obscure if geitonogamy makes a significant contribution to self-fertilization.

The influence of enantiostyly (reciprocal segregation of anthers and stigmas to different sides of the flower) on the outcrossing rate was examined by Fenster (1995) in *Chamaecrista fasciculata* (Leguminosae). Observations of pollinator movement between flowers on the same plant were made to determine if pollinators discriminate between floral types. Although pollinators moved randomly between flower types, the outcrossing rate was only marginally affected by the presence of enantiostyly. Enantiostylous plants outcrossed at a slightly lower rate than in enantiostylous plants only when the opportunity for geitonogamy was great. These results suggest that the contribution of enantiostyly to selfing is minimal (Fenster, 1995).

Despite much recent activity in the field of pollination biology, the extent to which animal pollinators drive the formation of new angiosperm species remains unresolved. One problem has been identifying floral adaptations that promote reproductive isolation. The evolution of a bilaterally symmetrical corolla restricts the direction of approach and movement of pollinators on and between flowers. Restricting pollinators to approaching a flower from a single direction facilitates specific placement of pollen on the pollinator. When coupled with pollinator constancy, precise pollen placement can increase the probability that pollen grains reach a compatible stigma. This has the potential to generate reproductive isolation between species, because mutations that cause changes in the placement of pollen on the pollinator may decrease gene flow between incipient species. Sargent (2004) predicted that animal-pollinated lineages that possess bilaterally symmetrical flowers should have higher speciation rates than lineages possessing radially symmetrical flowers. Using sister-group comparisons, Sargent (2004) demonstrated that bilaterally symmetric lineages tend to be more species rich than their radially symmetrical sister lineages. This study supports an important role for pollinator-mediated speciation and demonstrates that floral morphology plays a key role in angiosperm speciation. The study by Lázaro and Totland (2014) suggests that the relationship between flower size variation and floral symmetry may be influenced by population-dependent factors, such as ecological generalization and species' dependence on pollinators.

Flowers are very diverse, with a large range in the size, number, color, and morphology of the floral organs, differences that seem to reflect the adaptation of individual species to their mechanism of pollination. Gong and Huang (2009) have discussed the pollinator-mediated evolution of floral symmetry. Pollinator-mediated selection has been considered as one key factor molding the evolution of floral traits (Stebbins, 1974). The shift in flower symmetry from radial to bilateral, important in the evolution of flowering plants, is thought to have occurred as a consequence of strong selection exerted by specialized pollinators (Neal et al., 1998; Endress, 2001). Bilateral flowers, which are often visited by one type of pollinator (Stebbins, 1951), may obtain higher pollination efficiency because of the accuracy of the physical fit between flower and pollinator. This specialized pollinator system might mediate stabilizing selection, and thus variation in flower size would be lower in species with bilateral flowers than radial flowers (Wolfe and Krstolic, 1999), as postulated by the pollinator-mediated stabilizing selection (PMSS) hypothesis (Wolfe and Krstolic, 1999). In contrast, radial flowers are often pollinated by a wider range of animal species because of their simpler structures (Cronk and Möller, 1997). The diversity of pollinators causes radial symmetry to be associated with unstable selection and correspondingly permits large variation in phenotypes (Lande and Arnold, 1983).

Ornduff and Dulberger (1978) observed that flowers in *Wachendorfia paniculata* are dimorphic. Some individuals in a population produce flowers in which the style and one stamen are deflected to the right, and the other two stamens are deflected to left; other individuals produce flowers with the reverse arrangement of style and stamens. Intermorph crosses produced more seeds than self-pollinations or intra-morph pollinations. The dimorphism in the position of anthers and stigmas probably increases the level of intermorph pollination.

Renshaw and Burgin (2008) studied the enantiomorphy in flowers of *Banksia*. Dissection of 2400 flowers in situ from four *Banksia* species has revealed the *Banksia* unit inflorescence to be enantiomorphic (having left-right asymmetry) due to the orientation of the ovaries. The asymmetry of the ovary and its pronounced development of the anterior side in fruit formation means that fruits are also readily distinguished as having arisen from either a left- or right-handed flower.

Corolla chirality differs from monomorphic enantiostyly by not having a reciprocal stamen to pistil arrangement between flowers. While enantiostyly increases outcrossing by the differential placement of pollen on the pollinator resulting from the alternate deviation of the style and stamen among flowers on the same individuals (Jesson and Barrett, 2002a), species with unfixed corolla chirality do not have reciprocal placement of anthers and stigmas. However, the reduction in geitonogamous self-pollination could still result if pollinators behave differently on left and right flower, resulting in differential pollen placement on the pollinator's body.

The study by Diller and Fenster (2016) provides evidence that unfixed corolla chirality, unlike mirror image enantiostyly, does not represent an adaptation associated with promoting disassortative mating between floral morphs, or directed movement of pollen between flowers. Instead, their findings demonstrate that unfixed corolla chirality may be similar to other radial symmetrical flowers with open and generalized pollination system and consequently does not direct pollen movement between flowers. Ren et al. (2013) reported mirror-image flowers without buzz pollination in the Asian endemic *Hiptage benghalensis*.

These results indicate that floral syndromes and pollination adaptation in *H. benghalensis* differ completely from the New World Malpighiaceae and may help to explain evolutionary adaptations of the family during its long-distance dispersals from the New World to Asia.

The arrangement of flowers on inflorescences is important for determining the movement of pollinators within the inflorescence and, consequently, the overall mating success and fruit set of a plant. *Spiranthes spiralis* is an orchid that has a spiralled inflorescence. The species has two chiral forms that show opposite coiling directions (clockwise and counterclockwise). Scopece et al. (2017) tested if this arrangement of inflorescence influences pollinator attraction and behavior. Their finding suggests that pollinators visit the flowers sequentially from the lower part of the inflorescences and leave the anti-clockwise individuals more rapidly than the clockwise ones. However, this pollinator behavior is not detrimental for the pollination success of either of the two morphs.

The observations of Reyes et al. (2016) on various hypotheses of zygomorphic flower evolution are relevant They discussed three hypotheses. The **inflorescence type—flower orientation hypothesis** stipulates that pollinators do not land as well in horizontally oriented actinomorphic flowers as in flowers whose lower perianth margin is expanded in such way that it is usable as a landing platform. Selection would hence favour expansion of the lower margin of the perianth, which may also be used as advertising such as *Viola* (Violaceae) and *Lamium* (Lamiaceae). The **marginal flower landing—platform hypothesis** for pseudanthia (condensed inflorescence that look like flowers, many of which have peripheral flowers that look like petals). Provided that the inflorescence is rounded and flat-topped, an alternative solution is to have only the peripheral flowers expand into a platform while the others can remain actinomorphic and yet benefit from the pollinator landing on the inflorescence. More biomass tends to be allocated to zygomorphic flowers, so dedicating the "landing platform" role to a specific part of the inflorescence as in many Asteraceae. The **Dangerous lower margin hypothesis** is the complete opposite of the above two, once a species pollinated by flying vertebrate like birds and rarely bats or insects, which (nectar thieves) could benefit from having its lower margin smaller than the upper one and make pollination easier for larger pollinator that cannot land on the flower. Flowers with such configuration include *Chasmanthe* (Iridaceae) and *Musa* (Zingiberaceae).

17.5   GENETIC AND MOLECULAR BASIS OF FLORAL SYMMETRY

Twenty years ago, research groups led by Enrico Coen in the UK and Jorge Almeida in Portugal unraveled a network of key developmental genes controlling floral symmetry in the snapdragon (Luo et al., 1996; Almeida et al., 1997). In this species, unequal corolla and stamen development along the dorsoventral axis depends on the activity of four genes: CYCLOIDEA (CYC), DICHOTOMA (DICH), RADIALIS dorsoventral axis (RAD), and DIVARICATA (DIV). These four genes form two groups that act antagonistically to determine regional identities in the floral meristem. The first group (CYC-DICH-RAD) determines dorsal identity. CYC and DICH are close paralogues belonging to the TCP gene family of plant-specific transcription factors. Both genes are expressed in the dorsal region of the floral meristem prior to organogenesis; their expression becomes restricted to the dorsal petals and staminode (a non-functional and underdeveloped stamen) (CYC) and to the dorsal half of the dorsal petals (DICH) later in development. In the dorsal staminode, CYC has a negative effect on cell proliferation by repressing cell cycle genes. The activity of CYC and DICH is mediated by RAD, a MYB-class transcription factor. CYC and DICH can potentially bind directly to RAD, inducing its expression in the dorsal region of the developing flower. Ventral identity is specified by DIV, another MYB transcription factor. DIV activity is controlled posttranscriptionally in the dorsal and lateral floral organs by RAD, which outcompetes DIV for MYB-related DRIFs (DIV and RAD interacting factors). Whether this network, or some of its components, is conserved in other plant species has been a major focus of research over the past 20 years. It has been shown that some of these key molecular players, and in particular CYC-like genes, have been recruited repeatedly for the evolution of zygomorphy in diverse lineages.

Conspicuous asymmetries in forms that are polymorphic within a species can be genetically or environmentally determined. Jesson and Barrett (2002b) presented a genetic analysis of the inheritance of dimorphic enantiostyly, a sexual polymorphism in which all flowers on a plant have styles that are consistently deflected either to the left or the right side of the floral axis. Using *Heteranthera multiflora* (Pontederiaceae), they conducted crosses within and between left- and right-styled plants and scored progeny ratios of the style morphs in F1, F2, and F3 generations. Crosses conducted in the parental generation between morphs or right-styled plants resulted in right-styled progeny, whereas crosses between left-styled plants resulted in left-styled progeny. When putative heterozygous F1 plants were selfed, the resulting F2 segregation ratios were not significantly different from a 3:1 ratio for right- and left-styled plants. Crosses between left- and right-styled plants in the F2 generation yielded F3 progeny with either a 1:1 ratio of left- and right-styled plants or right-styled progeny. Their results are consistent with a model in which a single Mendelian locus with two alleles, with the right-styled allele (*R*) dominant to the left-styled allele (*r*), governs stylar deflection. The simple inheritance of dimorphic enantiostyly has implications for the evolution and maintenance of this unusual sexual polymorphism (Jesson and Barrett, 2002b).

Advances in flowering plant molecular phylogenetic research and studies of character evolution as well as detailed flower developmental genetic studies in a few model species (e.g., *Antirrhinum majus*) have provided a foundation for deep insights into flower symmetry evolution. From phylogenetic studies, it can be inferred that during flowering plant diversification transitions from radial to bilateral flower symmetry (and back to radial symmetry) have occurred. From developmental studies, we know that a genetic program largely dependent on the functional action of the *CYCLOIDEA* gene is necessary for differentiation along the snapdragon dorsoventral flower axis. Bringing these two lines of inquiry together has provided surprising insights into both the parallel recruitment of a CYC-dependent developmental program during independent transitions to bilateral flower symmetry, and the modifications to this program in transitions back to radial flower symmetry, during flowering plant evolution (Hileman, 2014).

The study of the genetic basis of the generation of flower form has been greatly aided by a handful of model plant species, including *Antirrhinum* (Snapdragon), *Arabidopsis*, and *Petunia*. These studies have, for instance, identified some of the genes that confer positional information in the floral meristem, such as the flower homeotic genes, mutations in which lead to the conversion of floral organs to types inappropriate for their position in the flower (Weigel and Meyerowitz, 1994). The sequences of floral organ identity genes and their expression patterns are largely conserved in the flowering plants examined to date, reflecting the nearly universal order of flower organ type: sepals, petals, stamens, and carpels, from the outer whorl to the center (Weigel and Meyerowitz, 1994. Reyes et al., 2016; Damerval et al., 2017).

Two genes, CYC and DICH, are expressed in dorsal domains of the *Antirrhinum* flower and determine its overall dorsoventral asymmetry and the asymmetries and shapes of individual floral organs, by influencing regional growth. Another gene, DIV, influences regional asymmetries and shapes in ventral regions of the flower through a quantitative effect on growth. However, DIV is not involved in determining the overall dorsoventral asymmetry of the flower, and its effects on regional asymmetries depend on interactions with CYC/DICH. These interactions illustrate how gene activity, symmetry, shape, and growth may be related (Almeida and Galego, 2005).

Nothing is known about the mechanism of assigning positional information along the dorsoventral axis of a flower. Luo et al. (1996) are the first to address this problem using molecular genetics, with their study of the *Antirrhinum* gene *cycloidea* (**cyc**). Plants mutant for *cyc* show a partial conversion of organs to ventral fates: lateral petals resemble the ventral petal, and dorsal petals are smaller and less lobed. The *cyc* mutation, in combination with a mutation in a second gene, *dichotoma* (**dich**), leads to a complete conversion of all petals to ventral fates, resulting in what is known as a peloric phenotype, a regular flower with multiple axes of symmetry.

The **cyc** and **dich** mutations also affect the fates of the sepals and stamens. In wild-type flowers, the two ventral stamens are shorter than the two lateral stamens, and the dorsal stamen is aborted to form a "staminode." The stamen filaments also twist to orient all the anthers toward the ventral direction. In peloric flowers, the dorsal stamens fail to abort, all stamens have ventral identity, and twisting of the stamen filament does not occur. In a wild-type flower, the dorsal sepal initiates later than the ventral and lateral sepals; in peloric mutants, all of the sepals arise about the same time, though the dorsal sepals are still smaller (Luo et al., 1996). A second effect of the **cyc** mutation is an increase in floral organ number in mutant plants. Wild-type flowers have five sepals, petals, and stamens (including the aborted dorsal stamen); peloric lines usually have six sepals, petals, and stamens, with no stamen abortion (Luo et al., 1996).

The cloning of the **cyc** gene was facilitated by the use of an allele generated by transposon insertion (Luo et al., 1996). The **cyc** sequence shows no homology to any gene of known function, but the encoded protein does contain sequences resembling a bipartite nuclear localization signal. The RNA expression pattern of the gene is more revealing; it is restricted to dorsal regions of the developing flower, where it presumably acts to promote a dorsal fate. Expression is first detectable in a small number of cells in the dorsal part of the early stage 1 flower, before dorsoventral asymmetry is visible. This expression domain increases as the floral meristem enlarges; by early stage 4, when the sepal primordia become clearly visible, expression is detected in the dorsal sepal, as well as dorsal regions of the floral meristem interior to the sepals, where the dorsal petals and stamen arise. By stage 6, when all organ primordia are visible, expression is seen only in the developing dorsal petals and staminode. Expression is maintained but becomes weaker in more mature flowers (Luo et al., 1996).

This expression pattern is consistent with at least two roles for **cyc**. The early expression in the dorsal portion of the flower primordium seems to control the number of sepals that initiate on the dorsal side, as well as their growth rate, as **cyc** mutants have an extra sepal and accelerated growth in the dorsal part of the first whorl. These two effects may be interrelated: increased floral meristem size at the time of organ initiation has been correlated with increased organ number in several mutants of *Arabidopsis* (Weigel and Clark, 1996). The late expression in the dorsal petals and staminode might be responsible for the distinctive dorsal petal morphology, including the generation of petal lobes and the growth retardation of the dorsal stamen primordium (Luo et al., 1996).

According to the "ABC" model (Weigel and Meyerowitz, 1994), floral organ identity is specified by the differential expression in each flower whorl of three types of activity, A, B, and C, which are provided by the products of the organ identity (homeotic) genes. Thus, A activity alone specifies sepals in the first whorl, A and B together specify petals in the second whorl, B and C together specify stamens in the third whorl, and C alone specifies carpels in the fourth whorl. In *Antirrhinum*, C activity is conferred by the *plena* gene, and B activity requires both *deficiens* and *globosa* (no *Antirrhinum* A function gene has yet been reported, though other flowering plants are known to contain A function genes). A activity negatively regulates C activity, and vice versa. In plants with loss-of-function *plena* alleles, A activity is present throughout the flower, leading to the conversion of stamens to petals in the third whorl, and carpels to sepals in the fourth whorl. In plants with gain-of-function *plena* alleles, such as the *ovulata* allele, *plena* is ectopically expressed throughout the flower, leading to the conversion of sepals to carpels in the first whorl and petals to stamens in the second whorl (Bradley et al., 1993). Recently, Krishnamurthy and Bahadur (2015) have also discussed the genetics of floral symmetry keeping in view the ABCDE model in considerable detail in this context.

CYCLOIDEA-like genes belong to the TCP family of transcriptional regulators and have been shown to control different aspects of shoot development in various angiosperm lineages, including flower monosymmetry in asterids and axillary meristem growth in monocots. Genes related to the CYC gene from *Antirrhinum* show independent duplications in both asterids and rosids. Kölsch and Gleissberg (2006) showed that CYC-like genes have also undergone duplications in two related Ranunculales families, Fumariaceae and Papaveraceae *s.str.* These families exhibit morphological diversity in flower symmetry and inflorescence architecture that is potentially related to functions of CYC-like genes.

To conclude, we concur with the observations of Damerval et al. (2017) whose remarks are appropriate for unlocking the secrets of symmetry: "New technologies will help us progress toward this goal and renew our approach of evolutionary developmental questions in the near future. Developmental studies are fundamental for identifying the crucial stages where the underlying molecular changes are predicted to take place. New non-destructive imaging methods such as the recently developed X-ray micro-computed tomography give access to the three-dimensional organization of developing flowers, enabling the observation of rare or difficult material. Additionally, live-imaging techniques such as light sheet fluorescence microscopy can be used in some species and organs." Further, these authors feel that, "with new generation sequencing techniques, genome and transcriptome sequencing has become feasible for virtually any species, challenging the candidate gene approach and giving access to potentially novel genes affecting the trait of interest. Techniques such as virus-induced gene silencing are feasible in a large panel of species, enabling functional validation. In addition, better resolved phylogenies and new theoretical and analytical approaches for reconstructing micro- and macroevolutionary patterns will help uncover the driving forces and constraints in phenotype evolution."

REFERENCES

Almeida, J. and L. Galego. 2005. Flower symmetry and shape in *Antirrhinum*. *Int. J. Dev. Biol.* 49(5–6): 527–537. doi:10.1387/ijdb.041967ja.

Almeida, J., M. Rocheta, and L. Galego. 1997. Genetic control of flower shape in *Antirrhinum majus*. *Development*. 124: 1387–1392.

Bahadur, B. and M.M. Rao. 1981. Vexillary aestivation in Fabaceae: A re-appraisal. *Curr. Sci.* 50(21): 950–952.

Barrett, S.C.H. 1988. The evolution, maintenance and loss of self-incompatibility systems. In: *Plant Reproductive Ecology, Patterns and Strategies*. (eds.), J. Lovett Doust and L. Lovett Doust. Oxford University Press, Oxford, UK. pp. 98–124.

Barrett, S.C.H. 2002. The evolution of plant sexual diversity. *Nat. Rev. Genet.* 3: 274–284. doi:10.1038/nrg776.

Barrett, S.C.H., L.K. Jesson and A.M. Baker. 2000. The evolution and function of stylar polymorphisms in flowering plants. *Annals of Botany* 85(SupplementA): 253–265 doi:10.10076/anbo.1999.1067

Bowers, K.A.W. 1975. The pollination ecology of *Solanum rostratum* (Solanaceae). *Amer. J. Bot.* 62: 633–638. doi:10.2307/2441943.

Bradley, D., R. Carpenter, H. Sommer, N. Hartley, and E. Coen. 1993. Complementary floral homeotic phenotypes result from opposite orientations of a transposon at the plena locus of *Antirrhinum*. *Cell*. 72: 85–95. doi:10.1016/0092-8674(93)90052-R.

Braun, A. 1835. Dr. Carl Schimper's Vorträge über dieMö glichkeit eines wissenschaftlichen Verständnisses der Blattstellung, nebst Andeutung der hauptsä chlichen Blattstellungsgesetze und insbesondere der neuentdeckten Gesetze der Aneinanderreihung von Cyclen verschiedener Maasse. *Flora*. 18: 145–192.

Braun, A. 1843. Uber Symmetrie in der Pflanzenbildung. *Verh Ges Dtsch Naturforsch Arzte*. 1842: 196–197.

Church, A.H. 1908. *Types of Floral Mechanism*. Clarendon, Oxford, UK.

Citerne, H.L., F. Jabbour, S. Nadot, and C. Demerval. 2010. The evolution of floral symmetry. *Adv. Bot. Res.* 54: 85–137. doi:10.1016/S0065-2296(10)54003-5.

Coen, E.S. 1996. Floral symmetry. *EMBO J.* 15: 6777–6788.

Crepet, W. 1996. Timing in the evolution of derived floral characters: Upper cretaceous (Turonian) taxa with tricolpate and tricolpate derived pollen. *Rev. Palaeobot. Palynol.* 90(3): 339–359. doi:10.1016/0034-6667(95)00091-7.

Crepet, W., K.C. Nixon, and M.A. Gandolfo. 2004. Fossil evidence and phylogeny: The age of major angiosperm clades based on mesofossil and macrofossil evidence from cretaceous deposits. *Am. J. Bot.* 91(10): 1666–1682.

Crepet, W.L. and K.J. Niklas. 2009. Darwin's second "abominable mystery": Why are there so many angiosperm species? *Am. J. Bot.* 96: 366–381.

Cronk, Q. and M. Möller. 1997. Genetics of floral symmetry revealed. *Trends Ecol. Evol.* 12: 85–86. doi:10.1016/S0169-5347(97)01028-8.

Damerval, C., F. Jabbour, S. Nadot, and L. Hélène. 2017. Evolution of symmetry in plants. In: *Evolutionary Developmental Biology*. (eds.), G.B. Müller, pp. 1–18, Springer International. doi:10.1007/978-3-319-33038-9_59-1.

Davis, T.A. 1964. Aestivation in Malvaceae. *Nature*. 201: 515–516. doi:10.1038/201515a0.

Davis, T.A. and C. Ramanujacharyulu. 1971. Statistical analysis of bilateral symmetry in plant organs. *Indian J. Stat.* 33: 259–290.

Delpino, F. 1887. Zigomorfia florale e sue cause. *Malpighia*. 1: 245–262.

Diller, C. and C.B. Fenster. 2014. Corolla chirality in *Hypericum irazuense* and *H. costaricense* (Hypericaceae): Parallels with monomorphic enantiostyly. *J. Torrey Bot. Soc.* 141: 109–114.

Diller, C. and C.B. Fenster. 2016. Corolla chirality does not contribute to directed pollen movement in *Hypericum peforatum* (Hypericaceae): Mirror image pinwheel flowers function as radially symmetric flowers in pollination. *Ecol. Evol.* 6(14): 5076–5086. doi:10.1002/ece3.2268.

Donoghue, M.J., R.H. Ree, and D.A. Baum. 1998. Phylogeny and the evolution of flower symmetry in the Asteridae. *Trends Plant Sci.* 3(8): 311–317. doi:10.1016/S1360-1385(98)01278-3.

Dulberger, R. 1981. The floral biology of *Cassia didymobotrya* and *C. auriculata* (Caesalpiniaceae). *Am. J. Bot.* 68: 1350–1360. doi:10.2307/2442734.

Dulberger, R. and R. Ornduff. 1980. Floral morphology and reproductive biology of four species of *Cyanella* (Tecophilaeaceae). *New Phytol.* 86: 45–56. doi:10.1111/j.1469-8137.1980.tb00778.x.

Endress, P.K. 1995. Floral structure and evolution in Ranunculanae. *Plant Syst. Evol.* Suppl 9: 49–61. doi:10.1007/978-3-7091-6612-3_5.

Endress, P.K. 1999. Symmetry in flower: Diversity and evolution. *Int. J. Plant Sci.* 160(6Suppl.): S3–S23. doi:10.1086/314211.

Endress, P.K. 2001. Evolution of floral symmetry. *Curr. Opin. Plant Biol.* 4(1): 86–91. doi:10.1016/S1369-5266(00)00140-0.

Endress, P.K. 2012. The immense diversity of floral monosymmetry and asymmetry across angiosperms. *Bot. Rev.* 78: 345–397. doi:10.1007/s12229-012-9106-3.

Endress, P.K. and J.A. Doyle. 2015. Ancestral traits and specializations in the flowers of the basal grade of living angiosperms. *Taxon.* 64: 1093–1116. doi:10.12705/646.1.

Faegri, K. and L. van der Pijl. 1979. *The Principles of Pollination Ecology.* 3rd ed. Pergamon Press, Oxford, UK.

Fenster, C.B. 1995. Mirror image flowers and their effect on outcrossing rate in *Chamaecrista fasciculata* (Leguminosae). *Am. J. Bot.* 82(1): 46–50. doi:10.2307/2445785.

Fenster, C.B., W.S. Armbruster, and M.R. Dudash. 2009. Specialization of flowers: Is floral orientation an overlooked step? *New Phytol.* 183: 497–501. doi:10.1111/j.1469-8137.2009.02852.x.

Fenster, C. B., W.S. Armbruster, P.M.R. Wilson, M.R. Dudash, and J.D. Thomson. 2004. Pollination syndromes and floral specialization. *Annu. Rev. Ecol. Evol. Syst.* 35: 375–403. doi:10.1146/annurev.ecolsys.34.011802.132347.

Ganders, F.R. 1979. The biology of heterostyly. *New Zealand J. Bot.* 17: 607–635. doi:10.1080/0028825X.1979.10432574.

Gao, J.-Y., P.-Y. Ren, Z.-H. Yang, and Q.-J. Li. 2006. The pollination ecology of *Paraboea rufescens* (Gesneriaceae): A buzz-pollinated tropical herb with mirror-image flowers. *Am. J. Bot.* 97: 371–376. doi:10.1093/aob/mcj044.

Giurfa, M., A.P. Dafni, and R. Neal. 1999. Floral symmetry and its role in plant-pollinator systems. *Int. J. Plant Sci.* 160: S41–S50. doi:10.1086/314214.

Giurfa, M., B. Eichmann, and R. Menzel. 1996. Symmetry perception in an insect. *Nature.* 382(6590): 458–461. doi:10.1038/382458a0.

Gomez, J.M., F. Perfectti, and J.P.M. Camacho. 2006. Natural selection on *Erysimum mediohispanicum* flower shape: Insights into the evolution of zygomorphy. *Am. Nat.* 168: 531–545. doi:10.1086/507048.

Gong, Y.-B. and S.Q.S. Huang. 2009. Floral symmetry: Pollinator mediated stabilizing selection on flower size in bilateral species. *Proc. Biol. Sci.* 276(1675): 4013–4020. doi:10.1098/rspb.2009.1254.

Hileman, L.C. 2014. Trends in flower symmetry evolution revealed through phylogenetic and developmental genetic advances. *Phil. Trans. R. Soc. Lond. B Biol. Sci.* 369(1648): 20130348. doi:10.1098/rstb.2013.0348.

Hutchinson, J. 1964. *The Genera of Flowering Plants.* Oxford University Press.

Jabbour, F., S. Nadot, and S. Damerval. 2009. Evolution of floral symmetry: A state of the art. *C.R. Biologies.* 332: 219–236. doi:10.1016/j.crvi.2008.07.011.

Jesson, L. K. 2002. The evolution and functional significance of enantiostyly in flowering plants. PhD Dissertation, University of Toronto, Toronto, Ontario, Canada.

Jesson, L.K. and S.C.H. Barrett. 2002a. Solving the puzzle of mirror-image flowers. *Nature* 417: 707. doi:10.1038/417707a.

Jesson, L.K. and S.C.H. Barrett. 2002b. The genetics of mirror-image flowers. *Proc. Roy. Soc. Lond. B.* 269: 1835–1839. doi:10.1098/rspb.2002.2068.

Jesson, L.K. and S.C.H. Barrett. 2003. The comparative biology of mirror-image flowers. *Intern. J. Plant Sci.* 164(6 Supplement): S237–S249. doi:10.1086/378537.

Jesson, L.K. and S.C.H. Barrett. 2005. Experimental tests of the function of mirror-image flowers. *Biol. J. Linn. Soc.* 85: 167–179. doi:10.3732/ajb.90.2.183.

Jesson, L.K., J. Kang, S.L. Wagner, S.C.H. Barrett, and N.G. Dengler. 2003. The development of enantiostyly. *Am. J. Bot.* 90: 183–195. doi:10.3732/ajb.90.2.183.

Johnson, S.D. and K.E. Steiner. 2000. Generalization versus specialization in plant pollination systems. *Trends Ecol. Evol.* 15: 140–143. doi:10.1016/S0169-5347(99)01811-X.

Kölsch, A. and S. Gleissberg. 2006. Diversification of CYCLOIDEA-like TCP genes in the basal eudicot families Fumariaceae and Papaveraceae s.str. *Plant Biol. (Stuttg).* 8(5): 680–687. doi:10.1055/s-2006-924286.

Krishnamurthy, K.V. and B. Bahadur. 2015. Genetics of flower development. In: *Plant Biology and Biotechnology.* (eds.), Bahadur, B. et al. Vol. 1. Springer India, pp. 385–407. doi:10.1007/978-81-322-2286-8_16.

Lande, R. and S.J. Arnold. 1983. The measurement of selection on correlated characters. *Evolution.* 37: 1210–1226. doi:10.2307/2408842.

Lázaro A. and O. Totland. 2014. The influence of floral symmetry, dependence on pollinators and pollination generalization on flower size variation. *Ann. Bot.* 114(1): 157–165. doi:10.1093/aob/mcu083.

Leppik, E.E. 1972. Origin and evolution of bilateral symmetry in flowers. In: *Evolutionary Biology.* (eds.), T. Dobzhansky, M.K. Hecht, and W.C. Steere. Appleton-Century-Crofts, New York.

Linnaeus, C. 1751. *Philosophia Botanica.* Kiesewetter, Stockholm.

Luo, D., R. Carpenter, C. Vincent, L. Copsey, and E. Coen. 1996. Origin of floral asymmetry in *Antirrhinum. Nature* 383: 794–799. doi:10.1038/383794a0.

Lyndon, R.F. 1998. Phyllotaxis in flowers and in flower reversion. In: *Symmetry in Plants.* (eds.), R.V. Jean and D. Barabé, pp. 109–124, World Scientific, Singapore.

Marazzi, B. and P.K. Endress. 2008. Patterns and development of floral asymmetry in *Senna* (Leguminosae, Cassinae). *Am. J. Bot.* 95(1): 22–40. doi:10.3732/ajb.95.1.22.

Neal, P.R., A. Dafni, and M. Giurfa. 1998. Floral symmetry and its role in plant-pollinator systems: Terminology, distribution, and hypotheses. *Annu. Rev. Ecol. Syst.* 29: 345–373. doi:10.1146/annurev.ecolsys.29.1.345.

Ornduff, R. and R. Dulberger. 1978. Floral enantiomorphy and the reproductive system of *Wachendorfia paniculata* (Haemodoraceae). *New Phytol.* 80: 427–434. doi:10.1111/j.1469-8137.1978.tb01577.x.

Palmer, A.R. 2004. Symmetry breaking and the evolution of development. *Science.* 306: 828–833.

Rao, M.M., K.L. Rao, and Bahadur, B. 1979. Corolla handedness in Papilionaceae. *Curr. Sci.* 48: 408–410.

Ren, M.X., Y.F. Zhong, and X.-Q. Song. 2013. Mirror image flowers without buzz pollination in the Asian endemic *Hiptage benghalensis* (Malpighiaceae). *Bot. J. Linn. Soc.* 173(4): 764–774. doi:10. 1111/boj.12101.

Renshaw, A. and S. Burgin. 2008. Enantiomorphy in *Banksia* (Proteaceae): Flowers and fruits. *Austral. J. Bot.* 56(4): 342–346. doi:10.1071/BT07073.

Reyes, E., H. Sauqet, and S. Nadot. 2016. Perianth symmetry changed at least 199 times in angiosperms evolutuion. *Taxon*, 12 October. 20 pages. doi:10.12705/8305/655.1.

Ronse De Craene, L.P., P.S. Soltis, and D.E. Soltis. 2003. Evolution of floral structures in basal angiosperms. *Int. J. Plant Sci.* 164(5 Suppl.): S329–S363. doi:10.1086/377063.

Rudall, P.J. and R.M. Bateman. 2004. Evolution of zygomorphy in monocot flowers: Iterative patterns and developmental constraints. *New Phytol.* 162: 25–44. doi:10.1111/j.1469-8137.2004.01032.x.

Running, M.P. 1997. Plant development: Making asymmetric flowers. *Current Biol.* 7(2): R89–R91. doi:10.1016/S0960-9822(06)00044-3.

Sargent, R.D. 2004. Floral symmetry affects speciation rates in angiosperms. *Proc. R. Soc. Lond. B.* 271: 603–608. doi:10.1098/rspb.2003.2644.

Sauquet, H., M. von Balthazar, S. Magallón, J.A. Doyle, P.K. Endress, E.J. Bailes, E.Barroso de Morais, et al. 2017. The ancestral flower of angiosperms and its early diversification. *Nat. Commun.* 8: 16047.

Schoute, J.C. 1935. On corolla aestivation and phyllotaxis of floral phyllomes. *Verh. Kon. akad. Wet. Amsterdam Afd. Natuurk.* 34(4): 1–77.

Scopece, G., B. Gravendeel, and S. Cozzolino. 2017. The effect of different chiral morphs on visitation rates and fruit set in the orchid *Spiranthes spiralis. Plant Ecol. Divers.* 10: 97–104. doi:10.1080/17550874.2017.1354093.

Scotland, R.W., P.K. Endress, and T.J. Lawrence. 1994. Corolla ontogeny and aestivation in the Acanthaceae. *Biol. J. Linn. Soc.* 114: 49–65. doi:10.1111/j.1095-8339.1994.tb01923.x.

Sokoloff, D.D., M.V.R. Remizowa, M. Bateman, and P.J. Rudall. 2018. Was the ancestral angiosperm flower whorled throughout? *Am. J. Bot.* 105: 5–15.

Soltis, D.E., S.A. Smith, N. Cellinense, K.J. Wurdack, D.C. Tank, S.F. Brockington et al. 2011. Angiosperm phylogeny: 17 genes, 640 taxa. *Am. J. Bot.* 98: 704–730. doi:10.3732/ajb.1000404.

Sprengel, C.K. 1793. *Das entdeckte Geheimniss der Natur im Bau und in der Befruchtung der Blumen.* Vieweg d. Ae., Berlin, Germany.

Stebbins, G.L. 1951. Natural selection and differentiation of angiosperm families. *Evolution.* 5: 299–324. doi:10.2307/2405676.

Stebbins, G.L. 1974. *Flowering Plants: Evolution above the Species Level.* Belknap Press, Cambridge, MA.

Todd, J.E. 1882. On the flowers of *Solanum rostratum* and *Cassia chamaecrista. Am. Nat.* 16: 281–287.

van der Niet, T. and S.D. Johnson. 2012. Phylogenetic evidence for pollinator-driven diversification of angiosperms. *Trends Ecol. Evol.* 27: 353–361. doi:10.1016/j.tree.2012.02.002.

Webb, C.J. and D.G. Lloyd. 1986. The avoidance of interference between the presentation of pollen and stigmas in angiosperms. I. Herkogamy. *New Zealand J. Bot.* 24: 163–178. doi:10.1080/0028825X.1986.10409725.

Weigel, D. and E.M. Meyerowitz. 1994. The ABCs of floral homeotic genes. *Cell.* 78: 203–209. doi:10.1016/0092-8674(94)90291-7.

Weigel, D. and S.E. Clark. 1996. Sizing up the floral meristem. *Plant Physiol.* 112: 5–10. doi:10.1104/pp.112.1.5.

West, E.L. and T.M. Laverty. 1998. Effect of floral symmetry on flower choice and foraging behaviour of bumble bees. *Can. J. Zool.* 76: 730–739. doi:10.1139/z97-246.

Wilson, J. 1887. On the dimorphism of the flowers of *Wachendorfia paniculata. Trans. Proc. Bot. Soc. Edinb.* 17: 73.

Wolfe L.M. and J.L. Krstolic. 1999. Floral symmetry and its influence on variance in flower size. *Am. Nat.* 154: 484–488. doi:10.1086/303249.

18 Transference of Positional Information from Bracteoles and Sepals to Petals in Species with Labile Handedness of Contort Corolla
Mechanical Forces or Prepatterning?

Polina V. Karpunina, Maxim S. Nuraliev,
Alexei A. Oskolski, and Dmitry D. Sokoloff

CONTENTS

18.1 INTRODUCTION

Certain patterns of corolla aestivation provide classic examples of asymmetry, or, more precisely, a rotational symmetry in plants (Schoute, 1935; Kaden and Urmantzev, 1971; Endress, 1999, 2001, 2012). The term *aestivation* means a mode of overlapping or contact between margins of adjacent organs of the same type in a flower bud. Different types of aestivation are recognized (e.g., Leins and Erbar, 2010; Ronse De Craene, 2010; Endress, 2011), and they are most commonly discussed with respect to whorled (rather than spiral) flowers. So-called valvate aestivation is an example of a polysymmetric aestivation. In valvate aestivation, adjacent organs contact by their margins but do not overlap each other. Of special interest is so-called contort aestivation, where each organ of a whorl has one margin overlapping adjacent organ and another margin overlapped by adjacent organ. It is commonly considered that this type of aestivation (like the valvate type) correlates with simultaneous initiation of all petals in a flower or at least with their sequential initiation with very short plastochrons (Ronse De Craene, 2010; Endress, 2011; Sokoloff et al., 2018a). The contort aestivation is conspicuous in corolla of various angiosperms belonging to different families and orders (Schoute, 1935; Kaden and Urmantzev, 1971; Bahadur et al., 1984; Scotland et al., 1994; Roels et al., 1997; Schönenberger and Endress, 1998; Endress, 1999, 2001, 2010, 2012; Schönenberger, 1999, 2009; Endress and Bruyns, 2000; Ma and Saunders, 2003; Ronse De Craene, 2010; Borg and Schönenberger, 2011; Sokoloff et al., 2018a). The contort aestivation can be found not only in petals but sometimes in stamens, sepals (Ronse De Craene, 2010), or even involucral bracts (Bahadur and Reddy, 1975).

The presence of a contort corolla is a constant feature of some plant species, whereas other species develop contort corolla with greater or lesser frequency. In a few taxa, flowers have a contort aestivation in two whorls, namely, in calyx and corolla (Ma and Saunders, 2003; Ronse De Craene, 2010) or in corolla and androecium (Roels et al., 1997; Sokoloff et al., 2018a).

Two mirror-shaped types of contort aestivation can be recognized. Their differences are explained below on the example of contort corolla. In a left-contort corolla, the left margin of each petal (observed from the center of a flower) occupies an inner position in the flower bud, being covered by the right margin of the adjacent petal. In a right-contort corolla, the right margin of each petal occupies an inner position in the flower bud, being covered by the left margin of the adjacent petal (Endress, 2001).

In some angiosperms, handedness of the contort corolla is fixed at the species level. In other words, there are angiosperm species in which the contortion of corolla is always left-handed and some other species with always right-handed contortion. Only rarely species with fixed left-handed and fixed right-handed corolla occur in the same genus (*Philadelphus*, Hydrangeaceae: Schoute, 1935). It is important that there are angiosperm species, genera, and even families (e.g., Polemoniaceae: Schoute, 1935; Kaden and Urmantzev, 1971; Schönenberger, 2009; Schönenberger et al., 2010) in which the *presence* of the contort corolla is not fixed, but its handedness is stable. In other words, their corolla is not always contorted but when contorted then its handedness is always fixed.

There are angiosperm species, in which direction of the corolla contortion is not fixed (Schoute, 1935; Bahadur et al., 1984; Endress, 1999). Both left- and right-contort corolla is present in different flowers, often in the same individual plant or even in the same inflorescence.

The presence of fixed versus unfixed handedness of the contort corolla is taxonomically significant. As pointed out by Endress (1999, 2001, 2012), the handedness is fixed in most asterids with contort corolla, whereas it is unfixed in many rosids with contort corolla. So far, the only member of asterids with well-documented unfixed handedness of the contort corolla is *Melanophylla* (Torricelliaceae: Apiales, Sokoloff et al., 2018a). This is apparently the only member of the large campanulid clade where the contort corolla is reported. The contort corolla is predominant in some genera and families (for example, Apocynaceae) of another large asterid clade, lamiids, where its handedness is always fixed at a level higher than species. Several well-known rosid families are characterized by frequent occurrence of the contort corolla, always with unfixed handedness, for example Oxalidaceae.

Given so high phylogenetic signal in characters related to handedness of the contort corolla, it is important to understand mechanisms of its developmental regulation. As recently highlighted by Ronse De Craene (2018), relative roles of genetic regulation and physico-dynamic aspects of flower development need further clarification, and this is clearly the case of the development of flowers with contort corolla.

Schoute (1935), in his detailed investigation of corolla aestivation patterns, discussed their correlations with calyx development. According to Schoute (1935), in taxa with fixed handedness of the contort corolla, the direction of petal contortion is independent of developmental patterns of the calyx, whereas in species with unfixed handedness of the contort corolla, the direction of contortion depends on the direction of the spiral of sepal initiation or aestivation (Schoute, 1935; Endress, 1999). It should be noted that the occurrence of this spiral in the calyx does not contradict the idea of whorled sepal arrangement. Rather, different members of the same whorl may initiate sequentially, and one common way of such initiation is so-called quincuncial pattern, where the angle between successively initiated organs (in this case, sepals) is 2/5 (of 360°). In angiosperms, the quincuncial *sequence of initiation* may (or may not) correlate with the quincuncial *aestivation* of sepals (with two external, two internal and one "intermediate" sepal). In a calyx with quincuncial aestivation (e.g., Ronse De Craene, 2010; Endress, 2011), sepals can be numbered sequentially starting from an outer one and ending with an inner one, with angles of 2/5 between successive numbers. As with the sequence of initiation, the direction of this spiral is either clockwise or anticlockwise. In species with unfixed handedness of the contorted corolla, when the calyx spiral (sequence of initiation and/or aestivation) is anticlockwise the corolla is right contort, and when the calyx spiral is clockwise the corolla is left contort (Schoute, 1935; Endress, 1999; Sokoloff et al., 2018a).

These observations suggest that, in species with unfixed handedness of the contort corolla, certain transference of positional information from calyx to corolla takes place. Direction of the calyx spiral is likely determined by the arrangement of bracteoles (see Eichler, 1875; Prenner, 2004; Ronse De Craene, 2004; Schönenberger and Grenhagen, 2005; Bachelier et al., 2011; Sokoloff et al., 2018a). The bracteoles are phyllomes situated on pedicel of lateral flowers (e.g., Prenner et al., 2009). Eudicots mostly possess two bracteoles that are inserted transversally, that is, in a plane passing between the flower subtending bract and the inflorescence axis. If the bracteoles are sequentially initiated, then the first sepal normally occurs closer to the first bracteole (continuing their initiation order), and this determines the direction of the calyx spiral. But how does the direction of the calyx spiral technically determine the handedness of the corolla? Broadly speaking, two possible mechanisms could be hypothesized.

1. The signal is transferred through mechanical forces in the developing flower. The petals appear simultaneously and the corolla is at first polysymmetric. During subsequent growth of petals, their shape is constrained by mechanical forces governed by the sepals so that a contortion of certain handedness appears.
2. The signal is transferred at an earlier stage, during prepatterning of petals on the floral meristem, so that the sites of petal initiation are *ab initio* more or less asymmetric.

Schoute (1935) hypothesized that petals tend to develop continuing the spiral of sepal initiation (even though we may not recognize their sequential appearance on practice). This mechanism is similar to that of transition between the initiation order of bracteoles and sepals described above. Thus, the first petal should appear, according to Schoute (1935), between sepals 1 and 3. Indeed, in the case of exact continuation of the sepal spiral into petals, the first petal should appear on the radius of sepal 1, but as the petals of core eudicots normally alternate with sepals, petal 1 appears between sepal 1 and adjacent sepal 3. In a calyx with quincuncial aestivation, the margin of sepal 1 is external to the margin of sepal 3. Therefore, the first petal develops in an asymmetric situation, and this is why its lateral side closest to sepal 3 is turned towards the center of the flower (Figure 18.1a,b). Petals 2 and 3 then develop in the same kind of asymmetric situation, and the kind of the resulting asymmetry in petals 1, 2, and 3 is exactly what we observe in a contort corolla (Figure 18.1a,b). Indeed, when the calyx spiral is clockwise, then the corolla is left contort (Figure 18.1a), and when the calyx spiral is anticlockwise then the corolla is right contort (Figure 18.1b). When corolla development proceeds to petals 4 and 5, the asymmetric situation created by the closest sepals should create an opposite kind of petal asymmetry that does not fit the pattern of contort corolla (for example, consider the pattern of overlapping between sepals 1 and 4 near petal 4, Figure 18.1a,b). As proposed by Schoute (1935), in species with contort corolla, development of petals 4 and 5 is canalized by the fact that the contort pattern is already established by the first three petals. In other words, the petals 4 and 5 use positional information of the petals 1, 2, and 3 rather than that of the sepals. According to the ideas of Schoute (1935), this is the way of appearance of contort corolla in species with unfixed direction of corolla contortion. Schoute thus assumed the crucial role of mechanical forces in this process, that is, the hypothetical mechanism (1).

Schoute's (1935) theory on developmental regulation of petal contortion is rather speculative. He did not conduct extensive original observations on initiation of primordia and further differentiation of organs in developing flowers. To our knowledge, the theory of Schoute (1935) was not subsequently tested using broad comparative developmental analysis. To make a step towards developmental analyses of these issues, we provide new data on early flower development in two species of *Melanophylla* (Torricelliaceae: Apiales). Developmental morphology

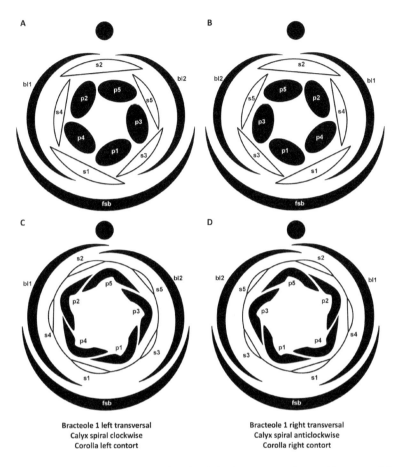

FIGURE 18.1 (a) and (b) Diagrams of perianth in the two enantiomorphic types of flowers with quincuncial aestivation of calyx and contort aestivation of corolla. These two types of flowers occur in species with unstable handedness of contort corolla; (c) and (d) diagrams of perianth in the two enantiomorphic types of flowers in *Melanophylla* (Torricelliaceae, Apiales), simplified from Sokoloff et al. (2018a). bl1, bl2, the first and the second bracteole; fsb, flower subtending bract; p1–p5, petals; s1–s5, sepals. The sepals are numbered according to their assumed sequence of initiation. The petals are numbered according the hypothesis of Schoute (1935) on their "cryptic initiation sequence." As suggested by Schoute (1935), though arising simultaneously to the eye, the petals may have been in fact laid down in a spiral order. This numbering of sepals and petals is adopted in subsequent figures with SEM images of *Melanophylla* flowers where we use it to individualize each organ. It does not imply that we have exhaustive empirical observations proving this sequence of initiation (though we did not get any data contradicting this sequence).

and anatomy of these species have been recently described by our team (Sokoloff et al., 2018a). Flowers of these species almost always possess a contort corolla (the exceptions are very rare). Both left- and right-handed corolla occur in each species, and patterns of arrangement of bracteoles and sepals correlate with handedness of the contort corolla (Figure 18.1c,d). Sokoloff et al. (2018a) provided some evidence of the occurrence of sequential sepal initiation according to a spiral of 2/5 in *M. alnifolia*, but these data were not complete, and there were no developmental data for the other species, *M. aucubifolia*. Here we present a more complete picture acquired through extensive search of different stages of early flower development, which is apparently quite rapid in these species. *Melanophylla* is of particular interest for testing ideas of Schoute (1935), because it actually has an open sepal aestivation (see Endress, 2011, for terminology). As outlined above, the quincuncial sepal aestivation plays an essential role in the hypothesis of Schoute (1935). Relatively small sepals of *Melanophylla*, however, are basally united in a tube (late congenital fusion), whereas margins of free sepal lobes do not overlap each other in the flower bud. As in many other Apiales, the corolla acts as the primary protecting structure of the flower bud in *Melanophylla*.

Material of *M. alnifolia* Baker and *M. aucubifolia* Baker was collected during an expedition to Marojejy National Park (Antsiranana Province, Madagascar) in October 2015, conducted in collaboration with Missouri Botanical Garden's Madagascar Research and Conservation Program. Voucher specimens are deposited in herbaria MW, P and TAN (*M. alnifolia*: *P. Karpunina, M. Nuraliev, A. Oskolski, D. Ravelonarivo, V. Razafindrahaja, J.H. Tonkaina PK 241*—MW0582177; *M. aucubifolia*: *P. Karpunina, M. Nuraliev, A. Oskolski, D. Ravelonarivo, V. Razafindrahaja, J.H. Tonkaina PK 263*—MW0582183). Flowers at different developmental stages were fixed and stored in 70% ethanol. For scanning electron microscopy (SEM), the material was dehydrated in 96% ethanol followed by 100% acetone. Dehydrated material was critical-point dried using a Hitachi HCP-2 critical point dryer, coated with gold and palladium using an Eiko IB-3 ion-coater (Tokyo, Japan) and observed using a CamScan 4 DV (CamScan, UK) at 20 kV at Moscow State University.

Throughout the chapter, we use numbering of sepals and petals according to ideas of Schoute (1935) for descriptive purpose, namely, to individualize each particular organ, as outlined in Figure 18.1. There is normally no problem to assign a flower at any developmental stage to one of the two mirror-shaped forms (Figure 18.1c,d), and then there is the only way of organ numbering. We discuss below to what extent this numbering fits our empirical data on the sequence of organ initiation.

18.2   FLOWER DEVELOPMENT IN *MELANOPHYLLA*

Both species of *Melanophylla* have flowers with five sepals, five petals, five stamens, and three carpels (see also Sokoloff et al., 2018a). Each flower is arranged in an axil of a subtending bract and possesses two bracteoles (Figures 18.2a,b,e; 18.3; 18.4; and 18.5c,e). The sepals are basally united in a tube. Their aestivation is open (Figures 18.2d,e and 18.3a,b). Petals are free, and their aestivation is (with very rare exceptions—Figure 18.2e) contort. Left or right position of bracteole 1 correlates with direction of petal contortion (Figure 18.1b,c,d). Stamens are free and their aestivation is usually contort, with direction of contortion always opposite to

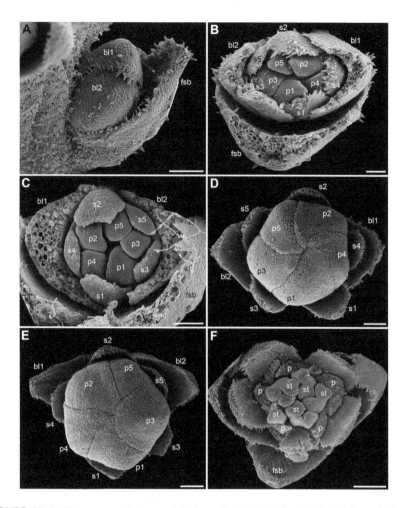

FIGURE 18.2 Flower morphology of *Melanophylla alnifolia* (SEM): (a) lateral view of flower bud enclosed by two bracteoles (note their overlapping margins); (b) and (c) top views of young flowers with bracteoles removed to show the contort corolla; (d) flower bud with contort aestivation of corolla; (e) rare example of flower bud with non-contort aestivation of corolla; and (f) flower bud with corolla removed to show the contort aestivation of androecium. bl1, bl2, the first and the second bracteole; fsb, flower subtending bract; p1–p5, petals; s1–s5, sepals; st, stamens. Organ numbering follows Figure 18.1c,d. Scale bars: 300 μm (a,d–f), 100 μm (b,c).

that of the petals of the same flower (Figures 18.2f and 18.3c,d). Stamen aestivation is out of the scope of the present chapter.

In both species, the two bracteoles differ considerably in size during early developmental stages (Figures 18.4a–c and 18.6a), and later in development a margin of the bracteole 1 clearly overlaps that with the bracteole 2 (Figure 18.2a). Sometimes

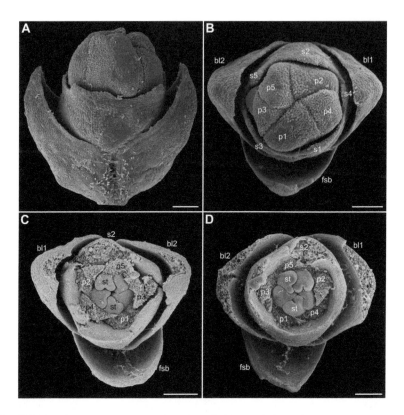

FIGURE 18.3 Flower morphology of *Melanophylla aucubifolia* (SEM): (a) flower bud with two bracteoles viewed from adaxial side; (b) top view of flower bud with contort aestivation of corolla; (c) and (d) flower buds with corolla removed to show contort aestivation of androecium. bl1, bl2, the first and the second bracteole; fsb, flower subtending bract; p1–p5, petals; s1–s5, sepals; st, stamens. Organ numbering follows Figure 18.1c,d. Scale bars: 300 μm (a,d), 100 μm (b,c).

they look like almost equal, but this may be due to the angle of view (Figure 18.6b). The two bracteoles are closer to the inflorescence axis than to the flower subtending bract (Figures 18.4c–e and 18.6), except probably during the earliest stages (Figure 18.4a). The flower meristem between the two bracteoles acquires a rounded-triangular shape, with two sides along the bracteoles and the third side along the subtending bract (Figures 18.4c, left flower, 18.6a,b). The sepal 1 seems to be indeed the first floral organ to initiate, at least in *M. alnifolia* (Figure 18.4b,d). We did not find this stage in our material of *M. aucubifolia*, but sepal 1 was larger than other sepals at some available subsequent stages (Figure 18.7b–d, but possibly not so in 18.7a,f). Sepals 2 and 3 initiate before sepals 4 and 5 in both species (Figures 18.4e,f and 18.6c). Sepal 3 is often smaller than sepal 2 in young flowers (Figures 18.4f; 18.5b,d,f; 18.6d; and 18.7c–f), but we found no direct evidence of their sequential

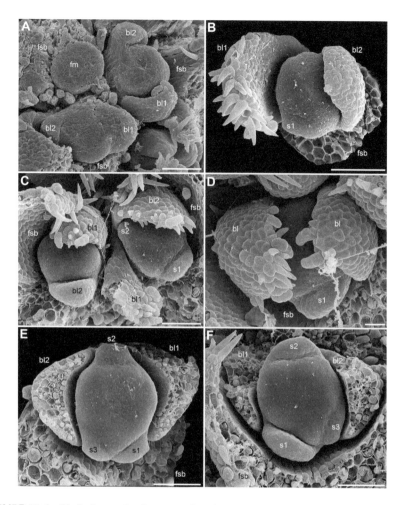

FIGURE 18.4 Early flower development of *Melanophylla alnifolia* (SEM): (a) stages before calyx initiation; (b) through (d) flowers with first evidence of calyx initiation; (e) and (f) flowers with three sepals initiated. bl1, bl2, the first and the second bracteole; floral meristem; fm, floral apex before sepal initiation; fsb, flower subtending bract; s1–s3, sepals. Organ numbering follows Figure 18.1c,d. Scale bars: 30 μm (a,b), 100 μm (c–f).

initiation in the present study. Sepal 5 is clearly the smallest one in the calyx of some young flowers (Figures 18.5c,d and 18.7d), and at least in one observed flower of *M. alnifolia*, sepals 1–4 were present and sepal 5 was yet absent (Figure 18.5a).

In both species, petal primordia (or first petal primordia) appear before sepals 4 and 5 (Figures 18.4f and 18.6c,d). Petal 1 appears to be the only clearly recognizable petal in the young flower of *M. aucubifolia* in Figure 18.6c (one can suggest that petal 2 is also visible here). In another young flower, only petals 1 and 2 are

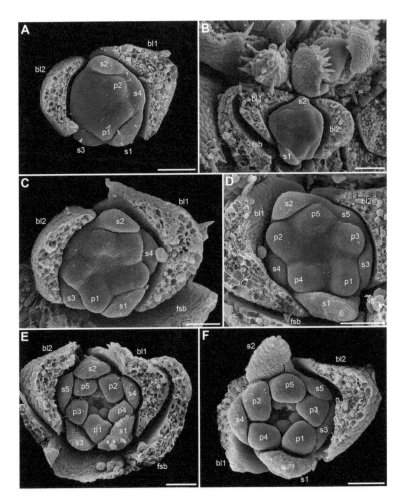

FIGURE 18.5 Later stages of flower development of *Melanophylla alnifolia* (SEM): (a), (c), and (e) flowers with anticlockwise arrangement of sepals 1–5; (b), (d), and (f) flowers with clockwise arrangement of sepals 1–5; (a) and (b) early development of corolla; (c) and (d) petals already asymmetric but their margins are yet not overlapping; (e) and (f) petal aestivation is recognizable, stamen primordia are present. bl1, bl2, the first and the second bracteole; fsb, flower subtending bract; p1–p5, petals; s1–s5, sepals. Organ numbering follows Figure 18.1c,d. Scale bars: 100 μm (a–f).

visible, and the shape of petal 1 is already slightly asymmetric (Figure 18.6d). In both species, petal asymmetry characteristic of left- or right-contort corolla appears early in development, when sepals 3–5 are yet very small (Figures 18.5c,d and 18.7a,d). Later in development, only sepals 1 and 2 cover the entire corolla in *M. aucubifolia* (Figures 18.7b,f) and *M. alnifolia* (Figure 18.2c, see also

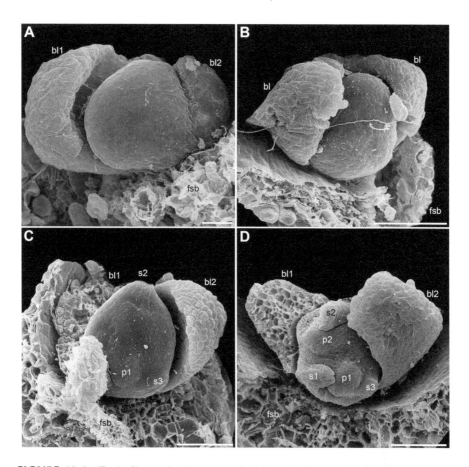

FIGURE 18.6 Early flower development of *Melanophylla aucubifolia* (SEM): (a) and (b) stages before calyx initiation; (c) and (d) flowers with three sepals and some petals initiated. bl1, bl2, the first and the second bracteole; fsb, flower subtending bract; p1, p2, petals; s1–s3, sepals. Organ numbering follows Figure 18.1c,d. Scale bars: 30 μm (a,b), 100 μm (c,d).

Figure 18.6d in Sokoloff et al., 2018a), but ultimately the corolla exceeds the calyx (Figures 18.2d,e and 18.3a,b).

At late developmental stages, petal 1 sometimes appears to be larger than other petals (Figures 18.2c,d and 18.3b). Petals 4 (Figure 18.7e) or 5 (Figure 18.7d) are smaller than other petals in some flowers. In the only flower with non-contort aestivation of corolla, the deviation from the contort condition is caused by altered overlapping between petals 3 and 5. Petal 5 occupies then an inner position to both adjacent petals (Figure 18.2e).

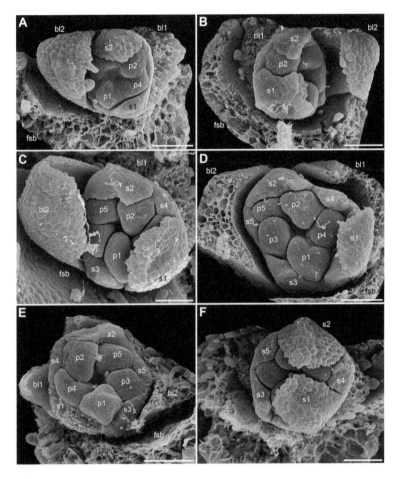

FIGURE 18.7 Later stages of flower development of *Melanophylla aucubifolia* (SEM): (a) and (b) flowers with sepals 1 and 2 well visible and much smaller other sepals. Sepal arrangement is anticlockwise in (a) and clockwise in (b); (c) through (e) young flowers with contort petal aestivation. Sepal arrangement is anticlockwise in (c, d) and clockwise in (e); (f) flower with relatively large sepals 1 and 2 partially enclosing the petals. bl1, bl2, the first and the second bracteole; fsb, flower subtending bract; p1–p5, petals; s1–s5, sepals. Organ numbering follows Figure 18.1c,d. Scale bars: 100 μm (a–f).

18.3   DISCUSSION

Here we provided first data on early flower development in *M. aucubifolia*. These agree with earlier data on *M. alnifolia* (Sokoloff et al., 2018a) that are also extended in this study. In general, our observations are congruent with many ideas of Schoute (1935). Sepal initiation is not synchronous. Sepals numbered 1 and 2 on theoretical grounds are indeed the first sepals to initiate. Moreover, some evidence of non-simultaneous petal initiation is obtained for *M. aucubifolia*. Again, petals numbered 1 and 2 according to the theory of Schoute (1935) were the first to initiate in this species. The fact that we did not find clear evidence of sequential petal initiation in *M. alnifolia* does not disprove the ideas of Schoute, because the differences in timing should be very small and difficult to observe.

However, our observations do not support the hypothesis on mechanical forces as a primary way of transference of positional information from the calyx to corolla in *Melanophylla*. We did not find any overlapping between sepal margins at any stages of development in *Melanophylla*. In the absence of such overlapping, an asymmetric situation for the developing petal cannot appear in the way proposed by Schoute (1935). Also, sepals are generally small by the time when petals show first signs of asymmetry. Thus, it looks like development of their asymmetry is autonomous of the surrounding sepals, just like in the species with fixed handedness of contort corolla (e.g., in *Asclepias*, Endress, 1994). Our observations on the development of petals 2 and 3 are instructive. According to Schoute (1935), these petals attain their asymmetry because they develop near sepals 4 and 5, respectively. As his theory suggests, the asymmetric arrangement of these petals appears because the margin of sepal 4 is overlapped by the margin of sepal 2, and the margin of the sepal 5 is overlapped by the margin of sepal 3. In *Melanophylla*, sepals 4 and 5 are yet absent at the time when petals 2 and 3 first appear and remain small when these petals show first evidence of asymmetry. If mechanical forces do really matter, one would expect asymmetric growth of petals 2 and 3 in the reverse direction, because the largest force they have (if any) is from sepals 1 and 2.

In our view, Schoute's (1935) theory can predict the interactions occurring between sites of future organs during their prepatterning rather than the effects of mechanical forces. Instead of mechanical forces, the inhibitory fields determining the prepatterned sites of organ initiation will play a role here. We suggest that the prepatterning of sepals 4 and 5 takes place before that of the first petals (in concordance with the common sequential prepatterning of floral whorls), even though their visible primordia appear in another sequence.

Prepatterning (e.g., Rudall, 2010) or spatial pattern formation (Choob and Penin, 2004) of organs during development of vegetative shoots and flowers was a subject of mathematical modeling and direct observations (Reinhardt et al., 2003; Alexeev et al., 2005; Smith et al., 2006; Choob and Yurtseva, 2007; Bayer et al., 2009; Choob, 2010; Chandler and Werr, 2014). It is commonly assumed that the prepatterning is related to details of polar transport of auxin (e.g., Reinhardt et al., 2003; Berleth et al., 2007). As tools for direct visualization of prepatterning are available, the ideas of Schoute (1935) could be tested by direct studies of petal prepatterning in one of model organisms. *Gossypium* could be one of models for such a study as a genetically well-explored plant with contort corolla of unfixed handedness (Tan et al., 2016). The theory predicts that sites of petal prepatterning should appear in a rapid sequence, the first one between sepals 1 and 3 and—what is probably easier to record—each site should be slightly asymmetric. Furthermore, it is important to employ a possibility of asymmetry of organ sites in mathematical modeling of prepatterning.

The way of transference of positional information should not be necessarily the same in all angiosperms with unfixed handedness of the contort corolla. *Melanophylla* is so far the only asterid genus with well-documented contort corolla of unstable handedness, whereas all other taxa with this type of petal aestivation belong to rosids. It may well be possible that mechanical forces are important in establishment of handedness of contort corolla in some rosids. Extensive comparative studies should answer this question.

The questions discussed here are related to the problem of criteria for distinguishing spiral and whorled organ arrangement in flowers (Sauquet et al., 2017, 2018; Sokoloff et al., 2018b). Even though the sepals and petals may be numbered continuously in a spiral sequence (s1–s5 followed by p1–p5), as proposed by Schoute (1935), this does not mean the perianth phyllotaxis should be scored spiral for analyses of morphological evolution. This pattern of initiation should be viewed as sequential organ initiation within sepal and petal whorls (Erbar and Leins, 1997; Endress, 2011). In our example, it is important that the actual sequence of organ initiation does not follow the idea of a continuous spiral, because the first petal(s) appear before the last sepal(s). The angles between successive organs (s1–s5 followed by p1–p5) also slightly deviate from the Fibonacci angle.

ACKNOWLEDGMENTS

We are grateful to Prof. Bir Bahadur for kind invitation to contribute to this volume. We thank the staff of the Department of Electron Microscopy at the Biological Faculty of Moscow University (G.N. Davidovich, A.G. Bogdanov, S.V. Polevova, A.N. Sungatullina) for their assistance and for providing SEM facilities. The staff of Missouri Botanical Garden's Madagascar Research and Conservation Program, specifically P.P. Lowry II, are gratefully acknowledged for organization of the fieldwork in Marojejy National Park. We thank D. Ravelonarivo, V. Razafindrahaja, and J.H. Tonkaina for field assistance. The field work was partially supported by the Russian Foundation for Basic Research, project 15-04-05836. The present study is conducted in the framework of the government order for the Lomonosov Moscow State University, projects No. AAAA-A16-116021660045-2 and AAAA-A16-116021660105-3, projects of the National Research Foundation of South Africa (incentive grant 109531), the University of Johannesburg, and the institutional research project no. AAAA-A18-118030690081-1 of the Komarov Botanical Institute.

REFERENCES

Alexeev DV, Ezhova TA, Kozlov VN, Kudryavtsev VB, Nosov MV, Penin AA, Skryabin KG, Choob VV, Shulga OA, Shestakov SV. 2005. Spatial pattern formation in the flower of *Arabidopsis thaliana*: Mathematical modelling. *Doklady Biological Sciences* 401: 133–135.

Bachelier JB, Endress PK, Ronse De Craene LP. 2011. Comparative floral structure and development of Nitrariaceae (Sapindales) and systematic implications. In: Wanntorp L, Ronse De Craene LP, eds. *Flowers on the Tree of Life*. Cambridge, UK: Cambridge University Press, 181–217.

Bahadur B, Reddy NP. 1975. Types of vernation in the cyathia of *Euphorbia milli* des Moulins. *New Phytologist* 75: 131–134.

Bahadur B, Reddy NP, Rao MM, Farooqui SM. 1984. Corolla handedness in Oxalidaceae, Linaceae and Plumbaginaceae. *Journal of Indian Botanical Society* 63: 408–411.

Bayer EM, Smith RS, Mandel T, Nakayama N, Sauer M, Prusinkiewicz P, Kuhlemeier C. 2009. Integration of transport-based models for phyllotaxis and midvein formation. *Genes and Development* 23: 373–384.

Berleth T, Scarpella E, Prusinkiewicz P. 2007. Towards the systems biology of auxin-transport-mediated patterning. *Trends in Plant Science* 12: 151–159.

Borg AJ, Schönenberger J. 2011. Comparative floral development and structure of the black mangrove genus *Avicennia* L. and related taxa in the Acanthaceae. *International Journal of Plant Sciences* 172: 330–344.

Chandler JW, Werr W. 2014. *Arabidopsis* floral phytomer development: Auxin response relative to biphasic modes of organ initiation. *Journal of Experimental Botany* 65: 3097–3110.

Choob VV. 2010. *The Role of Positional Information in Regulation of Development of Floral Organs and Leaf Series of Shoots*. Moscow, Russia: Binom.

Choob VV, Penin AA. 2004. Structure of flower in *Arabidopsis thaliana*: Spatial pattern formation. *Russian Journal of Developmental Biology* 35: 280–284.

Choob VV, Yurtseva OV. 2007. Mathematical modeling of flower structure in the family Polygonaceae. *Botanichesky Zhurnal* 92: 114–134.

Eichler AW. 1875. *Blüthendiagramme*. Leipzig, Germany: W. Engelmann.

Endress ME, Bruyns PV. 2000. A revised classification of the Apocynaceae *s.l.Botanical Review* 66: 1–56.

Endress PK. 1994. *Diversity and Evolutionary Biology of Tropical Flowers*. Cambridge, UK: Cambridge University Press.

Endress PK. 1999. Symmetry in flowers: Diversity and evolution. *International Journal of Plant Sciences* 160(Suppl.): S3–S23.

Endress PK. 2001. Evolution of floral symmetry. *Current Opinion in Plant Biology* 4: 86–91.

Endress PK. 2010. Flower structure and trends of evolution in eudicots and their major subclades. *Annals of the Missouri Botanical Garden* 97: 541–583.

Endress PK. 2011. Evolutionary diversification of the flowers in angiosperms. *American Journal of Botany* 98: 370–396.

Endress PK. 2012. The immense diversity of floral monosymmetry and asymmetry across angiosperms. *Botanical Review* 78: 345–397.

Erbar C, Leins P. 1997. Different patterns of floral development in whorled flowers, exemplified by Apiaceae and Brassicaceae. *International Journal of Plant Sciences* 158: S49–S64.

Kaden NN, Urmantzev YA. 1971. Isomery in live nature. II. Results of investigations. *Botanichesky Zhurnal* 56: 161–174.

Leins P, Erbar C. 2010. *Flower and Fruit. Morphology, Ontogeny, Phylogeny, Function and Ecology*. Stuttgart, Germany: Schweizerbart.

Ma OSW, Saunders RMK. 2003. Comparative floral ontogeny of *Maesa* (Maesaceae), *Aegiceras* (Myrsinaceae) and *Embelia* (Myrsinaceae): taxonomic and phylogenetic implications. *Plant Systematics and Evolution* 243: 39–58.

Prenner G. 2004. Floral development in *Polygala myrtifolia* (Polygalaceae) and its similarities with Leguminosae. *Plant Systematics and Evolution* 249: 67–76.

Prenner G, Vergara-Silva F, Rudall PJ. 2009. The key role of morphology in modelling inflorescence architecture. *Trends in Plant Science* 14: 302–309.

Reinhardt D, Pesce ER, Stieger P, Mandel T, Baltensperger K, Bennett M, Traas J, Friml J, Kuhlemeier C. 2003. Regulation of phyllotaxis by polar auxin transport. *Nature* 426: 255–260.

Roels P, Ronse De Craene LP, Smets EF. 1997. A floral ontogenetic investigation of the Hydrangeaceae. *Nordic Journal of Botany* 17: 235–254.

Ronse De Craene LP. 2004. Floral development of *Berberidopsis corallina*: A crucial link in the evolution of flowers in the core eudicots. *Annals of Botany* 94: 741–751.

Ronse De Craene LP. 2010. *Floral Diagrams: An Aid to Understanding Flower Morphology and Evolution.* Cambridge, UK: Cambridge University Press.

Ronse De Craene LP. 2018. Understanding the role of floral development in the evolution of angiosperm flowers: Clarifications from a historical and physicodynamic perspective. *Journal of Plant Research.* 131: 367–393.

Rudall PJ. 2010. All in a spin: Centrifugal organ formation and floral patterning. *Current Opinion in Plant Biology* 13: 108–114.

Sauquet H, von Balthazar M, Doyle JA, Endress PK, Magallón S, Staedler Y, Schönenberger J. 2018. Challenges and questions in reconstructing the ancestral flower of angiosperms: A reply to Sokoloff et al. *American Journal of Botany* 105: 127–135.

Sauquet H, von Balthazar M, Magallón S, Doyle JA, Endress PK, Bailes EJ, Barroso de Morais E et al. 2017. The ancestral flower of angiosperms and its early diversification. *Nature Communications* 8: 16047.

Schönenberger J. 1999. Floral structure, development and diversity in *Thunbergia* (Acanthaceae). *Botanical Journal of the Linnean Sociey* 130: 1–36.

Schönenberger J. 2009. Comparative floral structure and systematics of Fouquieriaceae and Polemoniaceae (Ericales). *International Journal of Plant Sciences* 170: 1132–1167.

Schönenberger J, Endress PK. 1998. Structure and development of the flowers in *Mendoncia, Pseudocalyx,* and *Thunbergia* (Acanthaceae) and their systematic implications. *International Journal of Plant Sciences* 159: 446–465.

Schönenberger J, Grenhagen A. 2005. Early flower development and androecium organization in Fouquieriaceae (Ericales). *Plant Systematics and Evolution* 254: 233–249.

Schönenberger J, von Balthazar M, Sytsma KJ. 2010. Diversity and evolution of floral structure among early diverging lineages in the Ericales. *Philosophical Transactions of the Royal Society B* 365: 437–448.

Schoute JC. 1935. On corolla aestivation and phyllotaxis of floral phyllomes. *Verhandeling der Koninklijke Akademie van Wetenschappen te Amsterdam, Afdeeling Natuurkunde* 34(4): 1–77.

Scotland RW, Endress PK, Lawrence TJ. 1994. Corolla ontogeny and aestivation in the Acanthaceae. *Botanical Journal of the Linnean Society* 114: 49–65.

Smith RS, Guyomarc'h S, Mandel T, Reinhardt D, Kuhlemeier C, Prusinkiewicz P. 2006. A plausible model of phyllotaxis. *Proceedings of the National Academy of Sciences* 103: 1301–1306.

Sokoloff DD, Karpunina PV, Nuraliev MS, Oskolski AA. 2018a. Flower structure and development in *Melanophylla* (Torricelliaceae: Apiales): Lability in direction of corolla contortion and orientation of pseudomonomerous gynoecium in a campanulid eudicot. *Botanical Journal of the Linnean Society* 187: 247–271.

Sokoloff DD, Remizowa MV, Bateman RM, Rudall PJ. 2018b. Was the ancestral angiosperm flower whorled throughout? *American Journal of Botany* 105: 5–15.

Tan JF, Walford S-A, Dennis ES, Llewellin D. 2016. Trichomes control flower bud shape by linking together young petals. *Nature Plants* 2: 16093.

19 Stylar Polymorphisms in Flowering Plants
An Overview

A. J. Solomon Raju

CONTENTS

19.1 INTRODUCTION

Hermaphroditism is the predominant sexual system in flowering plants. It is characterized by the presence of male and female sex organs in the same flower. This sexual system is advantageous for the plant to economize the resources for pollinator attraction and provide reproductive assurance if mates or pollinators are not available in the habitat (Lloyd 1987). In this sexual system, the synchrony in the expression of male and female sexual organs can cause intra-floral self-fertilization (Barrett 2003). Further, there is a tremendous potential for physical interference between male and female sexual functions during pollination and mating, especially in animal-pollinated plants (Barrett 2002). Since this sexual system provides scope for self-pollination and subsequent inbreeding depression, there is a need for plant species that suffer from inbreeding depression for the evolution of floral traits that contribute to diverse sexual systems. Barrett (2003) stated that inbreeding depression in hermaphroditic species is a major driving force to shape floral traits and evolve mating systems. Lloyd and Webb (1986) stated that physical interference between male and female parts is an important driver for the evolution of floral traits. Different workers stated that the size of floral parts displays, and the spatial and temporal deployment of anthers and stigmas are particularly important in influencing self-pollination, especially in plant species pollinated by animals (Darwin 1877;

Bertin 1993; Barrett 2002). The segregation of male and female sex organs in space or in time is viewed as an adaptation to avoid interference and increase reproductive success (Bertin and Newman 1993).

Dulberger (1992) stated that assessment of flower morphology is the prerequisite to understand the pollination process. The arrangement and position of the sexual organs together with the structure of the corolla play a key role for success or failure of fertilization and the attendant chances for genetic recombination in plants that depend on pollen vectors. Ganders (1979) observed that small differences in the positioning of stigmas and anthers can significantly change the rate of disassortative pollination. Webb and Lloyd (1986) reported various forms of spatial separation of male and female sex organs as floral adaptations that reduce pollen-stigma interference, especially when these adaptations occur in self-incompatible species where out-crossing benefits are assured through physiological mechanisms. Barrett et al. (2000) reported that diversifying selection of floral traits played a prominent role in floral evolution, and pervasive convergence in pollination mechanisms characterized many unrelated animal-pollinated groups. Barrett et al. (2000) reported that stylar polymorphisms are among the most interesting classes of sexual systems known in flowering plants. Individual plants are bisexual, but populations consist of two or three morphological types with differences in the length or orientation of the sexes. These differences are often combined with genetic incompatibility that limits intra-morph fertilization and promotes inter-morph mating. Further, stylar polymorphisms enhance the precision of cross-pollination and reduce lost mating opportunities associated with self-pollen interference. In this chapter, the importance of stylar polymorphisms, especially heterostyly, inversostyly, and flexistyly with reference to their adaptations to reduce self-pollination and promote cross-pollination has been explained based on the literature. Further, recent reports on the occurrence of "resupinate dimorphy" in flowering plants have also been reviewed to understand its importance to promote cross-pollination.

19.2   STYLAR POLYMORPHISM

Different types of stylar polymorphisms have been reported to date in different genera of different families of flowering plants. The important ones include heterostyly, stigma and anther height dimorphism in reciprocal positions, enantiostyly, inversostyly, resupinate dimorphy, and flexistyly. All these stylar polymorphisms are evolved in hermaphroditic species, which are adapted for pollination by insects. Of these, heterostyly is widespread while enantiostyly, resupinate dimorphy, and flexistyly are reported in certain genera. Inversostyly is reported only in *Hemimeris racemosa* (Scrophulariaceae). The plant species that demonstrate enantiostyly produce nectarless flowers with poricidal anthers, which are utilized as a pollen source by pollen-collecting bees, while the plant species that display other forms of stylar polymorphisms are nectariferous with longitudinally dehiscent anthers, which are a pollen and/or nectar source for different insects (Barrett et al. 2000; Pauw 2005; Armbruster et al. 2006; Harley et al. 2017). These stylar polymorphisms have been reviewed and briefly explained about their evolution, importance, and adaptation for ensuring reproductive success in flowering plants.

19.2.1 Heterostyly

Historically, Darwin (1862) first conducted experiments on *Primula*, and later Darwin (1877) in his book *The Different Forms of Flowers in Plants of the Same Species* pioneered the work on heterostyly. Subsequently, the work on this subject gathered momentum by Prof. H. G. Baker and his associates at University of California, Berkeley and later by Bahadur (1968, 2009) in India. Ganders (1979) and Barrett (1992) have reviewed the literature on heterostyly in flowering plants. Heterostyly is a genetic polymorphism known to occur in several angiospermic families of flowering plants, such as Acanthaceae, Boraginaceae, Clusiaceae, Connaraceae, Erythroxylaceae, Fabaceae, Gentianaceae, Iridaceae, Linaceae, Loganiaceae, Lythraceae, Menyanthaceae, Oleaceae, Nyctanthaceae, Ericaceae, Polemoniaceae, Oxalidaceae, Plumbaginaceae, Polygonaceae, Pontederiaceae, Primulaceae, Rubiaceae, Santalaceae, Saxifragaceae, Sterculiaceae, and Turneraceae. Most heterostylous species are distylous with two flower morphs, each one produced by different plants in equal ratio in a population, while a few species are tristylous with three types of flower morphs, each one produced by different plants that occur only in Lythraceae, Oxalidaceae, and Pontederiaceae, once again, in equal ratio in natural populations. In essence, the stigma-stamen morphism differing in the heights promotes effective cross-pollination. Bahadur (1968) first surveyed the literature on heterostyly in Rubiaceae and reported that this family has three subfamilies, Ixoroideae, Cinchonoideae, and Rubioideae, with the last one consisting of more heterostylous species, which is supported by Robbrecht (1988). Puff et al. (2005) stated that there are more heterostylous species in Rubiaceae than the heterostylous ones from all other plant families combined (Darwin 1865; Bahadur 1968). Heterostyly is more common in herbaceous members than the woody members of Rubiaceae. Bahadur (1963, 2009), Naiki and Nagamasu (2004), and Quinet et al. (2004) stated that heterostyly is often accompanied by differences in corolla tube length, filament length, pollen size, shape, surface sculpture, and stigmatic papillae size. Castro et al. (2004) reported that the majority of distylous species possess larger floral parts in thrum than in pin morph. Bahadur (1963, 2009) and Barrett (1992) reported that in distylous species, anthers, and stigmas are presented at equivalent/reciprocal heights in two alternate positions constituting long-styled form (pin-eyed) and short-styled (thrum-eyed) flower morphs. In tristylous species, sexual organ arrangements result in three distinct flower morphs, with one set of stigmas and two sets of anthers alternating at three levels at equivalent heights constituting long, mid, and short flower morphs. In both distylous and tristylous species, the long morph presents stigmas in the highest position, and anthers at the mid and short levels. The short morph has its stigmas in the lowest position and the two sets of anthers above them. In tristylous species, the mid-morph has stigmas at the mid level between the two sets of anthers. Bahadur (1968), Bahadur et al. (1984), and Ganders (1979) stated that heterostylous species often differ in pollen grain size, pollen production, pollen exine, pollen apertures, exine sculpturing, pollen color, presence of starch in pollen, stigmatic papillae size, corolla size, or morphology orientation of hairs in the corolla tube. Distylous species represent two compatibility groups at population level because long-styled and short-styled flower morphs are compatible with each other, while self-pollination and pollination between plants of the same form are incompatible. Tristylous species have three floral morphs in panmictic populations. Each morph has two types of stamens. In one morph, the pistil is short, and the stamens are long and intermediate; in the second morph, the pistil is intermediate, and the stamens are short and long; in the third morph, the pistil is long, and the stamens are short and intermediate. The lengths of stamens and pistils are in reciprocal position and are well adapted for pollination by pollinators, or different body parts of the same pollinator. Different authors reported that distyly and tristyly facilitate to optimize both male and female components of a plant's reproductive fitness. Further, reciprocal herkogamy improves the efficiency of pollen movement between long- and short-styled flowers via the transfer of pollen grains onto distinct positions of the pollinator's body that correspond to the heights of the receiving stigmas in the reciprocal morphs, a process known as disassortative pollination. The function of reciprocal herkogamy improves primarily male function by reducing the wastage of male gametes and increasing effective pollen transfer. The physiological incompatibility system evolved in heterostylous species mainly enhances female function by decreasing the wastage of female gametes (Darwin 1862, 1877; Ganders 1979; Ramaswamy and Bahadur 1984; Barrett 1992, 2002; Richards 1997; Barrett and Shore 2008; Bahadur 2009). Keller et al. (2014) reported that herkogamy in each flower morph reduces the probability of pollen transfer within the same flower and between flowers of the same morph, thus reducing the waste of incompatible pollen grains on nonreciprocal stigmas, and such incompatible pollen grains may also cause stigma or stylar clogging if the incompatibility reaction is weak, which in turn may diminish the probability of ovules to be fertilized by compatible pollen. Barrett (1992) reported that heterostyly is distinct among other types of sexual polymorphisms. It is a cross-pollination mechanism evolved in hermaphroditic species to facilitate pollination and reduce pollen waste by insects. Reciprocal positions of sex organs in the flower morphs of heterostylous species function to increase male fertility by actively promoting more precise pollen transfer and dispersal among plants that occur in populations with uniform sexual organs displaying herkogamy. These species achieve increased rates of male fertility by limiting functional interference between male and female sex organs and also by reducing pollen wastage through self-pollination. The floral design in heterostylous species is a perfect adaptation to assist them to exchange genetic material and maximize parental fitness by limiting the possibilities of inbreeding depression (Darwin 1877). Heterostyly functional through distyly and tristyly is a floral polymorphism that is usually genetically linked with a super gene "S" with two alleles, Ss and ss, with a sporophytic self-incompatibility system. Tristyly is under the control of two genes S and M, with several genotypes. As a breeding system, heterostyly is polyphyletic in origin and probably evolved several times within the same genus or family or in different families. Most of the families in which heterostyly is reported also have homostylous species (Ganders 1979; Bahadur 2009). The evolution of the heterostyly form of floral polymorphism appears to be related to multiple factors, such as habitat conditions, pollinator availability, and inbreeding depression, or even mutation in the S gene complex. The evolution of heteromorphic self-incompatibility expressed through stylar polymorphism is the first step in the evolution of this reproductive system and subsequently accompanied by the evolution of the reciprocal features of certain other floral traits. Weller (2009) stated that heterostyly has evolved on multiple occasions, and self-incompatibility associated with this sexual system is not related to the common multi-allelic self-incompatibility systems found in homomorphic species. An increased understanding of the relation between the reciprocal floral features of the breeding system and the nature of self-incompatibility is required for understanding the intrinsic and extrinsic conditions that favored the evolution of heterostyly. Phylogenetic approaches, combined with studies on the physiological and molecular genetic basis of heterostyly, offer promise in helping to resolve the questions about the origin of heterostyly. Therefore, further studies are necessary to understand the conditions that led to the evolution of heterostyly in plant species that display it.

19.2.1.1   Stigma-Height Dimorphism

Stigma-height dimorphism is another form of stylar polymorphism and quite distinct from heterostyly in hermaphroditic species. In this flower dimorphism, the anthers occupy the same position in both flower morphs, while stigmas occupy different positions in each flower morph. In one morph, the stigmas are long and occupy the position above the level of the anthers, while in the other morph, the stigmas are short and occupy the position below the level of the anthers (Barrett 2000). In *Narcissus*, the expression of stigma-height dimorphism is slightly different. The flowers possess two stamen levels in each morph that occupy slightly different positions within the floral tube. In one morph, the style is long and the stigma is located within or slightly above the upper-level stamens, while in the other morph, the style is short and the stigma is placed well below the lower-level stamens. The two flower morphs do not show any difference in pollen production rate and pollen grain size. Therefore, the stigma-height dimorphism is different and distinct from heterostyly because the reciprocity of stigma and anther positions in the two flower morphs is only weakly developed (Baker et al. 2000; Barrett 2000). This stylar polymorphism may be a transitional form in the evolution of distyly in hermaphroditic species, and in this evolutionary step also, it may limit the occurrence of self-pollination to the extent possible.

19.2.2   Enantiostyly

Enantiostyly is another form of stylar polymorphism reported in hermaphroditic species of at least 10 angiosperm families, including both dicotyledons and monocotyledons (Jesson et al. 2003). The dicotyledons include Solanaceae (Whalen 1979), Caesalpiniaceae (Bahadur et al. 1985, 1990/1991; Fenster 1995), Gesneriaceae (Harrison et al. 1999), and Gentianaceae (Lloyd and Webb 1992), while monocotyledons include Pontederiaceae (Wang et al. 1995), Philydraceae (Graham and Barrett 1995), and Haemodoraceae (Jesson and Barrett 2002). It is different from heterostyly and stigma-height dimorphism. The flowers possess the style, which is displaced away from floral axis and positioned either to the left or right, resulting in mirror-image flowers. This sexual polymorphism has been known since the late 19th century, but only recently, studies have been initiated to understand its evolution and functional importance (Barrett 2000, 2002). The studies indicated the occurrence of monomorphic and dimorphic enantiostyly with the first one as most common (Jesson and Barrett 2003). In monomorphic enantiostyly, individual plants produce floral forms, either mixed within an inflorescence or segregated between left- and right-styled inflorescences, but this condition is not a genetic polymorphism. Most of the species exhibit monomorphic enantiostyly (Jesson and Barrett 2003). The genus *Cassia* is a classical example for monomorphic enantiostyly (Subba Reddi et al. 1997). In dimorphic enantiostyly, individual plants are genetically uniform for style orientation and produce either left- or right-styled flowers (Jesson et al. 2003; Yu 2007). This is a rare form of stylar polymorphism evolved independently in only three closely related monocotyledonous families: Haemodoraceae, Pontederiaceae, Tecophilaeaceae (Jesson and Barrett 2003). The dimorphic enantiostyly is believed to be evolved from monomorphic enantiostyly due to its restricted occurrence and confinement to only three monocotyledonous families (Barrett et al. 2000). In enantiostylous species, the timing of stylar bending or displacement away from the floral axis is either in the bud as in *Wachendorfia paniculata*, *Dilatris corymbosa*, and *Philydrum lanuginosum* or at the beginning of anthesis as in *Heteranthera* spp., *Monochoria australasica*, *Cyanella lutea*, and *Solanum rostratum* (Jesson et al. 2003). Based on patterns of arrangement of style and stamens in individual flowers, Yu (2007) classified enantiostyly into reciprocal and non-reciprocal enantiostyly. Reciprocal enantiostyly is usually associated with the reciprocal deflection of a pollinating anther while non-reciprocal enantiostyly does not present any pollinating anther deflections. Most monomorphic enantiostylous species possess reciprocal enantiostyly, while dimorphic enantiostylous species characteristically possess reciprocal enantiostyly (Jesson and Barrett 2003). Richman and Lawrence (2018) reported a new form of enantiostyly called "aggregate enantiostyly" in *Allionia incarnata*, a species of Nyctaginaceae. In this species, a three-flowered capitulate umbel known as actinomorphic "compound blossom" displays enantiostyly. Each flower is zygomorphic and has a cluster of stamens, which are not deflected left or right, and a single style tipped with a capitate stigma; the styles are enantiomorphic and arranged in a "Y" shape, with two parallel "upper" styles pointing toward each other and away from the stamens, and one "lower" style pointing away from stamens and the upper styles. The style is slightly deflected toward each other and away from the stamens in the upper two flowers, while the style in the lower flower is deflected left or right and away from the stamens. Since enantiostyly is displayed in the capitulate umbel, it is referred to as "aggregate enantiostyly" and is a form of monomorphic enantiostyly. The flower visitors contact either anthers or stigmas but not both during flowering probing; this facilitates increased pollen export and effective out-crossing at the flower level. But these authors advocated further studies to determine the level of pollen export and receipt, and the extent of reduction of geitonogamy.

Dulberger (1981) documented that enantiostyly is probably a part of "floral syndrome" associated with certain floral traits, such as nectarless flowers with upper "feeding anthers" and lower "pollinating anthers," poricidal anther dehiscence, orientation of anthers leading to pollen deposition from the feeding anthers on the ventral side of the pollinator and of the "pollinating" anthers on its back or side, curved styles and stigmas touching the zone on which pollen has been deposited from the "pollinating" anthers, and minute stigmas. Graham and Barrett (1995) found association between dimorphic anthers, zygomorphy, and an outward floral orientation in monocotyledonous enantiostylous species. Such an association between these floral traits might be related to the presence of a pollination syndrome in which consistent positioning of pollinators is important for effective transfer of cross-pollen. Jesson and Barrett (2003) noted that the association of enantiostyly with heteranthery, the loss of nectaries, and the reciprocal placement of a pollinating anther indicate that the stylar polymorphism functions to increase the precision of cross-pollen transfer and to reduce interference of stigmas and anthers within or between flowers on the same plant. Since nectar is absent in enantiostylous species, the reward for pollinators is only pollen, and the flowers of these species are useful only for pollen-collecting pollinators. Jesson and Barrett (2002) also reported that most pollinators of enantiostylous species are large-bodied, pollen-collecting bees. Further, most of the species produce poricidal anthers, and pollen collection from these anthers requires special skills by way of vibrations by wings and squeezing of the anther sacs from the base to the top to release pollen from the apical pore. Therefore, enantiostyly is important for the promotion of cross-pollination by insects and in the evolution of breeding systems of plants with a hermaphroditic sexual system.

19.2.2.1 Inversostyly

Pauw (2005) described "inversostyly" for the first time in *Hemimeris racemosa*, a species of Scrophulariaceae. This is yet another stylar polymorphism with morphs displaying an either upward or downward orientation of style and the stamens in opposite positions. Here the species has zygomorphic flowers and presents a reciprocal arrangement of anthers and stigmas on a vertical plane without any alternation in the height of sex organs. Most populations of this species are inversostylous, dimorphic, and herkogamous while a few populations are homostylous, monomorphic, and non-herkogamous. In inversostylous populations, individual plants produce flowers with the same style orientation, either up or down on a vertical plane. The flowers have a pouch in which oil is secreted; this oil is the only reward offered to oil-collected bee pollinators. In the style-down morph, the style is positioned away from the oil-secreting pouch while anthers are positioned close to the pouch. The position of both style and anthers is reversed in the style-up morph. In homostylous populations, all individual plants produce flowers with the stigma and anthers clustered together in a down position, and both the sex organs contact each other resulting in autogamous self-pollination. Most inversostylous populations exhibit a slight bias in favor of style-down flower morph, and this bias increases with a decrease in pollinator abundance, while homostylous populations occur in areas where oil-collecting bees are less abundant and produce high levels of autogamous seed. The flowers of homostylous populations of *H. racemosa* resemble self-pollinating homostylous species of the same genus such as *Hemimeris sabulosa*, *H. gracilis* and *H. centrodes*. These species occur where their specialized pollinators are either rare or absent. The non-herkogamous homostylous populations of these species use autogamy as "fail-safe" pollination mode for reproductive assurance in areas where pollinators are not available. Barrett et al. (2000) noted that inversostyly has genetic basis because populations consist of equal ratios of the two morphs, and individual plants produce flowers of the same stylar orientation. This stylar polymorphism is distinct from enantiostyly in which flowers are asymmetrical with the style deflected either to the left or right. Jesson and Barrett (2003) stated that inversostyly is the complete reciprocity of all sex organs in flower morphs, while enantiostyly is the alternation of style with one of the stamens and the position of all other stamens remain unchanged. Barrett et al. (2000) noted that stylar polymorphism is not known from other species of Scrophulariaceae. Therefore, inversostyly is a specialized stylar polymorphism associated with a specialized pollination system. It is evolved to promote cross-pollination in areas where pollinators are abundant. Its flowers are evolved to produce and offer oil as a reward instead of nectar or pollen for specialized oil-collecting bees. Since inversostyly is not reported from other families thus far, further studies are required to record the occurrence and evolution of this stylar polymorphism in different families of flowering plants.

19.2.2.2 Resupinate Dimorphy

Harley et al. (2017) reported a novel floral dimorphism termed as "resupinate dimorphy" in three species of Hyptidinae genus *Eplingiella*: *E. brightoniae*, *E. cuniloides*, and *E. fruticosa* of the family Lamiaceae. In these species, all flowers present the style almost upright, closer to the posterior lip of the corolla, and occupy a more central position. But the populations of these studied by these authors showed that individual plants produced either resupinate or non-resupinate flower morphs, and the ratio of these individuals occurred in similar proportions. Half of the population produced resupinate flowers with stamens above the style-facilitating nototribic pollination while the other half of the population produced non-resupinate flowers with stamens below the style-facilitating sternotribic pollination. In both resupinate and non-resupinate flowers, the boat-shaped median lobe of the anterior corolla lip held the stamens inside under tension and released pollen similarly by explosion. In these species, both resupinate and non-resupinate flowers are structurally identical with the same stylar deflection, but only the presence or absence of resupination provides the dimorphism. The resupinate dimorphism functional in *Eplingiella* species is very close to "inversostyly," which is described in *H. racemosa* (Pauw 2005). This flower dimorphism is speculated to be a twisting of the short pedicel in these species. Resupination of the flowers by torsion of the pedicel has been reported in other Lamiaceae such as *Salvia gravida*, *S. vazquezii*, and *Hypenia* spp. (Atkinson 1998; Classen-Bockhoff et al. 2004). It is well-known in orchids and in a number of families of the Lamiales (Clark et al. 2006; Endress 2012). Therefore, resupinate dimorphy is a novel means of promoting cross-pollination and reducing self-pollination.

19.2.2.3 Flexistyly

Flexistyly is another form of polymorphism and is more advanced than the enantiostyly form of polymorphism in flowering plants. It is more proficient in promoting cross-pollination in plants that display hermaphroditic sexual system and are adapted for pollination by animals. It is a single stylar polymorphism, which combines both reciprocal herkogamy via stigma movement and heterodichogamy via temporal differences in sexual functions. Flexistyly was first reported in a tropical ginger species belonging to an Asian *Alpinia* genus (Zingiberaceae) distributed in China. This genus contains more than 250 species of perennials with terminal inflorescences that produce 2–10 flowers daily, and each flower is a hermaphrodite and lasts for only a day. The flowers are zygomorphic, and the stigma orientation alternates reciprocally by curving either upward away from the anthers (protandrous morph) or downward towards the anthers (protogynous morph). Individual plants express either protandrous or protogynous morph. Populations of these species consist of equal frequencies of both flower morphs. In both, the style position changes according to the time of the day; the flower morph that is protandrous in the morning becomes protogynous in the afternoon, while the flower morph that is protogynous in the morning becomes protandrous in the afternoon during the one-day flowering period (Li et al. 2001; Ren et al. 2007). Later, Sun et al. (2011) described the function of flexistyly in *Alpinia* in detail. In the protandrous morph known as cataflexistylous morph, stigmas begin in an unreceptive upward position and move downward into a receptive position where the stigmas contact pollinators. In a protogynous morph known as anaflexistylous morph, the stigmas begin in a receptive downward position and move upward into an unreceptive position where the pollinators contact only with the pollen. Stigmas of the both protandrous and protogynous morphs reciprocally move in the middle of the one-day flowering period. Anthers are held in the same position throughout flowering but disperse pollen only when stigmas are in the upward position where they cannot contact pollinators, promoting disassortative mating between floral morphs. Zhang et al. (2003) reported that flexistyly is functional in *Alpinia blepharocalyx*. In this species, the cataflexistylous morph shows the stigma in the erect position above the dehiscent anther when anthesis begins in the morning and in the curved position under the anther in the afternoon. In anaflexistylous morph, the receptive stigma is curved under the indehiscent anther in the morning and takes a reflexed superior position above the anthers as the latter begin to shed pollen in the afternoon. The style movement in both flower morphs is synchronous. These authors also noted certain traits that differentiate the two morphs in this species. Cataflexistylous flowers are larger, produce more pollen and nectar, and fewer ovules than in anaflexistylous flowers. This species is a self-compatible and primarily insect-pollinated species, but selfing is largely prevented and inter-morph pollen transfer is promoted by flexistyly. Sun et al. (2007) reported that in *Alpinia* species, the anaflexistylous morph with protogyny provides opportunities for the receipt of out-cross pollen before self-pollen is shed. These authors stated that heterodichogamy insulates from self-pollination, while herkogamy reduces interference between male and female functions in self-compatible flexistylous plant species. Different authors reported flexistyly in different genera of Zingiberaceae. Kress et al. (2005) reported flexistyly in *Alpinia*, *Amomum*, and *Etlingera* species. They speculated that flexistyly is functional in the genera *Plagiostachys* and *Paramomum* citing Cui et al. (1996) as source information. Flexistyly is reported in *Amomum tsaoko* (Zhang and Li 2002; Zhang et al. 2003), *A. maximum* (Ren et al. 2007), *Alpinia blepharocalyx* var. *glabrior* (Cui et al. 1996), *A. roxburghii* (Zhang et al. 2003), *A. nieuwenhuizii* (Takano et al. 2005), *A. mutica* (Sun et al. 2010), *A. havilandii*, and *Plagiostachys strobilifera* (Takano et al. 2009). In *A. nieuwenhuizii*, the pollinators show a bimodal pattern of foraging activity of pollinators, which matches well with the functions of the sex organs of flowers (Takano et al. 2005). *Alpinia hainanensis* is flexistylous, and the stylar movement behavior in cataflexistylous and anaflexistylous morphs is similar to that found in other *Alpinia* species. But he reported that the stylar movement behavior of the cataflexistylous morph is two hours later than that of anaflexistylous morph. Further, he also reported certain differences in anaflexistylous plants—asynchrony in stylar movement in different anaflexistlyous morphs of the same inflorescence or plant; anther dehiscence only after stigma takes the position above the anthers in each flower; more pollen output per flower; higher pollen/ovule ratio; and larger nectar volume than in cataflexistylous flower morphs. This plant species is self-compatible and obligately cross-pollinating, but anaflexistylous flowers show significant differences in fruit set rate while cataflexistylous flowers do not show such variation in fruit set rate (Wang et al. 2005). In *A. mutica*, which is cultivated as an ornamental species in Asia, the protandry in cataflexistylous flowers and protogyny in anaflexistylous flowers occur synchronously in the morning period, but anther dehiscence occurs early by noon in anaflexistylous flowers, indicating that the female phase is shortened while the exact time of commencement of stigma receptivity in cataflexistylous flowers is not documented. This species is reported to be self-incompatible, obligately vector-dependent, and fruit through cross-pollinations between the flower morphs only (Aswani and Sabu 2015). Wang et al. (2004) reported that *Alpinia* species with flexistyly and obligate xenogamy have low pollen: ovule ratios that are lower than the ratio provided for obligate xenogamy by Cruden (1977). The low pollen-ovule ratios in these species are attributed to much specialized, sophisticated, and very efficient pollinating mechanisms, and to the large stigma area (Cruden and Miller-Ward 1981; Jurgens and Gottsberger 2002; Wang et al. 2004).

Ren et al. (2007) reported that the genus *Amomum* has both flexistylous and non-flexistylous species. Yang et al. (2016) reported that *A. tsaoko* shows gender differentiation between cataflexistylous and anaflexistylous flower morphs. Cataflexistyled flowers with protandry display short duration of the female phase by the late downward movement of the style, indicating that they act more as the male gender, while anaflexistyled flowers with protogyny display short duration of male phase by the late upward movement of style, indicating that they act more as the female gender. Further, cataflexistylous plants with more flower numbers produce a lower fruit set rate, while anaflexistylous plants with a lower flower number produce a higher fruit set rate, indicating that cataflexistylous flower morphs function more as males while anaflexistylous flower morphs function more as females. In cataflexistylous plants, the flower number and fruit set percentage are not constant between years, while in anaflexistylous plants, the flower number and fruit set percentage are relatively constant between years. The changes in flower number and fruit set rate in cataflexistylous plants are attributed to the influence of resource environment and the possibility of existence of trade-off between male and female function, such as between stamen and pistil, pollen and ovule, flower and fruit (Campbell 2000), and also between number and size of organs, such as flower number and size, pollen number per flower, and size and seed number per fruit (Lloyd 1987; Thomson 1989). The functional gender differentiation between cataflexistylous and anaflexistylous morphs in *A. tsaoko* appears to be a function of resource allocation strategy to enhance sexual reproduction or accomplish reproductive assurance.

Wei et al. (2018) reported that Liliaceae species, *Eremurus altaicus* is hermaphroditic, self-compatible, and strikingly protandrous, but it shows style movement during flower life. The style is upright and very close to anthers, but it curves down before anther dehiscence and the stigma is not receptive. After complete pollen dispersal, the style returns to the upright state, and then the stigma is receptive, which extends for three days, facilitating the possibility of geitonogamy. These authors did not mention that this species is flexistylous, but the function of both herkogamy and dichogamy in monomorphic flowers certainly indicates that *E. altaicus* is flexistylous.

The evolution and function of flexistyly in Zingiberaceae is variously explained by different authors. Renner (2001) stated that flexistyly is a special form of heterodichogamy that promotes out-crossing, and it is probably derived from synchronous dichogamy, which may become a transitional step for the evolution of dioecy. Li (2002) and Zhang et al. (2003) mentioned that flexistyly is probably an intermediate phase from hermaphroditism to dioecy. Zhang et al. (2003) noted that flexistyly contributes to a reduction in autogamy, geitonogamy, and xenogamy among individuals of the same morph. Griffin et al. (2000) felt that the functionality of self-avoidance in flexistylous plant species depends on the rates of pollen deposition and removal, which in turn depend on foraging behavior and pollinator visitation rates. These various explanations on the function of flexistyly indicate that it is evolved to decrease inbreeding and promote out-breeding by temporally and spatially separating the presentation of pollen and receptive stigmas through active stylar movement. Further, this stylar polymorphism does not insulate the plants from the occurrence of autogamy or geitonogamy since some species of Zingiberaceae are self-compatible and likely to receive self-pollen and end up with fructification. Flexistyly is a unique stylar polymorphism that represents some features of dichogamy, herkogamy, enantiostyly, and heterostyly. Sun et al. (2007) suggested that manipulative studies are required to demonstrate the functional importance of flexistyly in pollen-stigma interference.

19.2.2.4 Diplostigmaty

Differentiation of female sexual organs in flowering plants is a rare occurrence. It is displayed in the form of diplostigmaty, which involves the possession of spatially and temporally distinct stigmas by the same pistil. The single pistil within a flower has an apical stigma, as occurs in most flowering plants, but also a secondary or basal stigma that occurs midway down the style, which is physically discrete and receptive several days after the apical stigma. Diplostigmaty is restricted in its taxonomic occurrence, but it is widely distributed in *Sebaea* genus in Gentianaceae family. It is a genus of insect-pollinated mostly African species (Marloth 1909; Hill 1913; Kissling et al. 2009). Kissling and Barrett (2013) reported the details of functionality of diplostigmaty for reproductive assurance in *S. aurea*, an insect-pollinated species of Western Cape, South Africa. The rarity of diplostigmaty in flowering plants is attributed to unusual ontogenetic features, such as the post-genital fusion of the two carpels that occur within a flower in *Sebaea* and the occurrence of flat rather than folded upper carpels allowing the development of secondary basal stigmas. Kissling et al. (2009) proposed that diplostigmaty functions to provide reproductive assurance when pollinator service is not reliable. Apical stigmas may receive mostly out-crossed pollen when pollinator service is reliable, while self-pollen could be autonomously deposited on basal stigmas when pollinator service is insufficient to fertilize all ovules. Kissling and Barrett (2013) provided evidence that basal stigmas in *S. aurea* function to enable autonomous delayed self-pollination, without limiting opportunities for out-crossing. The delayed selfing serves as a mechanism of reproductive assurance in populations with low plant density. Therefore, diplostigmaty is a novel example of a flexible mixed-mating strategy in plants that are responsive to changing demographic conditions.

19.3   CONCLUSIONS

Heterostyly and stigma-height dimorphism are commonly found in hermaphroditic plant species with individual plants producing a single flower morph. It is associated with actinomorphy and nectariferous flowers with prominent floral tubes. The nectar reward is usually situated at the flower base, which enables the spatial segregation of pollen along the style length of flower morph and the proboscis and body of the pollinator during flowering probing for nectar.

Enantiostyly is usually found in plant species with hermaphroditic species associated with zygomorphy, nectarless non-tubular flowers, poricidal anthers, and even heteranthery. It is classified into monomorphic and dimorphic stylar polymorphism; in the former, individual plants produce left- and right-styled flowers within the same inflorescence or on separate inflorescences of the same plant, while in the latter, individual plants produce either left- or right-styled flowers but not both. The pollen is segregated sideways on the pollinator's body during flower probing.

Inversostyly is another stylar polymorphism reported only in *H. racemosa*, which offers oil as a floral reward for specialized oil-collecting bees. This plant species produces inversostylous and homostylous populations; the former occurs in areas where pollinators are abundant, while the latter occurs in areas where pollinators are rare or absent. Inversostyly is associated with dimorphic flowers produced on different plants and herkogamy, while homostyly is associated with monomorphic flower form and non-herkogamy. This stylar polymorphism is a specialized one, associated with specialized pollination system and adapted for fruit set through cross-pollination in habitats where its specialized pollinators are abundant, while homostylous populations provide reproductive assurance for the species to produce fruit through autogamy if pollinators are not available.

Resupinate dimorphy is reported in Orchidaceae and certain genera of different families of Lamiales. These species produce resupinate and non-resupinate flowers on different plants, and both flower morphs are structurally identical with the same stylar deflection, but resupination of flowers in another set of plants contributes to dimorphism. In resupinate flowers, the lower lip takes the upper position while the upper lip takes the lower position due to which the stamens are situated above the stigma. Sternotribic pollination occurs in non-resupinate flowers, and nototribic pollination occurs in resupinate flowers. Resupination of the flowers is attributed to the torsion of the pedicel, as it has been reported in other genera of Lamiaceae. Resupinate dimorphy is speculated to be close to inversostyly, and it is a novel floral dimorphy evolved to promote cross-pollination and reduce self-pollination.

Flexistyly is a single stylar polymorphism that combines both reciprocal herkogamy and heterodichogamy and is reported in certain genera of Zingiberaceae. The plants that exhibit flexistyly produce cataflexistylous and anaflexistylous flower morphs on different plants. Plants producing cataflexistylous flower morphs are protandrous, while those producing anaflexistylous flower morphs are protogynous at the same time during morning with the reverse dichogamy occurring from noon onward in these forms. This stylar polymorphism is functional in reducing self-pollination and cross-pollination among individuals of the same flower morph.

Diplostigmaty is a novel expression of a flexible mixed-mating system in the genus *Sebaea* in the Gentianaceae family. It involves a single pistil possessing an apical stigma, which is primarily meant for the promotion of cross-pollination when pollinators are reliable, and a basal stigma meant for autonomous delayed selfing when pollinators are unreliable. This sexual system functions to provide reproductive assurance when pollinators are unreliable.

In flowering plants, the hermaphroditic sexual system is adaptive for reproductive assurance in diverse habitats and facilitates colonization of new areas. Because both male and female sex organs are borne and functional in the same flower, and self-pollination is possible if self-compatibility is functional. Hermaphroditic plants with the production of both pollen and nectar attract pollinators, which in effect contribute to both self- and cross-pollination, ensuring genetic variation and adaptability to different habitats. However, they could suffer significant reproductive costs from sexual interference, resulting in selection of various floral traits that reduce these costs. Selection of the sex function, either by the spatial separation of stigmas and anthers (Webb and Lloyd 1986) or by the temporal separation of stigma receptivity and anther dehiscence (Bertin and Newman 1993), has been viewed as an adaptation that reduces interference and enhances reproductive success. Certain plant species across families have developed some floral strategies to reduce self-pollination and promote cross-pollination. Stylar polymorphisms evolved in different hermaphroditic plant species provide flexibility to promote out-breeding while reducing inbreeding. The stylar movements in all forms of polymorphism reported in the studied plant species indicate that they are adapted for pollination by insects for fructification through cross-pollination supplemented by fructification through self-pollination in self-compatible species. However, stylar movement far away from anthers could be disadvantageous because pollinators may contact either male or female sex organs with different body parts while probing the flowers. Then, the flowers are more likely to experience a reduction in the precision of cross-pollination and a decrease in both male and female fitness and associated adaptations.

REFERENCES

Armbruster, W.S., Perez-Barrales, R., Arroyo, J., Edwards, M.E., and Vargas, P. 2006. Three-dimensional reciprocity of floral morphs in wild flax (*Linum suffruticosum*): A new twist on heterostyly. *New Phytol.* 171: 581–590.

Aswani, K. and Sabu, M. 2015. Reproductive biology of *Alpinia mutica* Roxb.(Zinziberaceae) with special reference to flexistyly pollination mechanism. *Intl. J. Pl. Reprod. Biol.* 7: 48–58.

Atkinson, R. 1998. A taxonomic revision of *Hypenia* (Mart. ex Benth.) Harley. Ph.D. Thesis, University of St. Andrews, St. Andrews.

Bahadur, B. 1963. Heterostylism in *Oldenlandia umbellata* L. *J. Genet.* 66: 429–440.

Bahadur, B. 1968. Heterostyly in Rubiaceae: A Review. *Osmania. Univ. J. Sci. Golden Jubilee Spl.* 4: 207–238.

Bahadur, B. 2009. Genetics of heterostyly and homostyly in *Oldenlandia umbellata* L. (Rubiaceae). *Proc. Andhra Pradesh Acad. Sci.* 13: 98–110.

Bahadur, B., Arthi, C., and Swamy, N.R. 1990/1991. SEM studies of pollen in relation enantiostyly and heteranthery in *Cassia* (Caesalpinaceae). *J. Palynol. (Silver Jubilee Volume).* 22/27: 7–22.

Bahadur, B., Kumar, P.V., and Reddy, N.P. 1985. Enantiostyly in *Cassia auriculata* L. In: *Pollination Biology—An Analysis.* R.P. Kapil (Ed.), pp. 239–249, D.K. Publishers, New Delhi, India.

Bahadur, B., Laxmi, S.B., and Swamy, N.R. 1984. Pollen morphology and heterostyly—A systematic and critical account. *Adv. Pollen Spore Res.* 13: 79–126.

Baker, A.M., Thompson, J.D., and Barrett, S.C.H. 2000. Evolution and maintenance of stigma-height dimorphism in *Narcissus*. I. Floral variation and style-morph ratios. *Heredity.* 84: 502–513.

Barrett, S.C.H. (Ed.) 1992. *Evolution and Function of Heterostyly. Theoretical & Applied Genetics Monographs.* Springer-Verlag, Berlin, Germany.

Barrett, S.C.H. 2000. Micro evolutionary influences of global change on plant invasions. In: *The Impact of Global Change on Invasive Species.* H.A. Mooney and R.K. Hobbs (Ed.), pp. 115–139, Island Press, Washington, DC.

Barrett, S.C.H. 2002. The evolution of plant sexual diversity. *Nat. Rev. Genet.* 3: 274–284.

Barrett, S.C.H. 2003. Mating strategies in flowering plants: The outcrossing-selfing paradigm and beyond. *Phil. Trans. Royal Soc. Ser. B.* 358: 991–1004.

Barrett, S.C.H., Jesson, L.K., and Baker, A.M. 2000. The evolution and function of stylar polymorphisms in flowering plants. *Ann. Bot.* 85(Suppl.): 253–265.

Barrett, S.C.H. and Shore, J.S. 2008. New insights on heterostyly: Comparative biology, ecology and genetics. In: *Self-incompatibility in Flowering Plants—Evolution, Diversity, and Mechanisms.* V.E. Franklin-Tong (Ed), pp. 3–32, Springer-Verlag, Berlin, Germany.

Bertin, R.C. 1993. Incidence of monoecy and dichogamy in relation to self-fertilization. *Am. J. Bot.* 80: 557–583.

Bertin, R.I. and Newman, C.M. 1993. Dichogamy in angiosperms. *Bot. Rev.* 59: 112–152.

Campbell, D.R. 2000. Experimental tests of sex-allocation theory in plants. *Trends Ecol. Evol. (Amst.)* 15: 227–232.

Castro, C.C., Oliveira, P.E., and Alves, M.C. 2004. Breeding system and floral morphometry of distylous *Psychotria* L. species in the Atlantic rain forest, SE Brazil. *Plant Biol.* 6: 755–760.

Clark, J.L., Herendeen, P.S., Skog, L.E., and Zimmer, E.A. 2006. Phylogenetic relationships and genetic boundaries in the Episcieae (Gesneriaceae) inferred from nuclear, chloroplast, and morphological data. *Taxon.* 55: 313–336.

Classen-Bockhoff, R., Speck, T., Tweraser, E., Wester, P., Thimm, S., and Reith, M. 2004. The staminal lever mechanism in *Salvia* L. (Lamiaceae): A key innovation for adaptive radiation? *Org. Divers. Evol.* 4: 189–205.

Cruden, R.W. 1977. Pollen-ovule ratios: A conservative indicator of breeding systems in flowering plants. *Evolution.* 31: 32–46.

Cruden, R.W. and Miller-Ward, S. 1981. Pollen-ovule ratio, pollen size, and the ratio of stigmatic area to the pollen-bearing area of the pollinator: A hypothesis. *Evolution.* 35: 964–974.

Cui, X.-L., Wei, R.-C., and Huang, R-F. 1996. A study on the breeding system of *Amomum tsaoko*. In: *Proc. 2nd Symp. Fam. Zingiberaceae.* T.-L. Wu, Q.-G. Wu, and Z.-Y. Chen (Eds.), pp. 288–296, Guangzhou, China.

Darwin, C. 1862. On the two forms, or dimorphic condition, in the species of *Primula*, and on their remarkable sexual relations. *J. Linn. Soc. (Bot.)* 6: 77–96.

Darwin, C. 1865. On the movements and habits of climbing plants. *J. Linn. Soc. Lond. (Bot.)* 9: 1–118.

Darwin, C. 1877. *The Different Forms of Flowers on Plants of the Same Species.* John Murray, London, UK.

Dulberger, R. 1981. The floral biology of *Cassia didymobotrya* and *C. auriculata* (Caesalpiniaceae). *Am. J. Bot.* 68: 1350–1360.

Dulberger, R. 1992. Floral polymorphism and their functional significance in the heterostylous syndrome. In: *Evolution and Function of Heterostyly.* S.C.H. Barrett (Ed.), pp. 41–84, Springer, Berlin, Germany.

Endress, P.K. 2012. The immense diversity of floral monosymmetry and asymmetry across angiosperms. *Bot. Rev.* 78: 345–397.

Fenster, C.B. 1995. Mirror-image flowers and their effect on outcrossing rate in *Chamaecrista fasciculata* (Leguminosae). *Am. J. Bot.* 82(1): 46–50.

Ganders, F.R. 1979. The biology of heterostyly. *N. Z. J. Bot.* 17: 607–635.

Graham, S.W. and Barrett, S.C.H. 1995. Phytogenetic systematics of the Pontederiales: Implications for breeding system evolution. In: *Monocotyledons: Systematics and Evolution.* P.J. Rudall, P.J. Cribb, D.F. Cutler, and C.J. Humphries (Eds.), pp. 415–441, Royal Botanic Gardens, Kew, UK.

Griffin, S.R., Mavraganis, K., and Eckert, C.G. 2000. Experimental analysis of protogyny in *Aquilegia canadensis* (Ranunculaceae). *Am. J. Bot.* 87: 1246–1256.

Harley, R.M., Giulietti, A.M., Abreu, I.S., Bitencourt, C., de Oliveira, F.F., and Endress, P.K. 2017. Resupinate dimorphy, a novel pollination strategy in two-lipped flowers of *Eplingiella* (Lamiaceae). *Acta Bot. Bras.* 31: 102–107.

Harrison, C., Moller, M., and Cronk, Q. 1999. Evolution and development of floral diversity in *Streptocarpus* and *Saintpaulia.Ann. Bot.* 84: 49–60.

Hill, A.W. 1913. The floral morphology of the genus Sebaea. *Ann. Bot.* 27: 479–489.

Jesson, L.K. and Barrett, S.C.H. 2002. Enantiostyly in *Wachendorfia* (Haemodoraceae): The influence of reproductive systems on the maintenance of the polymorphism. *Am. J. Bot.* 89: 253–263.

Jesson, L.K. and Barrett, S.C.H. 2003. The comparative biology of mirror-image flowers. *Int. J. Plant Sci.* 164: S237–S249.

Jesson, L.K., Kang, J., Wagner, S.L., Barrett, S.C., and Dengler, N.G. 2003. The development of enantiostyly. *Am. J. Bot.* 90: 183–195.

Jurgens, A. and Gottsberger, T.W.G. 2002. Pollen grain numbers, ovule numbers and pollen-ovule ratios in Caryophylloideae: Correlation with breeding system, pollination, life form, style number, and sexual system. *Sex. Pl. Reprod.* 14: 279–289.

Keller, B., Thomson, J.D., and Elena, C. 2014. Heterostyly promotes disassortative pollination and reduces sexual interference in Darwin's primroses: Evidence from experimental studies. *Funct. Ecol.* 28: 1413–1425.

Kissling, J. and Barrett, S.C.H. 2013. Diplostigmaty in plants: A novel mechanism that provides reproductive assurance. *Biol. Lett.* 9(5): 20130495.

Kissling, J., Endress, P.K., and Bernasconi, G. 2009. Ancestral and monophyletic presence of diplostigmaty in *Sebaea* (Gentianaceae) and its potential role as a morphological mixed mating strategy. *New Phytol.* 184: 303–310.

Kress, W.J., Liu, A.Z., Newman, M., and Li, Q.J. 2005. The molecular phylogeny of *Alpinia* (Zingiberaceae): A complex and phylogenetic genus of gingers. *Am. J. Bot.* 92: 167–178.

Li, Q-J. 2002. Study on the flexistyle outcrossing mechanism in *Alpinia* plants (Zingiberaceae). Ph.D. Thesis, Kunming Institute of Botany, Chinese Academy of Sciences (original not seen).

Li, Q.-J., Xu, Z.-F., Xia, Y.-M., and Gao, J.-Y. 2001. Study on the flexistyly pollination mechanism in *Alpinia* plants (Zingiberaceae). *Acta Bot. Sin.* 43: 364–369.

Lloyd, D.G. 1987. Allocations to pollen, seeds, and pollination mechanisms in self-fertilizing plants. *Funct. Ecol.* 1: 83–89.

Lloyd, D.G. and Webb, C.J. 1986. The avoidance of interference between the presentation of pollen and stigmas in angiosperms I. Dichogamy. *N.Z. J. Bot.* 24: 135–162.

Lloyd, D.G. and Webb, C.J. 1992. The evolution of heterostyly. In: *Evolution and Function of Heterostyly.* S.C.H. Barrett (Ed.), pp. 151–178, Springer, Berlin, Germany.

Marloth, R. 1909. A diplostigmatic plant, *Sebaea exacoides* (L.) Schinz (*Belmontia cordata* L.). *Trans. R. Soc. South Africa.* 1: 311–314.

Naiki, A. and Nagamasu, H. 2004. Correlation between distyly and ploidy level in *Damnacanthus* (Rubiaceae). *Am. J. Bot.* 91: 664–671.

Pauw, A. 2005. Inversostyly: A new stylar polymorphism in an oil-secreting plant, *Hemimeris racemosa* (Scrophulariaceae). *Am. J. Bot.* 92: 1878–1886.

Puff, C., Chayamarit, K., and Chamchumroon, V. 2005. *Rubiaceae of Thailand. A Pictorial Guide to Indigenous and Cultivated Genera.* The Forest Herbarium, National Park, Wildlife and Conservation Department, Bangkok, Thailand, pp. 245.

Quinet, M., Cawoy, V., Lefevre, L., Miegroet, F.V., Jacquemart, A.L., and Kinet, J.M. 2004. Inflorescence structure and control of flowering time and duration by light in buckwheat (*Fagopyrum esculentum* Moench). *J. Exp. Bot.* 55: 1509–1517.

Ramaswamy, N. and Bahadur, B. 1984. Pollen flow in dimorphic *Turnera subulata* (Turneraceae). *New Phytologist (Oxford.)* 98: 205–209.

Ren, P.-Y., Liu, M., and Li, Q.-J. 2007. An example of flexistyly in a wild cardamom species (*Amomum maximum*, Zingiberaceae). *Pl. Syst. Evol.* 267: 147–154.

Renner, S.S. 2001. How common is heterodichogamy? *Trends Ecol. Evol.* 16: 595–597.

Richards, A.J. 1997. *Plant Breeding Systems.* Chapman & Hall, London, UK.

Richman, S.K. and Lawrence, V.D. 2018. Aggregate enantiostyly: Floral visitor interactions with a previously unreported form of floral display. *J. Pollination Ecol.* 22: 9–54.

Robbrecht, E. 1988. Tropical woody rubiaceae. *Opera Bot. Belg.* 1: 1–271.

Subba Reddi, C., Solomon Raju, A.J., Atluri, J.B., and Bhaskara Rao, C. 1997. Enantiostyly, heteranthery and carpenter bee pollination in *Cassia alata* L. (Caesalpiniaceae). *J. Palynol.* 33: 149–152.

Sun, S., Cao, G.X., Luo, Y.J., and Li, Q.J. 2010. Maintenance and functional gender specialization of flexistyly. *Chinese J. Plant Ecol.* 34: 826–838.

Sun, S., Gao, J.Y., Liao, W.J., Li, Q.J., and Zhang, D.Y. 2007. Adaptive significance of flexistyly in *Alpinia blepharocalyx* (Zingiberaceae): A hand-pollination experiment. *Ann. Bot.* 91: 616–666.

Sun, S., Zhang, D.Y., Ives, A.R., and Li, Q.J. 2011. Why do stigmas move in a flexistylous plant? *J. Evol. Biol.* 24: 497–504.

Takano, A., Gisil, J., Yusoff, M., and Tachi, T. 2005. Floral and pollinator behavior of flexistylous Bornean ginger, *Alpinia nieuwenhuizii* (Zingiberaceae). *Pl. Syst. Evol.* 252: 167–173.

Takano, A., Julius, A., and Mohamed, M. 2009. First report of flexistyly in *Plagiostachys* (Zingiberaceae). *Acta Phytotax. Geobot.* 60: 56–59.

Thomson, J.D. 1989. Deployment of ovules and pollen among flowers within inflorescences. *Evol. Trends Plants.* 3: 65–68.

Wang, G., Muira, R., and Kusanagi, T. 1995. The enantiostyly and the pollination biology of *Monochoria korsakowii* (Pontederiaceae). *Acta Phytotax. Geobot.* 46: 55–65.

Wang, Y.-Q., Zhang, D.-X., and Chen, Z.-Y. 2004. Pollen histochemistry and pollen: Ovule ratios in Zingiberaceae. *Ann. Bot.* 94: 583–591.

Wang, Y.-Q., Zhang, D.-X., and Chen, Z.-Y. 2005. Pollination biology of *Alpinia hainanensis* (Zingiberaceae). *Acta Phytotax. Sinica.* 43: 37–49.

Webb, C.J. and Lloyd, D.G. 1986. The avoidance of interference between the presentation of pollen and stigmas in angiosperms. II. Herkogamy. *N. Z. J. Boyt.* 24: 163–178.

Wei, G., Miao, M., and Jun-Feng, F. 2018. Dichogamy and style curvature avoid self-pollination in *Eremurus altaicus. Curr. Sci.* 114: 387–391.

Weller, S.G. 2009. The different forms of flowers—What have we learned since Darwin? *Bot. J. Linn. Soc.* 160: 249–261.

Whalen, M.D. 1979. Taxonomy of solanum section androceras. *Gentes Herb.* 11: 359–426.

Yang, Y.-W., Qian, Z.-G., Li, A.-R., Pu, C.-X., Liu, X.-L., and Guan, K.-Y. 2016. Differentiation in fructification percentage between two morphs of *Amomum tsaoko* (Zingiberaceae). *Breed. Sci.* 66: 391–395.

Yu, L. 2007. Enantiostyly in angiosperms and its evolutionary significance. *Acta Phytotax. Sinica.* 45: 901–916.

Zhang, L. and Li, Q.J. 2002. Flexistyly and its evolutionary-ecological significance. *Acta Phytoecol. Sinica.* 26: 385–390.

Zhang, L., Li, Q.J., Deng, X.B., Ren, P.Y., and Gao, J.Y. 2003. Reproductive biology of *Alpinia blepharocalyx* (Zingiberaceae): another example of flexistyly. *Plant Syst. Evol.* 241: 67–76.

20 Enantiostyly in Angiosperms

Natan Messias de Almeida and
Cibele Cardoso de Castro

CONTENTS

20.1 INTRODUCTION

Enantiostyly, an expression created by Knuth (1906), is a kind of reciprocal herkogamy in which floral morphs exhibit reciprocal arrangement of sexual organs [in some cases, of sexual organs and petals (Almeida et al. 2013a)] in a horizontal plane (Bahadur et al. 1979, 1985; Barrett 2002). Enantiostylous species occur independently in 11 Angiosperm families (*sensu* APG IV 2016) and has broad distribution. Aside the secular knowledge on enantiostyly (Todd 1882), and the fact that even Darwin have had the intention of studying this polymorphism (Darwin 1877; Barrett 2010), several aspects of enantiostyly remain unknown. In this chapter, we aim to present what is known on enantiostyly in terms of the main presentation forms, functioning, genetics and evolution, particular strategies exhibited by some species, and perspectives of future studies.

20.2 ENANTIOSTYLOUS PATTERN AND FLORAL MORPH CLASSIFICATION

Dulberger (1981) suggested a set of floral features that would be a pattern for enantiostylous species (the enantiostylous flower syndrome): (1) nectarless flowers; (2) heteranthery (division of labor between anthers, i.e., food and reproduction); (3) poricidal anthers; (4) pollen deposition of food anthers on the ventral portion of pollinators and pollen deposition of reproduction anthers on the dorsal or lateral portion of pollinators' bodies; (5) curved style; and (6) a tiny stigma that contacts pollen deposition regions on pollinators' bodies.

The first classifications of enantiostyly basically took into account the occurrence of heteranthery and intermorph crosses. According to this classification, species were considered as (1) without heteranthery; (2) with heteranthery and no preferential intermorph crosses; and (3) with both heteranthery and preferential intermorph crosses (Knuth 1906). Nowadays enantiostylous species are classified as monomorphic and dimorphic. Individuals of monomorphic enantiostylous species exhibit both floral morphs that are randomly distributed (random enantiostyly), alternately distributed in the inflorescence (non-random enantiostyly), or even being exclusive to the inflorescence (the monomorphic individual has right and left inflorescences). Individuals of dimorphic enantiostylous species have only one floral morph; thus, the population has right and left individuals (Barrett 2002). Within monocots and dicots, ten families (*sensu* APG IV 2016) and 25 genera exhibit monomorphic enantiostylous, and four families and five genera exhibit the dimorphic type (Barrett 2002, 2010; Bezerra 2008).

Besides the diversification on pistil positioning described above, enantiostylous species may vary in relation to androecium positioning. In reciprocal enantiostylous species, the androecium is opposite to the pistil; thus, the style is turned to the right in relation to the flower center, and the stamens are turned to the left (right flowers) and/or the inverse (left flowers), resulting in flowers that form specular images (Jesson and Barrett 2002a). Non-reciprocal enantiostylous species produce flowers whose androecium positioning does not vary (Jesson and Barrett 2003; Yu and Dun-Yan 2007). The absence of style deflection combined with stamen deflection in the same species is recorded only in *Murdannia* (Commelinaceae; Evans et al. 2000; Jesson and Barrett 2003).

Combining the variations on pistil and anthers' positioning described above, the following enantiostylous types are found: reciprocal monomorphic, reciprocal dimorphic, non-reciprocal monomorphic, and non-reciprocal dimorphic. The non-reciprocal monomorphic enantiostyly is the most commonly found (Jesson et al. 2003), especially in the subtribe Cassiinae (*Cassia*, *Senna*, and *Chamaecrista*); it was also recorded in some species of Commelinaceae, Gentianaceae, Gesneriaceae, Haemodoraceae, and Pontederiaceae (Simpson 1990; Lloyd and Webb 1992; Harrison et al. 1999; Evans et al. 2000). Variations on the distance between anthers and stigma of both floral morphs are rare (Jesson and Barrett 2003; Almeida et al. 2013b).

Deviations in relation to the general enantiostylous pattern were also recorded. *Chamaecrista flexuosa* (Fabaceae), for example, has a third floral morph, called central (Almeida et al. 2013b). Its flowers were classified using morphological (positioning of the falcate petal: right and left) and functional (style positioning: right, left, and central) parameters. Central flowers may occur on the right (morphologically right but functionally central) and left (morphologically left but functionally central). No morphologically central flowers were observed. Because of the presence of a third floral morph, *C. flexuosa* was considered as exhibiting atypical enantiostyly, a term similar to that used for heterostyly, another type of reciprocal herkogamy (Barrett 1992).

FIGURE 20.1 Flowers and enantiostylous types of Cassiinae species. Schemes show the relative position of stamens and gynoecium: (a) *Chamaecrista flexuosa* (flexuosa type); (b) *Senna cana* (cana type); (c) *S. macranthera* var. *pudibunda* (macrantheratype); (d) *S. martiana* (martianatype); (e) *C. amiciella* (amiciellatype); (f) *C. repens* (repenstype); and (g) *C. ramosa* (ramosa type). Lines indicate the route of pollen grains, and circles show the final area of contact of pollen grains with the petals. (From Almeida, N.M. et al., *Plant Biol.*, 15, 369–375, 2015a.)

The most recent classification of enantiostylous species was suggested by Almeida et al. (2015a) for a species of the subtribe Cassiinae (Caesalpidoidae-Fabaceae). Using flower morpho-functional characteristics (reciprocity, area, and mode of pollen deposition on pollinators' bodies, number of petals related to pollen deposition, pollen pathway) and the complexity of pollen capture/deposition mechanisms, the authors defined seven enantiostyly types (Figure 20.1).

20.3 POLLEN CAPTURE AND DEPOSITION IN ENANTIOSTYLOUS SPECIES

Reciprocal monomorphic enantiostylous species have a direct pollen deposition (i.e., the anthers contact the pollinators). Pollen deposition (by the anthers) and capture (by the stigmas) occur in opposite areas of pollinators' bodies by each floral morph. Considering the reciprocal positioning of stigma and anthers in the flowers, right flowers deposit pollen at the left and capture pollen at the right portion of pollinators' bodies, whereas left flowers perform the inverse (Jesson and Barrett 2002a; Bezerra 2008; Krieck et al. 2008).

Enantiostylous species that are non-reciprocal and monomorphic usually exhibit an indirect pollen deposition mechanism, in which floral structures other than the anthers are involved (Westerkamp 2004; Costa et al. 2007; Dutra et al. 2009; Almeida et al. 2015a, 2015b; Amorim et al. 2017), allowing pollen deposition and capture on opposite portions at the pollinators' bodies. In the ricochet mechanism, the vibration performed by the pollinators allows pollen release from the anthers and also plays a role in the pollen conduction to a petal surface. The pollen grains strike against a petal and move to pollinators' bodies (Westerkamp 2004; Almeida et al. 2015a, 2015b; Amorim et al. 2017). This deposition occurs following the laws of light incidence, where the incidence and reflection angles are the same (Westerkamp 2004). Another mechanism of indirect pollen deposition recorded in non-reciprocal, monomorphic enantiostylous species was described by Almeida et al. (2013a) in *Chamaecrista ramosa*: after being released from the anthers, the pollen grains go through the inner petals surface, making a looping movement and reaching the pollinators' bodies (Almeida et al. 2013a, 2015a).

20.4   GENETICS AND EVOLUTION OF ENANTIOSTYLY

Little is known about the genetics of enantiostyly (Barrett 2010). In a study on the dimorphic enantiostylous *Heteranthera multiflora* (Pontederiaceae), Jesson and Barrett (2002b) showed that enantiostyly is controlled by a pair of alleles and that the right deflection of the style is dominant upon the left. The genetic expression of floral morphs for monomorphic enantiostylous species remains unknown (Jesson and Barrett 2002b; Barrett 2010).

Some studies investigated the evolutionary significance of enantiostyly (Barrett et al. 2000a). On one hand, most studies suggest that enantiostyly favors intermorph pollen transfer by means of xenogamy (Todd 1822; Knuth 1906; Irwin and Barneby 1976; Almeida et al. 2013a) or geitonogamy (Dulberger 1981; Gottsberger and Silberbauer-Gottsberger 1988; Bahadur et al. 1990; Fenster 1995; Laporta 2005). On the other hand, Dulberger (1981) suggested that the evolutionary significance of enantiostyly is the protection of the fragile gynoecium against buzz pollination that is generally performed by large bees.

The evolution of monomorphic and dimorphic enantiostyly was a mystery until the studies of Jesson et al. (2003) and Jesson and Barrett (2003). The evolution of the monomorphic type as being the result of a transition from a floral type bearing a central style was elucidated by the theoretical and experimental studies of Jesson et al. (2003) and Jesson and Barrett (2002a, 2003, 2005). The evolution of dimorphic enantiostyly presumably occurred as a selective response to the higher rates of autogamy and geitonogamy found in monomorphic species, similarly to what is known about the evolution of dioecious from monoecious species by fixing one floral morph in the individuals (Barrett 2010). The reduction on autogamy and geitonogamy and the increase on xenogamy rates in monomorphic enantiostylous species when compared to species with a central style corroborated the evolutionary transition from the monomorphic to the dimorphic enantiostyly. The conditions that allowed this transition are the occurrence of large, broadly distributed populations, the presence of effective pollinators, and the predominance of intermorph crosses (Jesson and Barrett 2002a).

Although the evolutionary theories describe reproductive advantages of dimorphic upon monomorphic enantiostylous species, only seven species were so far recorded bearing the former (Fenster 1995; Jesson and Barrett 2003, 2005; Vallejo-Marín et al. 2013; Almeida et al. 2018). The main hypothesis explaining the prevalence of monomorphic species is related to reproductive assurance, since the presence of both morphs plus self-compatibility could guarantee fruit set in environments with a low number of individuals and an inefficient pollination services (Almeida et al. 2015b, 2018).

Almeida et al. (2013b) suggested that the enantiostyly of *C. flexuosa* (Fabaceae) seems to contribute with a higher efficiency of pollen capture on a large extension of pollinators' bodies. Other authors stated that the enhancement of the pollen capture area on the pollinators' bodies minimizes the competition for pollen between stigmas, by means of the opposite placement of stigmas in the floral morphs; therefore, the competition for pollen and the stigma clogging constitute the selective pressures that allowed the evolutionary emergence of the polymorphism (Webb and Lloyd 1986; Neal et al. 1998).

Jesson and Barrett (2003) associated several floral traces with enantiostyly in monocots. Although they assume that the results should be more deeply evaluated, the association with heteranthery and loss of floral nectaries were highly significant; poricidal anthers and zigomorphy were also associated with enantiostyly, however with a lower significant value.

20.5 MATING SYSTEM AND RECIPROCITY

Conversely to what is generally observed for heterostylous species, enantiostylous species are intramorph compatible (Barrett 2002; Almeida et al. 2013a, 2013b, 2015b), and no difference on pollen tube growth are observed between intramorph and intermorph pollinations (Almeida et al. 2015b). Self-incompatibility is observed in some enantiostylous species (Bahadur et al. 1979; Carvalho and Oliveira 2003; Leite and Machado 2010).

Considering that self- and intramorph incompatibility are rare, enantiostylous species rely on flower morphological characters to maximize the chances of intermorph crosses (Barrett 2002; Almeida 2013a; Almeida et al. 2015b). According to what was discussed above, studies regarding intermorph pollen flow recorded that, in fact, pollen deposition and capture occur on opposite areas of pollinators' bodies by each floral morph (Jesson and Barrett 2005; Almeida et al. 2013a, 2013b, 2015b). Possibly this is the reason why the reciprocity is much more commonly recorded.

20.6 FLORAL MORPH PROPORTION

The high dependence of intermorph crosses to produce fruits and seeds led researchers to investigate floral morph proportions in monomorphic (by counting number of flowers of each morph in each individual) and dimorphic (by counting the number of individuals of each floral morph) enantiostylous species (Carvalho and Oliveira 2003; Almeida et al. 2013a, 2013b, 2018). The results showed similar proportion of floral morphs (also called as isoplethy), similarly to what is generally observed in heterostylous species (Teixeira and Machado 2004; Pereira and Barbosa 2006; Lenza et al. 2008; Novo et al. 2018). Almeida et al. (2013b) investigated both morphological (falcate petal turned to the right or to the left) and functional (style turned to the right, to the left, or central) floral morphs of *C. ramosa* (Fabaceae). The species was isoplethic only when the morphology was taken into account; the proportions of the functional morphs functional were different. This is the only case of anisoplethy in enantiostylous species.

20.7 FUNCTIONALLY DIMORPHIC INDIVIDUALS IN MONOMORPHIC SPECIES

According to what was commented above, studies show lower rates of xenogamy in monomorphic enantiostylous species (Jesson and Barrett 2002a, 2003, 2005; Jesson et al. 2003). Some studies found that monomorphic species may exhibit strategies that may favor xenogamy. For example, *Monochoria korsakowa* and *M. vaginalis* (Tang and Huang 2005), *C. flexuosa* (Almeida et al. 2013b), *C. amiciella*, *C. repens*, *Senna macranthera*, and *S. macranthera* var. *pudibunda* (Almeida et al. 2018) produce a low number of flowers daily (around three; Tang and Huang 2005; Almeida et al. 2013b, 2018), a strategy that enhances the chances of producing only one floral morph and, consequently, function as a dimorphic species (Barrett et al. 2000; Almeida et al. 2018).

The tendency of production of only one floral morph (independently on the number of flowers), may be observed in monomorphic enantiostylous species (Tang and Huang 2005; Almeida et al. 2018). Almeida et al. (2018) classified individuals of enantiostylous species of the tribe Cassiinae in relation to the daily production of right and left flowers and established the following types: functionally right (daily production of right flowers was more than twice the production of left ones), functionally left (daily production of left flowers was more than twice the production of right ones), and reciprocal (similar daily production of right and left flowers). The authors observed that individuals may maintain the right/left functional condition for six days, and the reciprocal condition for seven days. However, most individuals maintained the right/left functional condition for approximately three days, therefore varying in relation to the predominant morph that was produced daily. These results indicated that individuals exhibit an alternation of functional conditions, equilibrating the number of floral morphs produced in the populations.

20.8 AGGREGATE ENANTIOSTYLY

Richman and Venable (2018) described a new enantiostylous type in *Allionia incarnata* (Nyctaginaceae) called aggregate enantiostyly. The pollination unit is constituted by three zygomorphic, bisexual flowers that are aggregate in a capitulate umbel inflorescence. The stamens are located at the center of that pollination unit and have no deflection. The three styles are generally arranged like an "Y" (two styles above and

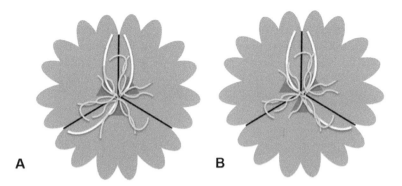

FIGURE 20.2 Typical display of flowers with aggregate enantiostyly of *Allionia incarnata* L., anthers (yellow), and stigmas (white): (a) left and (b) right. (From Richman, S.K., and Venable, D.L., *J. Pollinat. Ecol.*, 22, 49–54, 2018.)

one below), all of them far from the stamens and have a deflection. The lower style may be deflected to the right ("right flowers") or to the left ("left flowers"), in similar proportions in the population (Figure 20.2). The amount of pollen deposited on pollinators' bodies depends on the area in which pollinators land, and no differences were found regarding pollen capture for different floral morphs.

20.9 PERSPECTIVES FOR FUTURE STUDIES

There are several aspects on enantiostyly that deserve attention. The genetics driving floral morphs expression in monomorphic species (since a genetic basis for enantiostyly has been established so far only one taxon; Jesson and Barrett 2002b), similar proportion of floral morphs in the populations and tendency of production of only one floral morph by distinct individuals are quite important.

The predominance and evolutionary maintenance of monomorphic enantiostyly may be investigated by means of experiments concerning invading potential of dimorphic individuals in monomorphic populations. The simulation of distinct evolutionary scenarios is necessary to the comprehension of the real advantages assigned to dimorphic upon monomorphic enantiostyly and vice versa.

Checking morphometrically the exact reciprocity in enantiostyly is difficult when compared to heterostylous species, such as the calculation of the inaccuracy index (Armbruster et al. 2017). Conversely to heterostyly, enantiostyly exhibits several aspects influencing pollen deposition (on pollinators' bodies) and capture (by the stigma) that must be taken into account to establish a measurement pattern for the reciprocity of floral parts, such as the involvement of the commonly curved petals in pollen deposition, style deflection, vibratory movements of bees, and variations of enantiostyly (such as the enantiostylous types, Almeida et al. 2015a).

The variations of enantiostyly are something quite interesting that should be better investigated. The records in this decade of atypical enantiostyly, enantiostylous types, and aggregated enantiostyly reinforce this affirmation. On one hand, the investigation of the enantiostyly in typically enantiostylous taxa is a good option;

the tribe Cassiinae, for example, exhibits a broad morphological variation related to enantiostyly. On the other hand, taxa with still no enantiostylous records may present unexpected diversification. Finally, the investigation on the details concerning the functionality of enantiostyly is a challenge that would help the understanding of this interesting floral polymorphism.

REFERENCES

Almeida, N.M., Bezerra, T.T., Oliveira, C.R.S., Novo, R.R., Siqueira-Filho, J.A., Oliveira, P.E., Castro, C.C. 2015b. Breeding systems of enantiostylous cassiinae species (Fabaceae, Caesalpinioideae). *Flora* 215: 9–15. doi:10.1016/j.flora.2015.06.003.

Almeida, N.M., Castro, C.C., Leite, A.V.L., Novo, R.R., Machado, I.C. 2013a. Enantiostyly in *Chamaecrista ramosa* (Fabaceae-Caesalpinioideae): Floral morphology, pollen transfer dynamics and breeding system. *Plant Biology* 15: 369–375. doi:0.1111/j.1438-8677.2012.00651.x.

Almeida, N.M., Castro, C.C., Novo, R.R., Leite, A.V., Machado, I.C. 2013b. Floral polymorphism in *Chamaecrista flexuosa* (Fabaceae-Caesalpinioideae): A possible case of atypical enantiostyly? *Annals of Botany* 112: 1117–1123. doi:10.1093/aob/mct188.

Almeida, N.M., Cotarelli, V.M., Souza, D.P., Novo, R.R., Siqueira Filho, J.A., Oliveira, P.E., Castro, C.C. 2015a. Enantiostylous types of Cassiinae species (Fabaceae-Caesalpinioideae). *Plant Biology* 17(3): 740–745. doi:10.1111/plb.12283.

Almeida, N.M., Souza, J.T., Oliveira, C.R.S., Bezerra, T.T., Novo, R.R., Siqueira Filho, J.A., Oliveira, P.E., Castro, C.C. 2018. Functional dimorphic enantiostyly in monomorphic enantiostylous species of the subtribe Cassiinae (Fabaceae-Caesalpinioideae). *Plant Biology*. doi:10.1111/plb.12718.

Amorim, T., Marazzi, B., Soares, A.A., Forni-Martins, E.R., Muniz, C.R., Westerkamp, C. 2017. Ricochet pollination in Senna (Fabaceae)—Petals deflect pollen jets and promote division of labor among flower organs. *Plant Biology* 19: 951–962. doi:10.1111/plb.12607.

APG IV. 2016. An update of the Angiosperm Phylogeny Group classification for the orders and families of flowering plants: APG IV. *Botanical Journal of the Linnaean Society* 181: 1–20.

Armbruster, W.S., Bolstad, G.H., Hansen, T.F., Keller, B., Conti, E., Pelabon, C. 2017. The measure and mismeasure of reciprocity in heterostylous flowers. *New Phytologist* 215(2): 906–917. doi:10.1111/nph.14604.

Bahadur, B., Chaturvedi, A., Rama Swary, N. 1990. S.E.M. studies of pollen in relation to enantiostyly and heteranthery in *Cassia* (Caesalpinaceae). Current Perspectives in Palynology Research. *Journal of Palynology* 26: 7–22.

Bahadur, B., Kumar, P.V., Reddy, N. P. 1979. Enantiostyly in *Cassia* L. (Caesalpinioideae) together with SEM studies of pollen and its possible significance. In Reproduction in flowering plants. Abstr. Int. Symp. Christchurch, New Zealand.

Bahadur, B., Kumar, V.K., Reddy, N.P. 1985. Enantiostyly in *Cassia auriculata* L. (Caesalpinaceae). In: Pollination biology: An analysis. Kapil, R. N. (ed). Inter India Publication, New Delhi. pp. 239–24.

Barrett, S.C.H. 1992. *Evolution and Function of Heterostyly (S.C.H Barrett, ed.). Monographs on Theorical and Applied Genetics.* Springer-Verlag, Berlin, Germany.

Barrett, S.C.H. 2002. The evolution of plant sexual diversity. *Nature Reviews Genetics* 3: 274–284. doi:10.1038/nrg776.

Barrett, S.C.H. 2010. Darwin's legacy: The forms, function and sexual diversity of flowers. *Philosophical Transactions of the Royal Society B* 365: 351–368. doi:10.1098/rstb.2009.0212.

Barrett, S.C.H., Jesson, L.K., Baker, A.M. 2000a. The evolution and function of stylar polymorphisms in flowering plants. *Annals of Botany* 85: 253–265. doi:0.1006/anbo.1999.1067.

Barrett, S.C.H., Wilken, D.H., Cole, W.W. 2000b. Heterostyly in the Lamiaceae: The case of *Salvia brandegeei*. *Plant Systematics and Evolution* 223: 211–219. doi:10.1007/BF00985280.

Bezerra, E.L.S. 2008. Guilda de flores de óleo do Parque Nacional do Catimbau: fenologia, polinização e sistema reprodutivo. Tese de Doutorado, Universidade Federal de Pernambuco, Recife.

Carvalho, D.A., Oliveira, P.E. 2003. Biologia reprodutiva e polinização de *Senna sylvestris* (Vell.) (Leguminosae-Caesalpinioideae). *Revista Brasileira de Botânica* 26: 319–328.

Costa, C.B.N., Lambert, S.M., Borba, E.L., Queiroz, L.P. 2007. Postzygotic reproductive isolation between sympatric taxa in the *Chamaecristadesvauxii* complex (Leguminosae-Caesalpinioideae). *Annals of Botany* 99: 625–635. doi:10.1093/aob/mcm012.

Darwin, C. 1877. *The Different forms of Flowers on Plants of the Same Species.* London, UK: John Murray.

Dulberger, R. 1981. The floral biology *of Cassia didymobotrya* and *C. auriculata* (Caesalpiniaceae). *American Journal of Botany* 68: 1350–1360.

Dutra, V.F., Vieira, M. F., Garcia, F.C.P., Lima, H.C. 2009. Fenologia reprodutiva, síndromes de polinização e dispersão em espécies de Leguminosae dos campos rupestres do Parque Estadual do Itacolomi, Minas Gerais, Brasil. *Rodriguésia* 60: 371–387. doi:1590/2175-7860200960210.

Evans, T.M., Faden, R.B., Simpson, M.G., Sytsma, K.J. 2000. Phylogenetic relationships in the Commelinaceae: I. A cladistic analysis of morphological data. *Systematic Botany* 25(4): 668–691. doi:10.1043/0363-6445-28.2.270.

Fenster, C.B. 1995. Mirror image and their effect on outcrossing rate in *Chamaecrista fasciculata* (Leguminosae). *American Journal of Botany* 82: 46–50. doi:10.1002/j.1537-2197.1995.tb15647.x.

Gottsberger, G., Silberbauer-Gottsberger, I. 1988. Evolution of flower structures and pollination in neotropical Cassiinae (Caesalpiniaceae) species. *Phyton* 28: 293–320.

Harrison, C.J., Möller, M., Cronk, Q.C. 1999. Evolution and development of floral diversity in *Streptocarpus* and *Saintpaulia*. *Annals of Botany* 84: 49–60. doi:0.1006/anbo.1999.0887.

Irwin, H.S., Barneby, R.C. 1976. Notes on the generic of *Chamaecrista* Moench (Leguminosae: Caesapinioideae). *Brittonia* 28: 28–36.

Jesson, L.K., Kang, J., Wagner, S.L., Barrett, S.C., Dengler, N.G. 2003. The development of enantiostyly. *American Journal of Botany* 90: 183–195.

Jesson, L.K., Barrett, S.C.H. 2002a. Enantiostyly: Solving the puzzle of mirror-image flowers. *Nature* 417: 707. doi:10.1038/417707a.

Jesson, L.K., Barrett, S.C.H. 2002b. The genetics of mirror-image flowers. *Proceedings of the Royal Society of London B* 269: 1835–1839. doi:10.1098/rspb.2002.2068.

Jesson, L.K., Barrett, S.C.H. 2003. The comparative biology of mirror-image flowers. *International Journal of Plant Sciences* 164: 237–249. doi:10.1086/378537.

Jesson, L.K., Barrett, S.C.H. 2005. Experimental tests of the function of mirror-image flowers. *Biological Journal Linnean Society* 85: 167–179. doi:10.1111/j.1095-8312.2005.00480.x.

Jesson, L.K., Barrett, S.C.H., Day, T. 2003. A theoretical investigation of the evolution and maintenance of mirror image flowers. *American Naturalist* 161: 916–930. doi:10.1086/375176.

Knuth, P. 1906. *Handbook of Flower Pollination.* Oxford, UK: Clarendon Press.

Krieck, C., Finatto, T., Müller, T.S., Guerra, M.P., Orth, A.I. 2008. Biologia reprodutiva de *Alpinia zerumbet* (Pers.) B. L. Burtt & R. M. Sm. (Zingiberaceae) em Florianópolis, Santa Catarina. *Revista Brasileira de Plantas Medicinais* 10(2): 103–110.

Laporta, C. 2005. Floral biology and reproductive system of enantiostylous *Senna corymbosa* (Caesalpiniaceae). *Revista de Biología Tropical* 53(1–2): 49–61. doi:10.15517/rbt. v53i1-2.14361.

Leite, A.V., Machado, I.C. 2010. Reproductive biology of woody species in Caatinga, a dry forest of northeastern Brazil. *Journal of Arid Environments* 74: 1374–1380. doi:10.1016/j. jaridenv.2010.05.029.

Lenza, E. Ferreira, J.N. Consolaro, H., Aquino, F.G. 2008. Biologia reprodutiva de *Rourea induta* Planch. (Connaraceae), uma espécie heterostílica de cerrado do Brasil Central. *Revista Brasileira de Botânica* 31(3): 389–398. doi:10.1590/S0100-84042008000300003.

Lin, Y., Tan, D. 2007. Enantiostyly in Angiosperms and its evolutionary significance. *Acta Phytotaxonomica Sinica* 6: 901–916. doi:10.1360/aps07042.

Lloyd, D.G., Webb, C.J. 1992. The evolution of heterostyly. In *Evolution and Function of Heterostyly* (ed. S. C. H. Barrett), pp. 151–178. Berlin, Germany: Springer-Verlag.

Neal, P.R., Dafni, A., Giurfa, M. 1998. Floral symmetry and its role in plant-pollinator systems: Terminology, distribution, and hypotheses. *Annual Review of Ecology, Evolution, and Systematics* 29: 345–73. doi:10.1146/annurev.ecolsys.29.1.345.

Novo, R.R., Consolaro, H., Almeida, N.M., Castro, C.C. 2018. Floral biology of the velvet-seed *Guettarda platypoda* DC. (Rubiaceae): A typical distyly or style dimorphism? *Flora* 239: 62–70. doi:10.1016/j.flora.2017.11.008.

Pereira, M.S., Barbosa, M.R.V. 2006. A família Rubiaceae na reserva biológica Guaribas, Paraiba, Brasil. Subfamilias Antirheoideae, Cinchonoideae e Ixoroideae. *Acta Botanica Brasilica* 18(2): 305–318. doi:10.1590/S0102-33062004000200010.

Richman, S.K., Venable, D.L. 2018. Aggregate enantiostyly: Floral visitor interactions with a previously unreported form of floral display. *Journal of Pollination Ecology* 22(5): 49–54.

Simpson, M.G. 1990. Phylogeny and classification of the Haemodoraceae. *Annals of the Missouri Botanical Garden* 77: 722–784. doi:10.2307/2399670.

Tang, L.L., Huang, S.Q. 2005. Variation in daily floral display and the potential for geitonogamous pollination in two monomorphic enantiostylous *Monochoria* species. *Plant Systematics and Evolution* 253: 201–207. doi:10.1007/s00606-005-0309-5.

Teixeira, L.A.G., Machado, I.C. 2004. *Sabicea cinerea* Aubl. (Rubiaceae): Distilia e polinização em um fragmento de floresta Atlântica em Pernambuco, Nordeste do Brasil. *Revista Brasileira de Botânica* 27: 193–204. doi:10.1590/S0100-84042004000100019.

Todd, J.E. 1882. On the flowers of *Solanum rostratum* and *Cassia chamaecrista*. *American Naturalist* 16: 281–287.

Vallejo-Marín, M., Solís-Montero, L., Souto Vilaros, D., Lee, M.Y. 2013. Mating system in Mexican populations of the annual herb *Solanum rostratum* Dunal (Solanaceae). *Plant Biology* 15(6): 948–954. doi:10.1111/j.1438-8677.2012.00715.x.

Webb, C.J., Lloyd, D.G. 1986. The avoidance of interference between the presentation of pollen and stigma in angiosperms II: Herkogamy. *New Zealand Journal of Botany* 24: 135–162. doi:10.1080/0028825X.1986.10409726.

Westerkamp, C. 2004. Ricochet pollination in cassias—and how bees explain enantiostyly. In: Freitas, B.M. Pereira J.O.P. (Eds), *Solitary Bees: Conservation, Rearing and Management for Pollination*. Imprensa Universitária, Fortaleza, Brazil, 225–230.

Yu, L., Dun-Yan, T. 2007. Enantiostyly in angiosperms and its evolutionary significance. *Journal of Systematics and Evolution* 45(6): 901–916.

21 Inversostyly
A Still Less Known Reproductive Mechanism

Túlio Freitas Filgueira Sá, Natan Messias de Almeida, and Cibele Cardoso de Castro

CONTENTS

21.1 INTRODUCTION

The patterns of variation on flower morphology constituted one of the main inspirations of Darwin's (1877) studies. The naturalist stated that heterostyly was the floral polymorphism prevalent in angiosperms, a fact that was confirmed all along. Other polymorphisms were also reported, such as the enantiostyly, flexistyly, stigma-height dimorphism, and inversostyly (Darwin 1877; Todd 1882; Wilson 1887; Cui et al. 1996; Barrett 2002; Pauw 2005).

Inversostyly is a floral polymorphism recorded in only one species, *Hemimeris racemosa* (Scrophulariaceae), whose floral resource is the oil (Pauw 2005). The two floral morphs, produced in distinct individuals, exhibit reciprocal differences in the vertical arrangement of male and female organs of the flower (Figure 21.1). Style-down flowers have the style turned to the flower base and anthers turned to the upper portion of the flower, and style-up flowers exhibit the reverse. This reciprocal arrangement maximizes intermorph pollinations because it results in pollen deposition and capture at specific portions of pollinators' bodies; thus, it can be considered a reproductive mechanism that favors cross-pollination and minimizes the chances of autogamy. Considering its morphological and functional features, inversostyly is therefore similar to heterostyly (reciprocal differences of anthers and stigma height Bahadur [1968], Ganders [1979]) and enantiostyly (reciprocal differences in the horizontal arrangement of anthers and stigma, [Barrett 2002]). *H. racemosa* is mainly pollinated by specialized oil-collecting bees of the genus *Rediviva*. It is interesting to note that the proportion of the style-down morph is slightly higher in the populations, and this bias increases with decreasing abundance of pollinators (Pauw 2005).

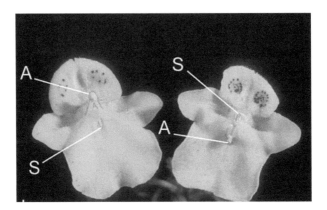

FIGURE 21.1 Style down (left) and style up (right) flowers collected in two adjacent plants of the inversostylous species *Hemimeris racemosa* (Scrophulariaceae). The dorsal petal lobe has been hinged open. A: anthers; S: stamen. (From Pauw, A., *Am. J. Bot.* 92, 1878–1886, 2005.)

Some populations of *H. racemosa* exhibit only a homostylous floral morph, whose male and female whorls are arranged in the down level within the flower and occur in environments with a low abundance of the oil-collecting bees, similarly to what was found for the homostylous *Hemimeris sabulosa*. Curiously, these bees more frequently visited flowers exhibiting inversostyly in relation to homostylous populations. It seems, therefore, that *Rediviva* bees constitute an important selective pressure maintaining the flower polymorphism of *H. racemosa* (Pauw 2005).

Specialized plant-pollinator systems may promote the evolution and the maintenance of flower polymorphisms (Arroyo and Barrett 2000; Pauw 2005; Pérez-Barrales and Arroyo 2010). Homostylous individuals are commonly observed in some populations of distylous species and sometimes is the only morph recorded in a species (Hamilton 1990; Consolaro et al. 2011). It is not known if the homostylous floral morph found in *H. racemosa* constitutes a derived character of inversostyly or if it represents an initial stage of sexual organ displacement that would result in the evolution of two floral morphs, similar to what is believed for the distylous species (Charlesworth and Charlesworth 1979).

Although homostylous populations of *H. racemosa* were less visited by the specialized *Rediviva* bees, they had a high fruit set, possibly due to lack of herkogamy and the presence of self-compatibility, which favors self-pollination and autogamy. Thus, it is possible that the autogamy constitutes a selective force favoring the maintenance of inversostyly in dimorphic populations of *H. racemosa*, similarly to what is believed for heterostylous (distylous and tristylous; Ganders 1979; Barrett 2002) and enantiostylous species (Barrett 2002).

A kind of inversostyly was recorded in two species of the genus *Eplingiella* [*E. cuniloides* and *E. brightoniae*, both (Epling) Harley and Pastore; Lamiaceae; Harley et al. 2017]. The arrangement of floral sexual organs is the same between morphs, but one floral morph is resupinate, thus inverting the arrangement of sexual organs in relation to the other (non-resupinate) floral morph. Floral resupination is a 180° twisting of the floral pedicel during anthesis (Schmucker 1929), and it is commonly observed in orchids and some botanical taxa such as *Salvia jurisicii* (Lamiaceae; Schmucker 1929; Classen-Bockhoff et al. 2004; Gibbs 2017). The presence of two floral morphs resulting from floral resupination is called as resupinate dimorphy and is considered a kind of functional inversostyly, since the sexual organs of the two floral morphs are vertically reciprocal (Gibbs 2017).

Floral symmetry is a particularly important aspect for species presenting some floral polymorphisms. Inversostyly, enantiostyly, and flexistyly are commonly recorded in species with zygomorphic, non-tubular flowers (Barrett 2010). Flower zygomorphy seems to maximize cross-pollinations because of the precision with which the sexual whorls contact specialized pollinators' bodies (Nilsson 1988; Citerne et al. 2010). The bilateral symmetry maximizes floral phenotypic integration in response to the selective pressures imposed by specialized pollinators (Berg 1959, 1960; Rosas-Guerrero et al. 2011; Pérez-Barrales et al. 2014). Therefore, zygomorphic flowers are more susceptible to the effects of a heterogeneous environment and to the positioning of those pollinators (Barrett 2010; Wolowski et al. 2014). Considering that floral symmetry influences the pollination process, it consequently impacts seed set in most plant species, independently of the mating system (from self-compatible to self-incompatible species, Lázaro and Totland 2014).

21.2 FUTURE STUDIES

Studies regarding the genetic control of floral morph production are necessary for the better understanding of inversostyly in *H. racemosa*. The relation between the proportional ratio of floral morphs and the disassortative pollination is not clear, as well as the effectiveness of *Rediviva* bees in maintaining *H. racemosa* polymorphism.

REFERENCES

Arroyo, J., Barrett, S. C. H. 2000. Discovery of distyly in *Narcissus* (Amaryllidaceae). *American Journal of Botany* 87: 748–751.

Bahadur, B. 1968. Heterostyly in Rubiaceae: A review. *Journal of Osmania University (Science). Golden Jubilee Special* 7: 207–238.

Barrett, S. C. 2010. Darwin's legacy: The forms, function and sexual diversity of flowers. *Philosophical Transactions of the Royal Society of London B: Biological Sciences* 365(1539): 351–368.

Barrett, S. C. H. 2002. The evolution of plant sexual diversity. *Nature Reviews Genetics* 3: 274–284.

Berg, R. L. 1959. A general evolutionary principle underlying the origin of developmental homeostasis. *The American Naturalist* 93: 103–105.

Berg, R. L. 1960. The ecological significance of correlation pleiades. *Evolution* 14: 171–180.

Charlesworth, D., Charlesworth, B. 1979. A model for the evolution of distyly. *The American Naturalist* 114: 467–498.

Citerne, H., Jabbour, F., Nadot, S., Damerval, C. 2010. The evolution of floral symmetry. *Advances in Botanical Research* 54: 85–137.

Classen-Bockhoff, R., Speck, T., Tweraser, E., Wester, P., Thimm, S., Reith, M. 2004. The staminal lever mechanism in *Salvia* L. (Lamiaceae): A key innovation for adaptive radiation? *Organisms Diversity and Evolution* 4: 189–205.

Consolaro, H., Silva, S. C. S., Oliveira, P. E. 2011. Breakdown of distyly and pin-monomorphism in *Psychotria carthagenensis* Jacq. (Rubiaceae). *Plant Species Biology* 26: 24–32.

Cui, X. L., Wei, R. C., Huang, R. F. 1996. A study on the breeding system of Amomumtsao-ko. In *Proceedings of the Second Symposium on the Family Zingiberaceae*, 288–296. Zhongshan, China: University Press, Guangzhou.

Darwin, C. 1877. *The Different forms of Flowers on Plants of the Same Species*. London, UK: John Murray.

Ganders, F. R. 1979. The biology of heterostyly. *New Zealand Journal of Botany* 17: 607–635.

Gibbs, P. E. 2017. Head over heels about floral polymorphism: A novel floral dimorphism based on resupination. *Plant Ecology and Diversity* 10(1): 1–3.

Hamilton, C. W. 1990. Variation on a distylous theme in Mesoamerican *Psychotria* subgenus *Psychotria* (Rubiaceae). *Memoirs of the New York Botanical Garden* 55: 62–75.

Harley, R. M., Giulietti, A. M., Abreu, I. S., Bitencourt, C., Oliveira, F. F. D., Endress, P. K. 2017. Resupinate dimorphy, a novel pollination strategy in two-lipped flowers of *Eplingiella* (Lamiaceae). *Acta Botanica Brasilica* 31(1): 102–107.

Lázaro, A., Totland, O. 2014. The influence of floral symmetry, dependence on pollinators and pollination generalization on flower size variation. *Annals of Botany* 114: 157–165.

Nilsson, L. A. 1988. The evolution of flowers with deep corolla tubes. *Nature* 334: 147–149.

Pauw, A. 2005. Inversostyly: A new stylar polymorphism in an oil-secreting plant, *Hemimeris racemosa* (Scrophulariaceae). *American Journal of Botany* 92: 1878–1886.

Pérez-Barrales, R., Arroyo, J. 2010. Pollinator shifts and the loss of style polymorphism in *Narcissus papyraceus* (Amaryllidaceae). *Journal of Evolutionary Biology* 23: 1117–1128.

Pérez-Barrales, R., Simón-Porcar, V. I., Santos-Gally, R., Arroyo, J. 2014. Phenotypic integration in style dimorphic daffodils (*Narcissus*, Amaryllidaceae) with different pollinators. *Philosophical Transactions of the Royal Society of London Series B* 369(1649): 20130258.

Rosas-Guerrero, V., Quesada, M., Armbruster, W. S., Pérez-Barrales, R., Smith, S. D. 2011. Influence of pollination specialization and breeding system on floral integration and phenotypic variation in *Ipomoea*. *Evolution* 65: 350–364.

Schmucker, T. 1929. Blütenbiologische und morphologische Beobachtungen. *Planta* 9: 718–747.

Todd, J. E. 1882. On the flowers of *Solanumrostratum* and *Cassia chamaecrista*. *The American Naturalist* 16: 281–287.

Wilson, J. 1887. On the dimorphism of the flowers of *Wachendorfia paniculata*. *Transactions and Proceedings of the Botanical Society of Edinburgh* 17: 73–77.

Wolowski, M., Ashman, T. L., Freitas, L. 2014. Meta-analysis of pollen limitation reveals the relevance of pollination generalization in the Atlantic forest of Brazil. *PLoS One* 9(2): e89498.

22 More on Buzz Pollination— Pollen Rebounds in Asymmetric Flowers

Thiago Magalhães Amorim, Arlete Aparecida Soares, and Christian Westerkamp

CONTENTS

Regarding the flowers of *Chamaecrista fasciculata*:

> "[…] *The points that are of special interest to us, are the sickle-shaped pistil, the stamens with long, rigid anthers, opening by terminal pores, and most of them pointed toward the incurved petal, which is always on the opposite side of the pistil* […]"
>
> **J. E. Todd**
>
> *American Naturalist*, 1882

"[...] *The pollen being thus forced out of the terminal anther-pores falls either directly upon the bee or upon the lateral petal which is pressed close against the bee's side. In this way the side of the bee which is next to the incurved petal receives the most pollen* [...]"

C. Robertson

Botanical Gazette, 1890

22.1 INTRODUCTION

Most flowering plants rely on animals for reproduction (Willmer, 2011), which visit flowers in search of several resources for their own survival (Proctor et al., 1996) and larvae development (Michener, 2007). In doing so, they inadvertently come into contact with the flower's reproductive organs. Anthers shed pollen on the pollinator's body, which in successive visits is transferred to conspecific stigmas. Thus, pollen and receptive stigma(s) usually occupy positions sufficiently similar to contact the same area on the visitor. This strategy occurs even in flowers with reciprocal reproductive organs, such as heterostyly (Barrett et al., 2000), enantiostyly (Jesson and Barrett, 2002, 2003), inversostyly (Pauw, 2005).

Going back to the mid-19th century, Hildebrand (1865) first reported bees collecting pollen through buzzing flowers of *Paeonia* (Paeoniaceae), with this pollination mode becoming more popularized after Michener (1962)'s description. In the 1980s, a survey by Buchmann (1983) showed that the "buzz-pollination mode" is expected for more than 20,000 angiosperm species. These "pollen-only flowers" (*sensu* Vogel, 1978) conceal the pollen inside tube-like poricidal anthers, in which the dry dusty grains are only accessible through a small pore at the anther tip (Buchmann, 1983) (Figure 22.1a and b). Sonicating bees are the major visitors of the extant flowers and have likely coevolved with these species (Vogel, 1978; Buchmann, 1983; Endress, 1994; De Luca and Vallejo-Marín, 2013; Cardinal et al., 2018). Buchmann and Hurley (1978) describe how bees collect pollen from buzz-pollinated flowers: (1) the bees cling to the anthers and sonicate the flowers using their indirect flight muscles; (2) when vibrated, the contact between the approaching anther walls and pollen within transfers energy to the grains; (3) the grains accumulate more and more energy; (4) by further collisions with the walls and other pollen grains, they finally accelerate towards the anther's pore and is ejected outwardly.

In most flowers, the anthers are the organs used to present (expose) pollen to floral visitors (Proctor et al., 1996; Willmer, 2011), including the poricidal ones (Buchmann, 1983). However, anthers do not always promote a precise placement of pollen on the pollinator body, making use of other floral organs to do so (Yeo, 1993). We have been studying a mechanism named "ricochet pollination" in flowers of genus *Senna* (Fabaceae; Cesalpinoideae) (Westerkamp, 2004; Amorim et al., 2017). Ricochet pollination and similar mechanisms do not seem to be restricted either to the genus *Senna* or to poricidal anther flowers. As indicated in the quotations above, *Chamaecrista* (Fabaceae; Cesalpinoideae), a genus related to *Senna* (Bruneau et al., 2001, 2008, 2013), a pollination role for the "cucullate" (incurved) petal was noticed

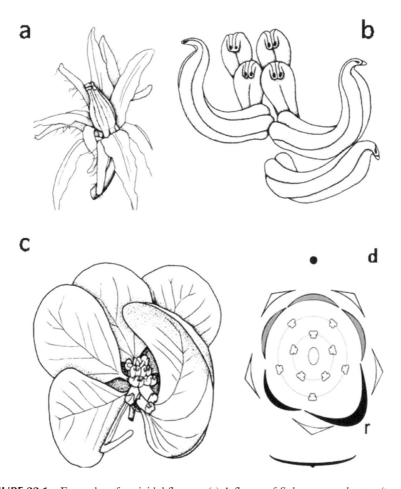

FIGURE 22.1 Examples of poricidal flowers: (a) A flower of *Solanum esculentum* (tomato) showing a central cone of anthers; (b) the asymmetric androecium of *Senna macranthera* showing a clear heteranthery (four middle short and three long lower stamens); (c) an asymmetric flower of *Senna spectabilis* showing short stamens, the long enantiostylous pistil and the asymmetric petal; and (d) a floral diagram representing the asymmetric flower of *Senna spectabilis* (note the asymmetric ricochet petal (r)).

in the late 19th century (Todd, 1882), as well as in more recent studies (Wolfe and Estes, 1992; Almeida et al., 2013). Additionally, many other buzz-pollinated species make use of floral organs other than the anthers to guide the pollen jets to the visitor: these floral organs can mimic the central anther cone of actinomorphic "solanoid flowers" (e.g., Bittrich et al., 1993) or form tube-like structures making the flower asymmetric (Huang and Shi, 2013). The biophysics of all of these systems are expected to function in accordance with the model proposed by Buchmann and Hurley (1978), since the structures providing additional pollen rebounds are expected to function as stamen prolongation (Amorim et al., 2017).

Here we provide a description of ricochet pollination and similar mechanisms in some species bearing asymmetric and symmetric flowers. We combine our own observations (*Senna* and *Chamaecrista*) with data from the literature to discuss these pollination mechanisms in such species. We also discuss and propose a biophysical approach that could allow ricochet pollination to take place.

22.1.1 RICOCHET POLLINATION IN GENUS *SENNA* (FABACEAE)

Although pollen and stigma are commonly leveled within a typical flower, this is not the case in many *Senna* flowers: the pistil can be twice as long as the stamens (Figure 22.1c). In most *sennas*, the pistil is deflected to the left or to the right from the middle plane of the flower (enantiostyly; Figure 22.1c) (Marazzi and Endress, 2008). As a consequence of a long, deflected pistil, the stigma touches the pollinating bees on their dorsal side (thorax or abdomen, depending on the *Senna* species), while the short stamens are all on the ventral side of the bee (Westerkamp, 2004; Amorim et al., 2017). This puzzling mismatch between anther and stylar tips is solved by (mostly asymmetric), that do reach the dorsal side of the pollinating bees and ricochet the pollen jets from the short stamens towards the bee's back (Delgado-Salinas and Sousa-Sánchez, 1977; Westerkamp, 2004; Amorim et al., 2017). After leaving the anthers, the pollen jets strike these petals, and the grains rebound multiple times, on their way to the dorsal side of the bees, allowing pollination.

Marazzi et al. (2006) proposed seven major clades for genus *Senna*. A description of floral architecture of *Senna* flowers is available in the studies by Marazzi et al. (2007) and Marazzi and Endress (2008), and is summarized in Figure 22.3. Species of clade I retain the ancestral traits of flower morphology, showing monosymmetric (zygomorphic) flowers similar to those of genus *Cassia*, the sister group of *Senna* (Marazzi et al., 2006). Monosymmetric flowers are also found in the most derived clade VII, but it is likely a case of reversal (Marazzi et al., 2006, 2007, 2008). Asymmetric (enantiomorphic) flowers are found mostly between clades II and VI. Thus, approximately 70% of genus *Senna* is composed of asymmetric flowers, this trait being affected by calyx, corolla, androecium, and gynoecium. In clade II, represented here by *S. alata*, only the calyx and the deflected pistil contribute to floral asymmetry while androecium and corolla are monosymmetric (Figures 22.2b and 22.3). Between clades III and VI, floral asymmetry is most evident because of the presence of an asummetric (ricochet) petal strongly modified in shape and size relative to the remaining petals (Figures 22.2a and 22.3). Its three-dimensional concave blade curves upwards or downwards depending on the clade or subclade. Asymmetry in the androecium, especially in the lower stamens, is also evident. Commonly, between species of clades III and VI, two stamens are deflected sideways, coming closer to the ricochet petal (Figure 22.3). Species of clade VI show the highest degree of floral asymmetry, in which the three lower stamens are completely deflected towards the ricochet petal (Figure 22.3).

Below, we describe the morphology and the ricochet pollination mechanism in three species of *Senna* with increasing degree of floral asymmetry: *Senna alata*, *S. rugosa*, and *S. splendida*. Further information on the functioning of flowers of *S. alata*, *S. spectabilis* and *S. acuruensis* is described in Amorim et al. (2017).

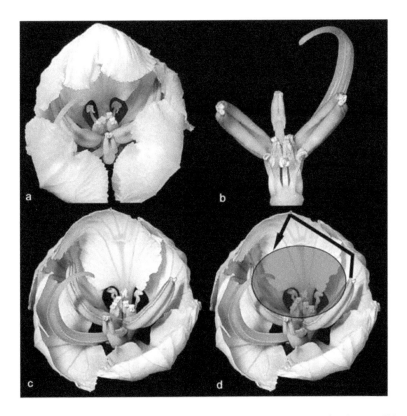

FIGURE 22.2 Flower of *Senna alata*: (a) entire flower; note two staminodes possibly acting as optical guides; (b) incurved style and whole set of stamens: 3 + 4 staminodes, two lateral pollinating anthers, and one central anther offering pollen for bee collection; (c) flower opened to reveal reproductive organs; and (d) same as before, with position of a vibrating bee (oval) and idealized trajectory (arrow) of the pollen jet from one anther.

22.1.1.1 *Senna alata*

In *S. alata* (Figure 22.3a), the 10 stamens are differentiated morphologically (Figure 22.3b) reflecting a division of labor (Müller, 1881, 1882, 1883; Luo et al., 2008; Vallejo-Marín et al., 2009; Vallejo-Marín et al., 2010): pollen for pollination and pollen available as a resource for bee collection. Moreover, only the three lower stamens provide these two functions, with the seven upper stamens reduced to staminodia, some possibly serving as optical guides (Figure 22.3a), and others providing a landing platform or means to secure a grip on the flower. Of the three active stamens (Figure 22.3b), the two laterals have greatly enlarged and curved anthers, whilst the median one is straight and smaller in size (Figure 22.3b and c). All three open via small, distal pores, and pollen is liberated from them by sonication. The ejected pollen forms a cloud, which is focused in a straight jet if the anther is elongated and the pore extended in a narrow tunnel.

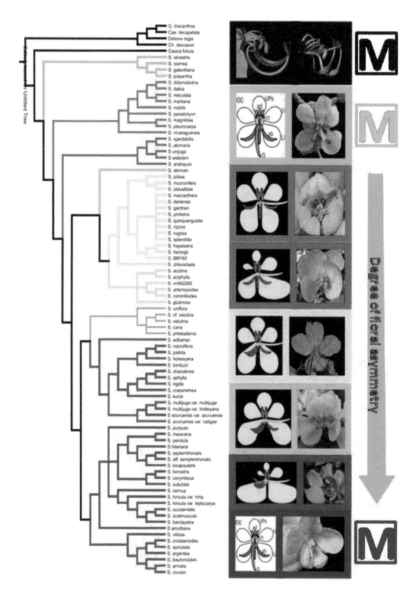

FIGURE 22.3 Part of the floral architecture variation in *Senna*. Phylogenetic three of *Senna* showing the seven major clades proposed by Marazzi et al. (2006) and updated by Marazzi and Sanderson (2010). The outgroups (black branches). The sister group of *Senna* represented by *Cassia* spp. Some representative species for the major clades of *Senna*. **Clade I**: shaded into magenta—*S. siamea*. **Clade II**: shaded into orange—*S. alata*. **Clade III**: shaded into blue—*S. spectabilis*. **Clade IV**: shaded into yellow—*S. macranthera*. **Clade V**: shaded into gray—*S. cearensis*. **Clade VI**: shaded into red—*S. acuruensis*. **Clade VII**: shaded into purple—*S. occidentalis*. M = monosymmetric flowers.

The five petals of the flower restrict access to the stamens (Figure 22.3a) since the distal parts of the two lowermost petals are folded upwards and thus limit the space below the functional stamens, whilst the median upper petal is hooded with the two lateral petals shielding its flanks. This arrangement leaves an aperture, which is about the diameter of the legitimate pollinators, females of large carpenter bees (*Xylocopa*, Hymenoptera: Apidae; compare Figure 22.3a and c).

Depending on the size of the species, these bees introduce the head and thorax, or head only, and secure the stamens as a perch. In this position, all stamens are below the bee's body, and the curved style allows the stigma to touch its back. Since the bees are slender relative to the inner diameter of the flower, there is a space around their body within the floral chamber (Figure 22.3d).

To collect pollen, the bee vibrates the flower. Only a small cloud of pollen from the lowermost, median anther is ejected directly onto the bee's body, effectively between the legs and lower thorax. A larger amount of pollen, however, is deposited on the back of the bee, arriving there by an astonishingly circuitous route (Figure 22.3d): the pollen cloud ejected from each of the two lateral anthers first hits the lateral (flank) petal on its own side, and from here it is ricocheted towards the upper hood petal, from where it is again ricocheted, this time in the direction of the bee's upper surface—at the side different from the anther of origin. Since the three reflectors involved in this process are three-dimensionally curved petals, neither the diameter of the pollen jets or their exact trajectory is predictable. Moreover, there are two pollen jets from the two large lateral anthers, and these will interfere with each other above the bee, resulting in a deposition also in the midline of the bee's back. When *S. alata* flowers are visited by the large bee *Xylocopa grisescens*, pollen is only deposited in a small triangular area on the top of the insect's head. With visits by the somewhat smaller *X. frontalis*, pollen is dusted in an area extending from the head over the first abdominal segment and reaching the wing bases on both sides. On the back of the bee, the pollen is well-hidden by the hood petal and thus unavailable for mid-leg gleaning while sitting in the flower. Thus, much pollen for pollination is separated from a minor amount for bee collection that is on the ventral side and within reach of the bee's legs.

Since the bee effectively fills the floral entrance, it is, however, impossible to observe the flux of pollen within the flower during visits. Our experiments revealed that when the anthers were artificially sonicated with the adjacent flank petal removed, the jet of pollen was shot through this gap. When the upper hood petal was removed, the pollen jet initially ricocheted from the flank petal now exited from the flower via this space. Only with both petals in situ was the pollen able to bounce twice and achieve a trajectory analogous to that during bee visits.

22.1.1.2 *Senna rugosa* and *Senna splendida*

Senna rugosa (Figure 22.4a) and *S. splendida* (Figure 22.4b) have similar floral architecture. Here, the floral standard is made up of the three uppermost petals. One lower petal is asymmetric, spoon-shaped, grows downwardly, and is always opposite to the pistil (Figure 22.4a and b). The androecium is composed of three upper staminodes,

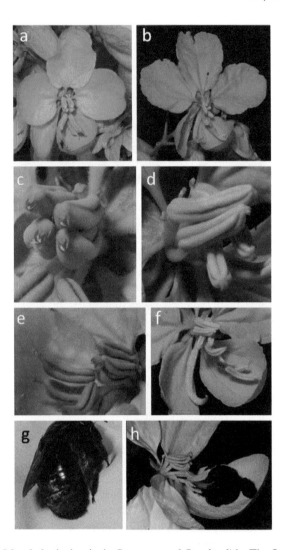

FIGURE 22.4 Morphological traits in *S. rugosa* and *S. splendida*. The flower of *S. rugosa* (a) and *S. splendida* (b) (note the lower deflector petal). The feeding stamens of *S. rugosa* (c) and *S. splendida* (d) with downwardly oriented apical pores. The pollinating stamens of *S. rugosa* (e) and *S. splendida* (f) approaching the deflector petal towards which they point their pores. In (g) is a large *Xylocopa* bee dusted on the dorsal abdomen after visiting a flower of *S. rugosa*, while in (h) is a mid-size *Bombus* bee sonicating a flower of *S. splendida*.

four central and three lower stamens. In both species, the pistil is displaced sideways, grows downwardly, curves upwardly, and levels with the tip of the asymmetric petal.

In *S. rugosa*, at the tip of the central stamens is a short, downward tube opening through a single pore resembling an open mouth (Figure 22.4c). The lower stamens are nearly the same size but half the size of the pistil (Figure 22.4a). Two of them are displaced to the inner blade of the asymmetric petal, while the remaining one is on the same size of the pistil, but the anther is twisted so its tip points towards the asymmetric petal (Figure 22.4a and e). At the tip of the anthers, two well-separated pores open on the dorsal side of a tube (Figure 22.4a and e).

In *S. splendida*, the tip of the central stamens is oriented downwardly through two short, fused tubes bearing a single pore each (Figure 22.4d). One lower stamen always exceeds the others, and together with the neighboring one, approaches and point their tips to the inner blade of the asymmetric petal. These two stamens are nearly the same size of the pistil. The remaining stamen, although in the same side of the pistil, is the smallest, but the anther can be twisted and have its tip pointed towards the asymmetrical petal. In all three lower stamens, two well-separated pores are elevated by a short tube (Figure 22.4f).

In both these species, the apical tubes of the anthers and stamen arrangements determine the direction of the released pollen jets. The short tubes on the tip of the central anthers drive the pollen jets downwards. As for the lower stamens, the two laterally displaced ones eject pollen to the asymmetric petal. The two laterally displaced stamens eject pollen to the tip of the asymmetric petal, while the opposing one ejects pollen to the center of this petal. The asymmetric petal deflected the pollen jets from the lower stamens. The pollen leaves the flower through the tip of the petal. The asymmetric petal was equally able to deposit pollen on the surrogate bee's back (Figure 22.4g). They are mainly visited by large female *Xylocopa* bees, which we considered as the potential legitimate pollinators (Figure 22.4h). The bees visit flowers of both morphs and seem not to discriminate between them. The bees cling to the central stamens but may use the asymmetric petal as a perch. The asymmetric petal comes close to the dorsal abdomen of the bees (Figure 22.4h), where the stigma touches, and the pollen load is deposited on this part of the bee's body (Figure 22.4g). Bees usually do not try to groom this area while perched on the flower. Pollen from the central stamens is deposited on the ventral parts of the bees.

22.1.2   RICOCHET TRAITS IN *SENNA*

In genus *Senna*, ricochet pollination seems to be strongly associated with the traits that lead to floral asymmetry. In most ricocheting *sennas*, a single highly asymmetrical petal stands out, and the pollinating stamens tend to be deflected to this petal (Figure 22.3). Ricochet traits involving corolla, androecium, and gynoecium (Amorim et al., 2017) seem to have evolved independently multiple times (>60% of genus *Senna*), as indicated by phylogenetic (Marazzi et al., 2006) and morphological studies (Marazzi et al., 2007; Marazzi and Endress, 2008) (Figure 22.3). If the flowers of different clades have dissimilar developmental pathways (e.g., Tucker, 1996), all allowing ricochet pollination at anthesis, then it is a candidate for floral convergence in genus *Senna*.

A step further is determining whether ricochet pollination is a key innovation in *Senna*. If so, this mechanism may have provided an accelerated speciation rate in the genus. Any key innovation must be supported by empirical studies showing adaptive functions of ricochet pollination. The biological roles of ricochet pollination are yet to be fully comprehended. We showed that symmetric and asymmetric deflector petals work as a functional extension of the pollinating stamens (Amorim et al., 2017). Along with the typical heteranthery (shorter feeding stamens versus longer pollinating ones) of mostly monosymmetric flowers, ricochet pollination seems to maintain the division of labor between floral organs in the whole genus. However, heteranthery alone in monosymmetric flowers seems to be efficient at segregating pollen into feeding and pollinating functions (Mesquita-Neto et al., 2017). Why the pollinating stamens are shortened in many ricocheting *sennas* while asymmetric deflector petals partially assume their role, is a puzzle yet to be solved.

22.1.3　RICOCHET POLLINATION IN GENUS *CHAMAECRISTA* (FABACEAE; CAESALPINOIDEAE)

Chamaecrista species were formerly included in the subtribe Cassiinae and considered a sister genus of *Cassia* and *Senna* (Irwin and Barneby, 1982). Recent phylogenetic studies suggest the *Cassia*-clade, in which *Batesia* is sister to *Chamaecrista*, while *Melanoxylon* is the sister group of *Cassia* and *Senna* (Bruneau et al., 2001, 2008, 2013). In *Chamaecrista*, enantiomorphic flowers achieve high levels of asymmetry, mainly caused by one petal (the cucullus) and an opposing pistil. Evidence of this petal taking part in pollen deposition on the bee body goes back to the late 19th century (Todd 1882; see the quotations above). Further studies show the role of this petal in guiding pollen to the bee body (Gottsberger and Silberbauer-Gottsberger, 1988; Wolfe and Estes, 1992; Westerkamp, 2004; Costa et al., 2007; Almeida et al., 2013; Costa et al., 2013).

In *C. flexuosa*, five upper stamens (Figure 22.5a) serve for bee foraging, while the remaining five and the pistil rest along the asymmetric petal (Figure 22.5b). Pollen is ejected towards the asymmetric petal and guided to the bee's back.

In *C. desvauxii*, the asymmetric petal is positioned to the left or to the right in opposition with the pistil (Figure 22.5c). Seven anthers point their apical pores to the asymmetric petal, and the remaining three point outwards, functioning as feeding anthers (Figure 22.5c). A very similar construction occurs in *C. ramosa*, in which the asymmetric petal works together with two upper petals to form a functional arch to guide pollen to the bee's back (Almeida et al., 2013).

In *C. fagonioides*, the asymmetric petal is a folded structure (Figure 22.5d). The 10 short stamens form a central cone and converge their apical pores to the base of the asymmetric petal. The vibrations cause pollen grains to pass through the folded petal and come out at its tip towards to bee body (Figure 22.5d). A similar floral architecture and flower functioning are described for *C. hispidula* and *C. campestrris* (Gottsberger and Silberbauer-Gottsberger, 1988). The most complex case occurs in *C. planifolia* (Figure 22.5g) and flowers with similar construction. Here, the asymmetric petal is completely twisted into a tube-like structure with a functional apical pore. At its base, all 10 stamens converge their apical pores. This tube-like petal guides the pollen jets up to the back of the bees. The pistil is on the opposite side with the stylar tip at the same height of the tube-like pore (Figure 22.5f).

22.1.4　POLLEN REBOUNDS IN SOME ANGIOSPERM FAMILIES

Up to this point, we have described the ricochet pollination in *Senna* and *Chamaecrista*. These are buzz-pollinated flowers bearing poricidal anthers. *Senna* and *Chamaecrista* were assigned by Vogel (1978) to the "solanum type" in his general

FIGURE 22.5 Part of the floral architecture variation in *Chamaecrista*: *C. flexuosa* in a front (a) and above (b) view; (c) right-styled flower of *C. desvauxii* showing the asymmetric petal towards which seven stamens are pointing their pores; (d) left-styled flower is being visited by a tiny Halictidae (note the asymmetric petal around the bee); (e) flower of *C. fagonioides*; (f) close look of the folded asymmetric petal and all stamens converging their pores to its base; (g) flower of *C. planifolia*; and (h) a detail of the complex involving the tube-like petal, androecium and gynoecium of *C. planifolia*.

classification of pollen-only flowers. A "solanoid flower" (*sensu* Faegri, 1986) has a central cone formed by poricidal anthers. In the case of *Senna* and *Chamaecrista*, flowers also show different degrees of asymmetry, in which part of the androecium are specialized into bee foraging, and asymmetric petals can ricochet pollen towards pollinating spots on bees, allowing pollination. Some poricidal species seem to use petals or petal-like structures to guide the pollen jet. However, some species bearing longicid anthers can not only mimic typical solanoid flowers (mostly monosymmetric) by using different floral organs to make a centrally pored cone, but also use organs to mimic the deflector petals described for *Senna* and *Chamaecrista* flowers (mostly asymmetric). Across the angiosperms, ricochet-like mechanisms seem to have evolved, but overlooked as pollination mechanisms. Thus, our aim here is to call attention to potential floral systems making use of pollen ricochet.

In the Old-World genus *Acridocarpus* (Malpighiaceae) (Figure 22.6a), the flowers are poricidal and resupinate. Thus, the typical standard petal for this family (Davis et al., 2014) does not exist, resulting in two upper and three lower petals (Vogel, 1990). For example, in *A. natalitius*, the two upper petals are folded, as in *S. alata*. At least the two uppermost stamens of the outer whorl are covered by these petals, towards which they point their pores. The remaining stamens point their pores upwards towards the bee ventral side. Here, the bees must land on all stamens, the feeding ones being mainly those of the inner whorl. The style is bifid and each of its branch curves upwards so that the two stigmas touch the back of the bees, both on the left and on the right side at once. Moreover, the height of the stylar tips is the same as the tips of the two upper, folded petals. Here, the two upper petals probably reflect pollen jets coming from the two adjoining anthers and guide them to the back of the bees.

In *Exacum affine* (Gentianaceae) (Figure 22.6b), the three lower petals are slightly concave. There are four bright yellow, downwardly curved, middle, feeding stamens that stay below the visiting bees. A single lower pollinating stamen is slightly displaced sideways opposite the pistil, pointing their pores to one lower petal lobe. Flowers are enantiostylous. The pistil is longer than the pollinating stamens and touches the flanks of the visiting bee (Russell et al., 2015). In laboratory experiments with a tuning fork by Dr. S. Buchmann and colleagues, when the pollen jet hit the lower petal, it was clearly deflected in the opposite direction (see a video at http://www.anneleonard.com/buzz-pollination/). Pollen from the displaced stamen, as well as some form the feeding stamens, is expect to rebound after hitting the petal towards the flanks of the bee, effecting pollination.

Petalostylis (Fabaceae; Caesalpinoideae) (Figure 22.6c) is an endemic Australian genus which, along with *Labichea*, forms the subtribe Labicheinae (Cassieae) (Lewis, 2005). A detailed study on the development and morphology of *Petalostylis* is provided by Tucker (1998). The most impressive structure is the incurved petal-like style, being its side broadened by inflated lateral ridges. The three fertile stamens have curved anthers, with V-shaped pores directed to the petal-like style. An amount of this pollen is expected to hit the petal-like style and ricochet in a way similar to the observed in flowers of *Senna* and *Chamaecrista*.

In the Central Amazonian *Ternstroemia* spp. (Pentaphylacaceae) (Figure 22.6d), five imbricate, basally fused, petals converge into a cone, enclosing up to 40 longitudinally dehiscing anthers. The tips of the petals are caniculate and create

FIGURE 22.6 Flowers showing ricochet-like mechanisms (arrows indicate distinct floral allowing rebounds of pollen): (a) *Acridocarpus natalitius*, source: http://www.efloras.org/object_page.aspx?object_id=106281&flora_id=800); (b) *Exacum affine*, source: http://www.ross.no/communicate/2015/02/08/exacum-affine/); (c) *Petalostylis cassioides*, source: https://www.flickr.com/photos/72842252@N04/11526668003/in/photostream/); (d) *Ternstroemia lineata*, source: https://www.inaturalist.org/taxa/273854-Ternstroemia-lineata); (e) *Pedicularis groenlandica*, source: https://alchetron.com/Pedicularis-groenlandica#demo); and (f) *Cochliostema odoratissimum*, source: http://commons.hortipedia.com/wiki/File:Cochliostema_odoratissimum_flower_photo_file_PDB_130KB.jpg.)

an extra space for pollen exiting around the protruding style. Pollen is shed into the cone, which guide the grains to the pollinator body during vibratory pollen collection (Bittrich et al., 1993). In many species of Ochnaceae (e.g., *Adenarake muriculata*, *Tyleria bicarpellata*, *Ty. pendula*, etc.), the cone is formed either by half-closed petals or by flat staminodes, which enclose poricidal anthers. The stigma(s) protrude through the apical pore of the cone. Under the vibrations of the bees, the pollen is expelled from the anthers and continues to bounce through the cone up to its apical pore, and eventually to the bee body (Kubitzki and Amaral, 1991; Matthews et al., 2012).

In many *Pedicularis* species (Orobanchaceae) (Figure 22.6e), petal-mediated pollen jets have also evolved, but in a different manner (Faegri and van der Pijl, 1979). The flowers have tubular corollas, with a trilobate lower lip and an upper lip folded into a hood-like or helmet structure through the fusion of the two upper petals (the galea). The galea has a basal opening and covers four introrse anthers and the style, in which the stigma protrudes through the apical pore of the galea. In buzz-pollinated flowers of *Pedicularis*, the pollen is early-shed inside the galea. When gathering pollen, bumblebees vibrate the flowers, causing the grains to accelerate, ricochet off along the tube-like structure, and eventually escape through the apical pore of the galea to the bee's body (Macior, 1968, 1983; Faegri and van der Pijl, 1979; Huang and Shi, 2013).

In *Cochliostema odoratissimum* (Commelinaceae) (Figure 22.6f), the filaments of two of the three stamens outgrow and form a tube-like structure (about the same length of the opposed style) enclosing the three helically twisted, longitudinal anthers and opening through a pore at the tip. Pollen falls inside the tube, and when the bees vibrate the flower, it is canalized through the tube to the insect's body (Hardy et al., 2000; Endress, 1994).

22.1.5 Ricochet Pollination Is a Special Type of Secondary Pollen Presentation

We interpret ricochet pollination as a type of transference of function in angiosperms (Corner, 1958), since the petals partially assume the role of delivering pollen onto the pollinator body. Particularly, the transference of function occurs when pollen ricochet becomes an example of secondary pollen presentation (2PP; Yeo, 1993), where pollen is not deposited on the visitor's body by direct contact with the anther (as in "normal" pollination). Here, however, pollen is not always (but see Almeida et al., 2013) statically exposed by another floral organ as in typical 2PP. Rather, the pollen arrives on the bee's back by a process that can be characterized as virtual 2PP (Westerkamp, 2004).

In at least 25 angiosperm families, pollen presentation and exportation occur on floral structures other than the anthers—a likely independently evolved mechanism called "secondary pollen presentation" (2PP; Yeo, 1993). Flowers can even present their pollen in tertiary organs (e.g., *Polygala myrtifolia*, Westerkamp and Weber, 1997). Howell et al. (1993) identify nine modes of 2PP according to which organ presents the pollen, whether the grains are exposed or concealed, and how pollen is loaded onto the presenter: (1) enveloping bract (pherophyll) presentation (e.g., Araceae); (2) perianth presentation (e.g., Sterculiaceae); (3) anther filament presentation (e.g., Santalaceae); and (4) style presentation (six different types in Marantaceae, Cannaceae, Lobeliaceae, Campanulaceae, Asteraceae, Calyceraceae, Proteaceae, Fabaceae, Myrtaceae, Rubiaceae, Goodeniaceae, Brunoniaceae, and Polygalaceae). Recently, Fan et al. (2015) have described a novel mode of 2PP in which pollen is presented on the anther crest of *Zingiber densissimum* (Zingiberaceae).

In most instances of 2PP, pollen is statically exposed by the secondary floral organs, even in those species with an explosive movement of the style, such as in some Fabaceae (Faboideae) (Castro et al., 2008) and many Marantaceae (Claßen-Bockhoff and Heller, 2008). 2PP mostly occurs in nectar flowers, while in pollen-only flowers (Vogel, 1978) it appears to be restricted to Fabaceae (Faboideae; e.g., *Coronilla*, *Ononis spinosa*, *Lupinus luteus*, tribe Genisteae) and Proteaceae (e.g., *Isopogon dubius*, *Petrophile biloba*, and *Grevillea pilulifera*) (Yeo, 1993).

In a broad sense, if the specialized organs work as a functional extension of the poricidal anthers, these are clear examples of 2PP in buzz-pollinated flowers. These findings raise the question of whether and why 2PP driven by ricochet-like mechanisms has evolved in many buzz-pollinated species in the angiosperms.

22.1.6   The Functioning of the Ricochet Pollination Is Still a Mystery—*Senna* as a Study Model

We combine our own observations with data provided by the literature to propose a model for the functioning of ricochet pollination in genus *Senna*. A biophysical model for buzz pollination was elegantly developed by Buchmann and Hurley (1978). The vibrations of the bee are transmitted to the whole flower and cause the anthers to shake. The anthers function as resonating chambers allowing mature pollen grains to elastically bounce around, detaching themselves from the anther walls and from other pollen grains. The grains gain more and more kinetic energy and eventually escape through the anther pores. In our studies, when the jets of pollen grains leave the pollinating anthers, they go straight towards the inner surface of a deflector petal. In *sennas* studied by us, according to the physical laws of reflection, the grains ricochet off the petal surface and strike an upper area of that same petal (or of another petal in the case of *S. alata*), where a new ricochet takes place, repeating this pattern until the grains eventually reach the bee's back (Figure 22.7a).

For a ricochet to take place, an oblique incidence (collision) angle is necessary, which coincides with the actual direction of the pollen jets with respect to the deflector petal. Assuming a total conservation of the kinetic energy, a specular reflection is expected (incidence angle equals reflection angle; Figure 22.7c). In this simplified model, based on the physical laws of collision, each pollen grain ricochets off the deflector petal similar to what happens to a billiard ball colliding multiple times with the billiard table rails. The inner surface of the deflector petals, however, is not smooth, but has a superficial ultrasculpturing (Figure 22.7b), commonly found in species of *Senna* (Ojeda et al., 2009). Such bumps introduce the variable "wall roughness" into the system, which may affect the final reflexion (ricochet) angle (Konan et al., 2009). The wall roughness component also excludes the possibility of a pollen sliding along the petal surface. Such wall roughness is also found within the anthers: a non-cellular ultrasculpturing made of sporopollenin known as orbicles or Übisch bodies (Buchmann et al., 1977; Kreunen and Osborn, 1999). Buchmann and Hurley (1978) proposed that these minute projections might "randomize" the movement of the colliding pollen grains. According to Knudsen's cosine law, the final reflexion angle actually falls within a distribution of probabilities that can go even beyond the theoretical specular ricochet angle (Figure 22.7d) (Knudsen, 1952;

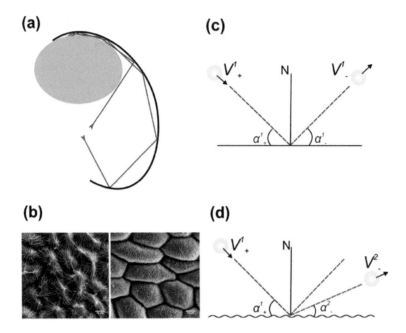

FIGURE 22.7 Schematic representation of the pollen ricochet mechanism in *Senna*: (a) scheme of *S. spectabilis* showing the trajectory of two pollen grains from the lowermost (red trajectory) and lateral (blue trajectory) stamens; (b) ultrasculpturing of the inner surface of the ricochet petal of *S. spectabilis* (left panel) and *S. alata* (right panel); (c) theoretical specular reflexion of the pollen grain on a smooth surface; and (d) theoretical non-specular reflexion of the pollen grains on rough-walled surface. N is the normal. V1+ and α1+ represent the directional velocity and the incidence angle before the collision. V1− and α1− represent the opposite directional velocity and the reflexive angle after the collision. V2− and α2− represent the opposite directional velocity and the reflexive angle after a collision in a non-specular system. The waved solid line represents the inner surface of the ricochet petal.

Morais, 2011). If this is true, the jet of pollen approximates more and more to the petal surface, and becomes more distant from the bee surface. This would likely prevent the pollen from being captured by the flanks of the bee before reaching their dorsal side, where pollination takes place. Knudsen's cosine law is based on the principles of fluid dynamics and considers the whole mass of pollen instead of a single pollen grain.

The velocity of the pollen grains is an important variable in buzz-pollinated systems (Corbet and Huang, 2014). The velocity of the grains must be sufficient to reach the pollinator body after leaving the anthers. However, some intrinsic factors can interfere in the velocity of the grains and must, therefore, be overcome. In Buchmann and Hurley's (1978) model, at the beginning of the vibrational pulse, still inside the anther, the grains ricochet elastically with the same opposite velocity and have the maximum amount of kinetic energy because the coefficient of restitution equals 1. After leaving the anthers, the pollen grains drastically lose their kinetic energy (Buchmann and Hurley, 1978). In the case of *Senna*, while the pollen is driven to the deflector petal(s), such energy loss is expected to diminish the capacity of pollen to ricochet off the petal surface. Additionally, the wall roughness component decelerates the pollen grain (Kussin and Sommerfeld, 2002). Since the flying bees become positively charged owing to the friction with the air, the negatively charged pollen grains are naturally attracted to the bee body (Erickson and Buchmman, 1983; Gan-Mor et al., 1995; Vaknin et al., 2000, 2001). Again, the velocity of the grains is expected to diminish, and so does their capacity to rebound elastically and reach pollinating area on the bee's back. Electrostatic forces are, however, essential in buzz-pollinated systems, since it ensures a pollen-bee-stigma movement, which is extremely adaptive for plants (Erickson and Buchmman, 1983; Vaknin et al., 2000, 2001). So, how do *Senna* flowers solve this problem?

Because in nature flowers of *Senna* are effectively pollinated, another source of energy must exist to overcome the decelerating effects of wall roughness and physical forces on the pollen grains. The vibrating force of the bee is sufficient to sonicate the whole flower that thus functions as a vibrating bee-flower-system in its entirety. This new energy input causes the deflector petal(s) to shake vigorously. When the decelerated grains strike the deflector petal(s), it is hit back (like a baseball bat), then restoring the velocity necessary for successive pollen rebounds. In a recent work using buzz-pollinated flowers of *Pedicularis*, Corbet and Huang (2014) showed that multiple collisions with the walls of the tube-like petal can increase the negative charge of the grains, helping the pollen to firmly adhere onto the bee body. Such triboelectric properties likely play a role in ricocheting flowers: after leaving the anthers the extra collisions on the petal surface must increase the negative charge of the grains, enabling them to firmly adhere on the pollinating area of the bee.

Another important issue is considering the resonant properties of both flowers and pollinators. Parameters such as mass, stiffness, tuning, and damping affect the natural frequencies of flowers (Harder and Barclay, 1994; King and Buchmann, 1996). Bees are capable of adjusting their frequency and amplitude, depending on the context, affecting the amount of pollen removed from the anthers (Burkart et al., 2011; De Luca et al., 2013, 2014; Morgan et al., 2016; Switzer and Combes, 2016). Amplitude differs from frequency since it has to do with the intensity of the vibrational pulse, and it is often measured in velocity or acceleration units (De Luca et al., 2013, 2014). For instance, if only the frequency is taken into account, then a tiny bee, which produces frequencies at higher pitches (Burkart et al., 2011) could steal all the pollen from a large flower. In our studies on *Senna*, we have considered *Xylocopa* bees as the proper pollinators. These large bees can produce a strong vibrational intensity high enough to dislodge a greater amount of available pollen (De Luca et al., 2013). An interesting question is: can the sonication produced by bees be adjusted to activate the ricochets in our studied *sennas*? Species of *Senna* bear flowers of contrasting sizes (Delgado-Salinas and Sousa-Sánchez, 1977) and are worked and pollinated by bees of the corresponding sizes. It would be interesting to test the hypothesis that the larger the flower, the greater must be the vibrational intensity to activate the ricochets, meaning that larger bees are the effective pollinators of larger flowers and smaller bees are better pollinators of smaller flowers. This would support the hypothesis of an adaptive radiation in current floral diversity of *Senna* (Delgado-Salinas and Sousa-Sánchez, 1977).

22.1.7 CONCLUDING REMARKS

Our survey on ricocheting flowers includes taxa from Monocotyledons and Dicotyledons: Commelinales (Commelinaceae), Malpighiales (Malpighiaceae, Ochnaceae and Orobanchaceae), Ericales (Pentaphylacaceae), Fabales (Fabaceae; Labicheineae), and Gentianales (Gentianaceae). In these species, flowers can take advantage of all floral structures, either alone or in combination: sepals, petals, staminodes, filaments, anther connectives, and styles. In three cases, concave petals (*Acridocarpus*), corolla lobes (*Exacum*), and a petal-like style (*Petalostylis*) seem to function similar to the deflector petals of *Senna* and some *Chamaecrista* species. Experimental studies are needed to determine whether: (1) pollen jets are ejected to and deflected by these structures; and (2) the flowers rely on such structures to deposit pollen on areas on the pollinator body touched by the stigma. Pollen can also be guided either by a porous cone formed by corolla or staminodes (*Adenarake, Salvagesia, Tyleria*) or by a tube-like structure formed by a petal (*Chamaecrista*), petal lobes (*Pedicularis*), or anther connectives (*Cochliostema*). In *Chamaecrista* and *Cochliostema*, the pistil is deflected either to the right or to the left (enantiostyly), while in the remaining species, the pistil is included within the cone, with the stigma protruding through the apical pore. Under the bee buzzing, pollen is canalized along the cone or the tube, escapes through the apical pore, and is eventually shed on the bee body. Distinct floral structures serving as resonating chambers and canalizing pollen in buzz-pollinated flowers from non-related angiosperm families is a clear example of floral convergence.

The biological meanings of ricochet mechanisms are still speculative. Buzz-pollinated flowers are, thus, open and challenging systems yet to be explored. What is the ecological and evolutionary significance of this pollination mechanism in buzz-pollinated flowers? Specifically, why should these species combine pollen jet with ricochet? We list some non-mutually exclusive hypotheses. By avoiding the formation of long stamens, flowers could allocate energy to produce more pollen grains. Stout stamens with large anthers could work as better resonating chambers. Flowers with short stamens could rely on other floral organs, such as petals, to promote division of labor. Pollen is, then, deviated towards "safe sites" on the bee's back. Concerning tube-like structures, they unite pollen from all anthers and direct the jets independently from the stamens. A single jet could deposit pollen more precisely on the bee body. Long tubes increase the number of collisions of pollen grains, which can change their electrostatic properties. The functioning of ricochet pollination is likely more complex than thought so far, since the hypotheses put forward above do not support a simplistic ricochet in a two-dimensional view. We hope to have set the stage to explore the functioning of ricocheting flowers in the angiosperms.

REFERENCES

Almeida, N. M., C. C. Castro, A. V. Lima Leite, R. R. Novo, and I. C. Machado. 2013. "Enantiostyly in *Chamaecrista ramosa* (Fabaceae-Caesalpinioideae): Floral Morphology, Pollen Transfer Dynamics and Breeding System." *Plant Biology* 15(2): 369–375.

Amorim, T., B. Marazzi, A. A. Soares, E. R. Forni-Martins, C. R. Muniz, and C. Westerkamp. 2017. "Ricochet Pollination in *Senna* (Fabaceae)—Petals Deflect Pollen Jets and Promote Division of Labour among Flower Organs." *Plant Biology* 19: 951–962.

Barrett, S. C., L. K. Jesson, and A. M. Baker. 2000. "The Evolution and Function of Stylar Polymorphisms in Flowering Plants." *Annals of Botany* 85: 253–265.

Bittrich, V., M. C. E. Amaral, and G. A. R. Melo. 1993. "Pollination Biology of *Ternstroemia laevigata* and *T. dentata* (Theaceae)." *Plant Systematics and Evolution* 185: 1–6.

Bruneau, A., F. Forest, P. S. Herendeen, B. B. Klitgaard, and G. P. Lewis. 2001. "Phylogenetic Relationships in the Caesalpinioideae (Leguminosae) as Inferred from Chloroplast TrnL Intron Sequences." *Systematic Botany* 26(3): 487–514.

Bruneau, A., J. J. Doyle, P. Herendeen, C. E. Hughes, G. Kenicer, G. Lewis, B. Mackinder et al. 2013. "Legume Phylogeny and Classification in the 21st Century: Progress, Prospects and Lessons for Other Species-Rich Clades Legume Phylogeny and Classification in the 21st Century: Progress, Prospects and Lessons for Other Species-Rich Clades." *Taxon* 62: 217–248.

Bruneau, A., M. Mercure, G. P. Lewis, and P. S. Herendeen. 2008. "Phylogenetic Patterns and Diversification in the Caesalpinioid Legumes." *Botany* 86: 697–718.

Buchmann, S. L. 1983. "Buzz Pollination in Angiosperms." In *Handbook of Experimental Pollination Biology*, edited by Jones, C. E. and Little, R. J., 73–113. New York: Van Nostrand Reinhold.

Buchmann, S. L, C. E. Jones, and L. J. Colin. 1977. "Vibratile Pollination of *Solanum douglasii* and *S. xanti* (Solanaceae) in Southern California." *Wasmann Journal of Biology* 35(1): 1–25.

Buchmann, S. L and J. P. Hurley. 1978. "A Biophysical Model for Buzz Pollination in Angiosperms." *Journal of Theoretical Biology* 72(4): 639–657.

Burkart, A., K. Lunau, and C. Schlindwein. 2011. "Comparative Bioacoustical Studies on Flight and Buzzing of Neotropical Bees." *Journal of Pollination Ecology* 6(16): 118–124.

Cardinal, S., S. L. Buchmann, and A.L. Russell. 2018. "The Evolution of Floral Sonication, a Pollen Foraging Behavior Used by Bees (Anthophila)." *Evolution* 72–73: 590–600.

Castro, S., P. Silveira, and L. Navarro. 2008. "How Does Secondary Pollen Presentation Affect the Fitness of *Polygala vayredae* (Polygalaceae)?" *American Journal of Botany* 95(6): 706–712.

Claßen-Bockhoff, R. and A. Heller. 2008. "Floral Synorganization and Secondary Pollen Presentation in Four Marantaceae from Costa Rica." *International Journal of Plant Science* 169(6): 745–760.

Corbet, S. A. and S.-Q. Huang. 2014. "Buzz Pollination in Eight Bumblebee-Pollinated *Pedicularis* Species: Does It Involve Vibration-Induced Triboelectric Charging of Pollen Grains?" *Annals of Botany* 114(8): 1–10.

Corner, E. J. H. 1958. "Transference of Function." *The Journal of the Linnean Society Botany* 56: 33–40.

Costa, C. B. N, J. A. S. Costa, L. P. de Queiroz, and E. L. Borba. 2013. "Self-Compatible Sympatric *Chamaecrista* (Leguminosae-Caesalpinioideae) Species Present Different Interspecific Isolation Mechanisms Depending on Their Phylogenetic Proximity." *Plant Systematics and Evolution* 299(4): 699–711.

Costa, C. B. N., S. M. Lambert, E. L. Borba, and L. P. De Queiroz. 2007. "Post-Zygotic Reproductive Isolation between Sympatric Taxa in the *Chamaecrista desvauxii* Complex (Leguminosae-Caesalpinioideae)." *Annals of Botany* 99(4): 625–635.

Davis, C. C., H. Schaefer, Z. Xi, D. A. Baum, M. J. Donoghue, and L. J. Harmon. 2014. "Long-Term Morphological Stasis Maintained by a Plant-Pollinator Mutualism." *Proceedings of the National Academy of Sciences of the United States of America* 111(16): 5914–5919.

De Luca, P. A., D. A. Cox, and M. Vallejo-marín. 2014. "Comparison of Pollination and Defensive Buzzes in Bumblebees Indicates Species-Specific and Context-Dependent Vibrations." *Naturwissenschaften* 101: 331–338.

De Luca, P. A., L. F. Bussière, D. Souto-Vilaros, D. Goulson, A. C. Mason, and M. Vallejo-Marín. 2013. "Variability in Bumblebee Pollination Buzzes Affects the Quantity of Pollen Released from Flowers." *Oecologia* 172(3): 805–816.

De Luca, P. D. and M. Vallejo-Marín. 2013. "What's the 'buzz' about? The Ecology and Evolutionary Significance of Buzz-Pollination." *Current Opinion in Plant Biology* 16(4): 429–435.

Delgado-Salinas, A. O. and M. Sousa-Sánchez. 1977. "Biología Floral Del Género *Cassia* En La Región de Los Tuxtlas, Veracruz." *Boletín de La Sociedad Botánica de México* 37: 5–52.

Endress, P. K. 1994. *Diversity and Evolutionary Biology of Tropical Flowers*. Cambridge, UK: Cambridge University Press.

Erickson, E. H., and S. L. Buchmann. 1983. "Electrostatics and Pollination." In *Handbook of Experimental Pollination Biology*, edited by Eugene Jones, C. and John Little, R., 173–184. New York: Van Nostrand Reinhold.

Faegri, K. 1986. "The Solanoid Flower." *Transactions of the Botanical Society of Edinburgh* 45(supl): 51–59.

Faegri, K, and L. van der Pijl. 1979. *The Principles of Pollination Ecology. The Principles of Pollination Ecology*. 3rd ed. Vol. 3. Pergamon Press.

Fan, Y. L., W. J. Kress, and Q. J. Li. 2015. "A New Secondary Pollen Presentation Mechanism from a Wild Ginger (*Zingiber densissimum*) and Its Functional Roles in Pollination Process." *PLoS ONE* 10(12): 1–13.

Gan-Mor, S., Y. Schwartz, A. Bechar, D. Eisikowitch, and G. Manor. 1995. "Relevance of Electrostatic Forces in Natural and Artificial Pollination." *Canadian Agricultural Engineering* 37(3): 189–194.

Gottsberger, G. and I. Silberbauer-Gottsberger. 1988. "Evolution of Flower Structures and Pollination in Neotropical Cassiinae." *Phyton* 293–320.

Harder, L. D. and R. M. R. Barclay. 1994. "The Functional Significance of Poricidal Anthers and Buzz Pollination: Controlled Pollen Removal from *Dodecatheon*." *Functional Ecology* 8: 509–517.

Hardy, C. R., D. W. Stevenson, and H. G. Kiss. 2000. "Development of the Gametophytes, Flower, and Floral Vasculature in *Cochliostema odoratissimum* (Commelinaceae)." *American Journal of Botany* 87: 131–157.

Hildebrand, F. 1865. "Über Die Befrunchtung Der *Salvia* (Lamiaceae) Mit Hilfe von Insekten." *Pringsheims Jahrbücher Für Wissenschaftliche Botanik* 4: 451–476.

Howell, G. J., A. T. Slater, and R. B. Knox. 1993. "Secondary Pollen Presentation in Angiosperms and its Biological Significance." *Australian Journal of Botany* 41(4–5): 417–438.

Huang, S.-Q. and X.-Q. Shi. 2013. "Floral Isolation in *Pedicularis*: How Do Congeners with Shared Pollinators Minimize Reproductive Interference?" *New Phytologist* 199(3): 858–865.

Irwin, H. S. and R. C. Barneby. 1982. "The American Cassiinae: A Synoptical Revision of Leguminosae Tribe Cassieae Subtribe Cassiinae in the New World." *Memories of the New York Botanical Garden* 35(1): 1982.

Jesson, L. K. and S. C. H. Barrett. 2002. "Solving the Puzzle of Mirror-Image Flowers." *Nature* 417(6890): 707.

Jesson, L. K. and S. C. H. Barrett. 2003. "The Comparative Biology of Mirror-Image Flowers." *International Journal of Plant Sciences* 164(5 Suppl.): S237–S249.

King, M. J. and S. L. Buchmann. 1996. "Sonication Dispensing of Pollen from *Solanum laciniatum* Flowers." *Source: Functional Ecology Functional Ecology* 10(10): 449–456.

Knudsen, M. 1952. *The Kinetic Theory of Gases*. London, UK: Methuen.

Konan, N. A., O. Kannengieser, and O. Simonin. 2009. "Stochastic Modeling of the Multiple Rebound Effects for Particle-Rough Wall Collisions." *International Journal of Multiphase Flow* 35(10): 933–945.

Kreunen, S. S. and J. M. Osborn. 1999. "Pollen and Anther Development in *Nelumbo* (Nelumbonaceae)." *American Journal of Botany* 86(12): 1662–1676.

Kubitzki, K. and M. C. E. Amaral. 1991. "Transference of Function in the Pollination System of the Ochnaceae." *Plant Systematics and Evolution* 177: 77–80.

Kussin, J. and M. Sommerfeld. 2002. "Experimental Studies on Particle Behaviour and Turbulence Modification in Horizontal Channel Flow with Different Wall Roughness." *Experiments in Fluids* 33(1): 143–159.

Lewis, G. P. 2005. "Tribe Cassieae." In *Legumes of the World*, edited by Lewis, G. P., Schrire, B., MacKinder, B. and Lock, M., 592. Kew, UK: Royal Botanic Gardens.

Luo, Z., D. Zhang, and S. S. Renner. 2008. "Why Two Kinds of Stamens in Buzz-Pollinated Flowers? Experimental Support for Darwin's Division-of-Labour Hypothesis." *Functional Ecology* 22(5): 794–800.

Macior, L. W. 1968. "Pollination Adaptation in *Pedicularis canadensis*." *American Journal of Botany* 55(9): 1031–1035.

Macior, L. W. 1983. "The Pollination Dynamics of Sympatric Species of *Pedicularis* (Scrophulariaceae)." *American Journal of Botany* 70(6): 844–853.

Marazzi, B. and P. K. Endress. 2008. "Patterns and Development of Floral Asymmetry in *Senna* (Leguminosae, Cassiinae)." *American Journal of Botany* 95(1): 22–40.

Marazzi, B, P. K. Endress, L. P. Queiroz, and E. Conti. 2006. "Relationships within *Senna* (Leguminosae, Cassiinae) Based on Three Chloroplast DNA Regions: Patterns in the Evolution of Floral Symmetry and Extrafloral Nectaries." *American Journal of Botany* 93(2): 288–303.

Marazzi, B., E. Conti, and P. K. Endress. 2007. "Diversity in Anthers and Stigmas in the Buzz-Pollinated Genus *Senna* (Leguminosae, Cassiinae)." *International Journal of Plant Sciences* 168(4): 371–391.

Matthews, M. L., M. D. C. E. Amaral, and P. K. Endress. 2012. "Comparative Floral Structure and Systematics in Ochnaceae s.l. (Ochnaceae, Quiinaceae and Medusagynaceae; Malpighiales)." *Botanical Journal of the Linnean Society* 170: 299–392.

Mesquita-Neto, J. N., B. K.P. Costa, and C. Schlindwein. 2017. "Heteranthery as a Solution to the Demand for Pollen as Food and for Pollination—Legitimate Flower Visitors Reject Flowers without Feeding Anthers." *Plant Biology* 19(6): 942–950.

Michener, C. D. 1962. "An Interesting Method of Pollen Collecting by Bees from Flowers with Tubular Anthers." *Revista de Biologia Tropical* 10(2): 167–175.

Michener, C. D. 2007. *The Bees of the World*. 2nd ed. Baltimore, MD: Johns Hopkins University Press.

Morais, A. F. 2011. "Escoamento de Fluidos Complexos e Transporte de Partículas Em Geometrias Irregulares." Universidade Federal do Ceará.

Morgan, T., P. Whitehorn, G. C. Lye, and M. Vallejo-marín. 2016. "Floral Sonication Is an Innate Behaviour in Bumblebees That Can Be Fine-Tuned with Experience in Manipulating Flowers." *Journal of Insect Behaviour* 29: 233–241.

Müller, H. 1881. "Two Kinds of Stamens with Different Functions in the Same Flower." *Nature* 24: 307–308.

Müller, H. 1882. "Two Kinds of Stamens with Different Functions in the Same Flower." *Nature* 27: 30.

Müller, F. 1883. "Two Kinds of Stamens with Different Functions in the Same Flower." *Nature* 27: 364–365.

Ojeda, I., J. Francisco-Ortega, and Q. C. B. Cronk. 2009. "Evolution of Petal Epidermal Micromorphology in Leguminosae and Its Use as a Marker of Petal Identity." *Annals of Botany* 104(6): 1099–1110.

Pauw, A. 2005. "Inversostyly: A New Stylar Polymorphism in an Oil-Secreting Plant, *Hemomeris racemosa* (Scrophulariaceae)." *America Journal of Botany* 92(11): 1878–1886.

Proctor, M., P. Yeo, and A. Lack. 1996. *The Natural History of Pollination*. Portland, OR: Timber Press.

Robertson, C. 1890. "Flowers and Insects. V." *Botanical Gazette* 15(8): 199–204.

Russell, A. L., R. E. Golden, A. S. Leonard, and D. R. Papaj. 2015. "Bees Learn Preferences for Plant Species That Offer Only Pollen as a Reward." *Behavioral Ecology* 27(3): 731–740.

Switzer, C. M. and S. A. Combes. 2016. "Bumblebee Sonication Behavior Changes with Plant Species and Environmental Conditions." *Apidologie* 48(2): 223–233.

Todd, J. E. 1882. "On the Flowers of *Solanum rostratum* and *Cassia chamaecrista*." *American Naturalist* 16(4): 281–287.

Tucker, S. C. 1996. "Trends in Evolution of Floral Ontogeny in *Cassia* Sensu Strictu, *Senna*, and *Chamaecrista* (Leguminosae: Caesalpinioideae: Cassieae: Cassiinae); a Study in Convergence." *American Journal of Botany* 83(6): 687–711.

Tucker, S. C. 1998. "Floral Ontogeny in Legume Genera *Petalostylis*, *Labichea*, and *Dialium* (Caesalpinioidea: Cassieae), a Series in Floral Reduction." *American Journal of Botany* 85(2): 184–208.

Vaknin, Y., S. Gan-Mor, A. Bechar, B. Ronen, and D. Eisikowitch. 2000. "The Role of Electrostatic Forces in Pollination." *Plant Systematics and Evolution* 222(1–4): 133–142.

Vaknin, Y., S. Gan-mor, A. Bechar, B. Ronen, and D. Eisikowitch. 2001. "Are Flowers Morphologically Adapted to Take Advantage of Electrostatic Forces in Pollination?" *New Phytologist* 152(2): 301–306.

Vallejo-Marín, M., E. M. Da Silva, R. D. Sargent, and S. C. H. Barrett. 2010. "Trait Correlates and Functional Significance of Heteranthery in Flowering Plants." *New Phytologist* 188(2): 418–425.

Vallejo-Marín, M., J. S. Manson, J. D. Thomson, and S. C. H. Barrett. 2009. "Division of Labour within Flowers: Heteranthery, a Floral Strategy to Reconcile Contrasting Pollen Fates." *Journal of Evolutionary Biology* 22(4): 828–839.

Vogel, S. 1978. "Evolutionary Shifts from Reward to Deception in Pollen Flowers." In *The Pollination of Flowers by Insects*, edited by A. J. Richards, 89–96. London, UK: Academic Press.

Vogel, S. 1990. "History of the Malpighiaceae in the Light of Pollination Ecology." *Memories of the New York Botanical Garden* 55: 130–142.

Westerkamp, C. 2004. "Ricochet Pollination in Cassias–and How Bees Explain Enantiostyly." In *Solitary Bees—Conservation, Rearing and Management for Pollination.*, edited by B. F. Freitas and J. P. Pereira, 285. Fortaleza, Brazil: Imprensa Universitária.

Westerkamp, C. and A. Weber. 1997. "Secondary and Tertiary Pollen Presentation in *Polygala myrtifolia* and Allies (Polygalaceae, South Africa)." *South African Journal of Botany* 63(5): 254–258.

Willmer, P. 2011. *Pollination and Floral Ecology*. Princeton, NJ: Princeton University Press.

Wolfe, A. D. and J. Estes. 1992. "Pollination and the Function of Floral Parts in *Chamaecrista fasciculata* (Fabaceae)." *American Journal of Botany* 79(3): 314–317.

Yeo, P. F. 1993. *Secondary Pollen Presentation. Form, Function and Evolution*. Wien, Austria: Springer-Verlag.

23 Handedness in Plants in Relation to Yield
A Review

Bir Bahadur and Monoranjan Ghose

CONTENTS

23.1 INTRODUCTION

According to Dixon (1983), the "green spiral" or the spiral phyllotaxis literally means, "leaf arrangement," but the term may be applied to the arrangement of seeds, florets, petals, scales, twigs, etc., which are all parts of green plants. Spiral phyllotaxis is an archetypal arrangement studied along with the patterns of sequential branching around the plant stems. However, the spirals seen in plants differ in number and may be clockwise or counterclockwise. Indeed, the numbers thus exhibited are consecutive pairs taken from the celebrated Fibonacci series 0, 1, 1, 2, 3, 5, 8, 13, 21, 34, 55 ... etc.

Phyllotaxis is the arrangement of leaves on a plant stem (term derived from the ancient Greek word *phýllon* leaf and *táxis* arrangement. Phyllotactic spirals (Fibonacci spirals) form a distinctive class of patterns commonly found in plant world. The basic leaf arrangement is alternate, opposite, or even whorled. In alternate phyllotaxis, the leaves are arranged in various types of spirals, which may be right (counterclockwise) or left (clockwise) with each leaf arising at a different node on the stem in ascending order.

Spirality in various plants has been studied in relation to phyllotaxy by Imai (1927) in *Pharbitiesnil*, Schroder (1953) in *Citrus*, Allard (1946) in *Nicotiana*, Kundu and Sharma (1965) in *Corchorus capsularis*, and Kanahama et al. (1989) in tomato and some Solanaceous species but not in relation to their yield. Hence, the present review is on the relationship between handedness and yield of plants.

Narayana (1942) was the first to examine the relationship between spirality and yield in coconuts based on observations of 70 trees selected at random, from different yield groups and concluded that the yield of two groups was not significant. Davis (1972) followed the work further at the Central Coconut Research Station, Kasaragod, Kerala, where details of yield parameters like number of spathes produced, female flowers produced, setting percentages, yield of nuts per bunch and copra were investigated over a period of time. With these background case studies, results of various studies on different crop plants are discussed here.

23.2   FOLIAR SPIRALS AND YIELD IN PALMS

The phyllotaxis of palms is usually alternate, and in most species two consecutive leaves are placed at an angular deflection of 137.5° as confirmed by most coconut workers. This leads to a crown of leaves arranged in five spirals, which run in clockwise in one palm or counterclockwise in another. The phyllotaxy of each spiral is nearly two-fifths, that is, the sixth leaf stands over the first, the seventh leaf stands over the second, and so on. The number of R/L palms in any plantation scored is more or less equal (Patel 1938), and most coconut workers agree on this. The angle 137.5°with the remaining angle 222.5° makes one full revolution, and the ratio forms a proportion (0.618), which is known as the golden proportion. The handedness of coconut palms can also be determined by assessing the direction of spadix supported by the subtending leaf. If the spadix arises from the right side of its supporting leaf then the palm is left-handed and vice versa (Toar et al. 1979).

The number of foliar spirals varies in different palm species, for example, *Areca catechu* has single foliar spiral, *Arenga saccharifera* has two spirals, *Borassus flabellifer* has three, *Cocos nucifera* has five, *Elaeis guineensis* has eight spirals, and so on. All these numbers follow the Fibonacci sequence, that is, 1, 1, 2, 3, 5, 8, 13, 21, etc.

23.2.1   *Cocos nucifera* L. (Coconut: Arecaceae)

From the base of the well-grown coconut tree, one can easily determine a left- or a right-handed palm viewing the position of the immediately younger leaf (Figure 23.1).

FIGURE 23.1   Coconut leaf crown showing left- and right-spiralled arrangement.

Although the spirality is not genetically determined, the yield of fruits depends on the foliar spirality as reported by Davis (1962a, 1962b, 1963, 1972). Davis (1962b) reported that frequency of left-spiralled plants among 3,028 coconut trees sampled in India was 52.05%, and among 13,842 trees sampled elsewhere it was 52.90%. He further reported that data collected from various countries and locations starting from Tonga Islands in the Pacific Ocean, going westward via the Indian Ocean, Africa, Atlantic Ocean, and the America showed that the sums of the lefts and rights are almost equal. The asymmetry is not inherited and has been regarded as trivial (Davis 1962a).

A study on coconut fruit yield data collected from Kayangulam, Kerala, during 1955 and 1960 showed that 58 left-spiralled healthy coconut palms produced 65.25 fruits on an average, and 70 right-spiralled palms produced 53.93 fruits per palm per year. These data indicate that left-spiralled palms significantly produced more fruits than those of the right-spiralled palms (Davis 1962b). In further extension of research at Central Coconut Research Station, Kayangulam, Davis (1963) selected 384 coconut trees of which 177 were left-spiralled and 207 right-spiralled. The excess of right-spiralled palms was not significant at the 5% level. The yield data of all 384 trees from 1949 to 1960 were collected. The final results are shown in graphs (Figures 23.2 through 23.6). In the differences of yield between left-spiralled and right-spiralled in each of the five comparisons, it is seen that the mean number of nuts on the left-spiralled exceeds that of the right-spiralled coconut palms.

Davis (1963) postulated that higher nut yield of the left-spiralled palms is due to the fact that they possess more leaves. He further showed that average green leaves per healthy left and right palms are 31.19 and 29.59, respectively. This indicates that lefts have an average 5.4% more leaves than the right-handed palms.

However, later, Davis (1972) determined the spirality of 177 of the 384 palms of Kayangulam, Kerala, India, as right-spiralled (according to single spiral) and found that right-spiralled coconuts produced significantly more fruits than left-spiralled palms. Probably his (Davis 1963) earlier determination of spirality of coconut palms was wrong, which he corrected in his later publications.

Satyabalan et al. (1964), however, did not find any association between direction of foliar spirality and yield in coconuts. On the other hand, coconut yield data collected

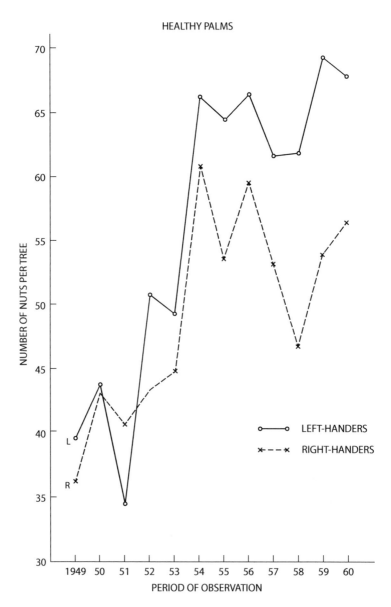

FIGURE 23.2 Yearly yields of healthy coconuts from 1949 to 1960. (From Davis, *J. Genet.*, 58, 186–215, 1963.)

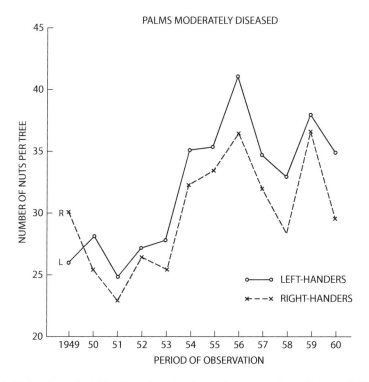

FIGURE 23.3   Annual yields of moderately diseased coconuts from 1949 to 1960. (From Davis, T.A., *J. Genet.*, 58, 186–215, 1963.)

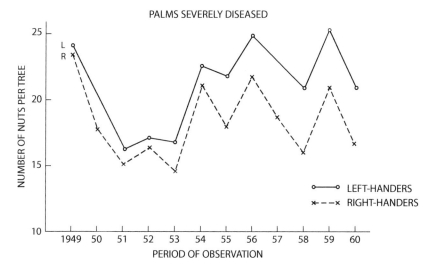

FIGURE 23.4   Annual yields of severely diseased coconuts trees from 1949 to 1960. (From Davis, T.A., *J. Genet.*, 58, 186–215, 1963.)

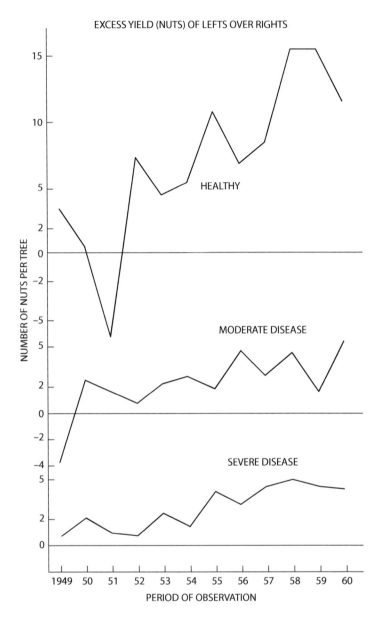

FIGURE 23.5 Excess nut yields of left-handed over right-handed coconuts. (From Davis, T.A., *J. Genet.*, 58, 186–215, 1963.)

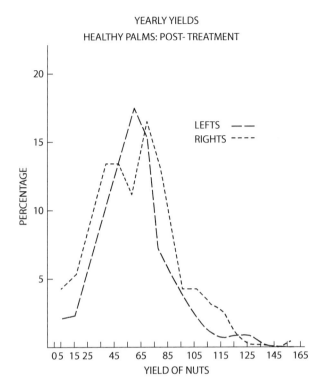

FIGURE 23.6   Annual nut yields of healthy left-spiralled and right-spiralled coconuts from 1955 to 1960. (From Davis, T.A., *J. Genet.*, 58, 186–215, 1963.)

from the Coast Agricultural Station, Government of Tanzania (East Africa), on 481 palms for 5 years (January 21, 1960 to December 22, 1965) revealed that a right-spiralled palm produced, on an average, 80.32 nuts per year, while a left-spiralled one produced only 64.77 nuts per palm, the difference being 24.01%, which is highly significant (Davis 1972).

Davis (1972) collected another set of data set on coconut yield on 130 palms, of which 60 were left-spiralled, from Boston, Jamaica, West Indies, relating to the period from 1961 to 1969. The result showed that the right-spiralled palms produced more fruits per tree during each year. On the aggregate, a right-spiralled palm produced 9.94% fruits in excess of a left-spiralled one.

Data collected from Trust Territory of Pacific Islands during the calendar year 1963 on kernel yield of 100 left-spiralled and 100 right-spiralled coconut plants showed that the left-right difference of kernel yield per tree per year was 13.48%.

The average yield of kernel per tree per year was 31.24 lbs. and 27.53 lbs. for right-spiralled and left-spiralled, respectively (Davis 1964a).

Bahadur et al. (1978a) studied some physiological aspects of left- and right-handed coconut palms and tried to correlate them with yield. They observed that the right-spiralled palms possess more chlorophyll, large stomata, high stomatal index, and low rate of transpiration than the left-spiralled ones. On the basis of these observations, it appears that right-handed palms are metabolically superior than the left-handed coconuts; hence, higher fruit production is probably correlated to their efficient metabolism.

23.2.2 *Borassus flabellifera* L. (Palmyra Palm)

Davis (1972) collected data from 10 female palmyra palms from a village 80 km north of Kolkata. Five palms were left-spiralled and the rest were right-spiralled. On an average, a right-spiralled palmyra produced 190.2 fruits, and a left-spiralled one produced 114.2 fruits, the difference being 68.3%.

Davis (1972) also collected yield data of another set of 31 palmyra palms from Srikakulam (Andhra Pradesh, India), of which 14 palms were left-spiralled and 17 were right-spiralled. The data indicated that a right-spiralled palm produced on an average 147.50 fruits, and a left-spiralled palm 108.47 fruits. The excess for the right-spiralled palms is 35%. The result of the two populations shows a similar trend, that is, right-spiralled palms produced more fruits.

23.2.3 *Areca catechu* (Betel Nut Palm)

A. catechu possesses only a single foliar spiral, which may be left-spiralled or right-spiralled. In a study on 54 areca palms in the vicinity of Kolkata, which comprised of 32 left-spiralled and 22 right-spiralled palms, a left-spiralled palm on an average produced 134.53 fruits per year while a right-spiralled palm produced 186.55 fruits. This indicates that the right-spiralled *Areca* palms produced 38.67% in excess of the left-spiralled palms (Davis 1972).

23.2.4 *Hibiscus cannabinus* L. (Deccan Hemp)

Deccan hemp is a tall erect plant, and the leaves are arranged in alternate phyllotaxis. This arrangement of leaves distinguishes the plants into left-spiralled and right-spiralled. The two kinds of plants are distributed almost equally in any large population. These plants are cultivated as a good bast fiber crop. There is an association between the foliar spirality and the aestivation of the flowers. A left-spiralled plant or shoot produces a greater number of flowers having counterclockwise contortion. Also, right-spiralled plants or shoots produce flowers with clockwise contortion (Davis 1972). The two types of flowers occur equally in large populations (Davis 1964b). Kundu (1968) studied 49 left-spiralled and 50 right-spiralled plants of Deccan hemp at the premises of the Indian Statistical Institute, Kolkata. She reported that a right-spiralled plants produced, on an average, 4.28 fruits (i.e., 15.92%) more than that of a left-spiralled plants.

Ghosh and Davis (1978) worked for 10 years on the foliar spirality and aestivation of flowers in *Hibiscus cannabinus*. They found that the left-spiralled plants produced 26%–42% left-twisting flowers and 57%–73% right-twisting flowers. On the other hand, right-spiralled plants produced 58%–70% left-twisting and 29%–41% right-twisting flowers and suggested that a negative association exists between the foliar spirality and the aestivation of corolla. They also mentioned that fruits developed from left-twisting flowers of the left-spiralled plants and those developed from right-twisting flowers of right-spiralled plants were better in quality. However, there is no significant difference in yield of seeds between left- and right-spiralled plants.

23.2.5 *DISOCOREA ESCULENTA* (DIOSCOREACEAE)

Davis (1972) experimented on two stem twiners namely, cow pea (*Vigna sinensis*) and Dioscorea (*Dioscorea esculenta*) to see the effect of coiling on yield. Cow pea normally twines counterclockwise (right-spiralled) and produces fruits and seeds. The experiment was conducted in three methods using bamboo poles as support: (1) plants allowed twining normally, that is, right-spiralled, counterclockwise; (2) plants forced to twine clockwise, that is, left-spiralled; and (3) plants forced to grow vertically. The treatments differed significantly from one another. The number of fruits produced by the left-spiralled and verticals were greater than the right-spiralled plants by 42.89% and 20.92%, respectively. In the production of seeds also, the left-spiralled and vertical plants were superior to the right-spiralled plants by 51.80% and 18.10%, respectively. The results of these experiments were in conformity with the result obtained by Reber (1960, 1964) on pole beans in Australia.

The same treatment that was followed in cow pea (*Vigna sinensis*) was employed in *D. esculenta*. The twiners were made to coil on bamboo poles with three sets of experiments as followed in cow pea. *D. esculenta* produces medium-sized tuberous roots and normally the stem always twines clockwise. The result of the experiments showed that on an average, a clockwise plant moving according to the normal way produced 814 grams of tubers, the reversed counterclockwise plant produced 756 grams, and the vertical ones produced the maximum of 1161 grams. This indicates that the left plants are superior to the rights by 7.67%, and the verticals are better than the lefts by 42.63% and the rights by 53.57% (Davis 1972).

23.2.6 *DISOCOREA ALATA*

Various morphophysiological, biochemical, and yield aspects of tubers were studied in the right-handed twiner (natural spiral, CCW), left, forced reversed, clockwise, and forced to grow vertically along with bamboo with no twining. In general, the vertically grown plants were found to be superior over the left- and right-twining plants (Figure 23.7a–c). Electrophoretic studies of fresh mature leaves and tuber extracts showed interesting results. The various amino acids of mature leaves showed that the vertically grown plants contained seven amino acids, *viz.*, Cystine, Ornithine, Lysine, Glycine, Glutamic acid, Alanine and Tyrosine, and hence considered as superior over the left- and right-handed twiners, which contain only 6 and 5 amino acids respectively, *viz.*, Cystine, Orninthine, Glycine, Glutamic acid,

FIGURE 23.7 Comparison of *Dioscorea alata* tubers of: (A) right-twining; (B) vertically grown; and (C) left-twining plants. (From Narsaiah, G., Unpublished PhD Thesis, 1984.)

Alanine, and Tyrosine in left-spiralled plants and Orninthine, Arginine, Glycine, Glutamic acid, and Methionine in right-spiralled plants. On the other hand, the compositions of amino acids in tubers of vertically grown plants are Lysine, Cystine, Glutamic acid, Alanine, Methionine, Valine, and Leucine. This is in contrast to left-twining plants that contain only four amino acids, *viz.*, Alanine, Valine, Isoleucine, and an unassigned amino acid, and in right-twining plants only five amino acids, *viz.*, Lysine, Glycine, Glutamic acid, Valine, and Tyrosine, were detected. From this it may be inferred that the vertically grown plants have more amino acids including Methionine, which enhances the food value of tubers. However, the stomatal index, stomatal size, and chlorophyll content of mature leaves are negatively correlated. From the yield point of view, the tubers weight of vertically grown plants of both S1 and S2 generations was found to be superior over the right- and left-twining plants by 15.5% and 5.35%, respectively (Narsaiah 1984).

23.2.7 *PHASEOLUS MULTIFLORUS*

Another interesting work was reported by Reber (1960) on reversed bean vines. He planted nine different kinds of pole beans, for example, Hawaiian, Kentucky wonder, Corn bean, Scarlet runner (*Phaseolus multiflorus*), etc. Normally, these bean plants twine around the poles in a right-hand direction. The vines on even numbered poles of three rows were unwound and twined backward. All vines and pods were allowed to ripen, wither, and dry on the poles and then harvested. The field data on each hill consists of: number and weight of pods, number and weight of beans, weight of shucks, number and weight of vines. The results show a better ratio of weight in ounces of beans/ounces of shucks, and ounces of beans/ounces of vines, for the reversed vines compared to the normal vines (Figures 23.8 and 23.9).

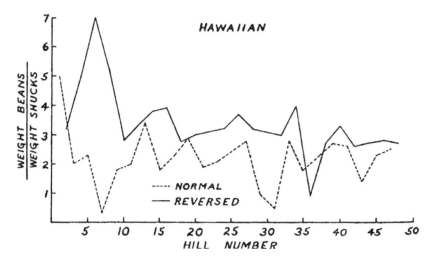

FIGURE 23.8   Weight of beans/weight of shucks ratios in normal and reversed vines of the Hawaiian bean. (From Reber, G., *Castanea*, 25, 122–124, 1960.)

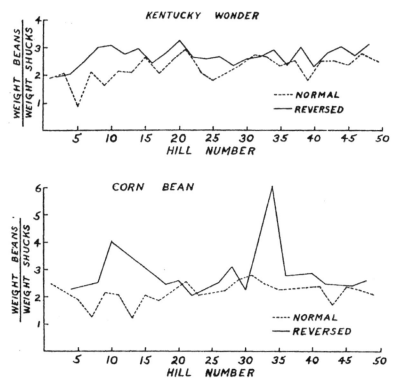

FIGURE 23.9   Weight of beans/weight of shucks ratios in normal and reversed vines of Kentucky wonder and Corn beans. (From Reber, G., *Castanea*, 25, 122–124,1960.)

Reber (1964) continued the experiment on two other varieties of the same bean. The artificial reversal of the direction of coiling in these two varieties of bean vines caused an increase in the weight ratio of beans to shucks.

23.2.8 *Psophocarpus tetragonolobus*: Winged Bean (Fabaceae)

Psophocarpus tetragonolobus L. (DC), the winged bean, is a tropical underutilized multi-use food crop grown as an annual to perennial and twines to the right side around the bamboo support. It is nutrient-rich, and all parts of the plant are edible. The species shows seedling handedness and also exhibits twining handedness. The left, right, and vertically grown plants were tested for yield like pods, pods/plant, mean number of seeds per pod, and mean seed weight per plant selected at random. It was shown that the yield of vertically grown plants was more than the left- and right-twining plants (Figure 23.10a–c). From the nutritional value point of view, the vertically grown plants showed more amounts of amino acids (namely, Orninthine, Histidine, Serine, Hydroxyproline, Alanine, Proline, Tyrosine, Methionine, Isoleucine, and Phenylalanine). This is followed by left-twining plants (namely, those rich in, Cystine, Orninthine, Histidine, Arginine. Glycine, Glutamic acid, Alanine, Tyrosine, Methionine, Tryptophan, and Isoleucine) and right-twining

FIGURE 23.10 *P. tetragonolobus* plants in fruiting condition: (A) right-twining; (B) reversed left-twining and vertically grown plants, right twining; and (C) reversed left-twining and vertically grown plants. (From Narsaiah, G., Unpublished PhD Thesis, 1984.)

plants (namely, those rich in Cystine, Orninthine, Arginine, Glycine, Glutamic acid, Alanine, Tyrosine, Methionine and Isoleucine).

The vertically grown plants had 10 amino acids, the left-twining plants contained 9 amino acids, and the right-twining plants had only 8 amino acids. The root nodules index is also higher in vertically grown plants than the left- and right-twining plants developed from right-handed seedlings.

23.2.9 *CAJANUS CAJAN*: PIGEON PEA

Bahadur et al. (1980) worked on handedness in 5,057 seedlings belonging to 15 cultivars of the pigeon pea (*Cajanus cajan*). Out of these, 52.34% were right-handed, 47.06% were left-handed, and 0.6% were neutral seedlings. In right-handed plants, the roots were profusely branched and were longer with a higher number of root nodules than the left-handed plants. Similarly, right-handed plants contained more chlorophyll and higher stomatal indices than the left-handed plants. The percentage difference of left-handed and right-handed seedlings is in conformity with the law of occurrence of bioenantiomorphs (Dubrov 1978). Rao (1980) also observed an excess of right-handed seedlings in *Phaseolus* and *Vigna*. In three cultivars of the pigeon pea, the number of pods and the weight of seeds per plant in right-handed plants were significantly more than the left-handed plants. In another study in *C. cajan*, Rao et al. (1985) reported that the total number of fruits and seeds produced and the seed weight was comparatively more in flowers with right-handed corolla than in the left-handed ones.

23.2.10 *VIGNA RADIATA* (GREEN GRAM) AND *VIGNA MUNGO* (BLACK GRAM)

Rao and Bahadur (1982) studied the root length and root nodule characteristics in left- and right-handed seedlings of green gram and black gram. They found that in green gram, the primary root lengths of left-handed seedlings were significantly greater than those of the right-handed seedlings. In black gram, the primary root length was significantly greater in the left-handed than the right-handed seedlings. The root nodule size, indices, and number were also more in left-handed than in the right-handed plants in both green gram and black gram. This indicated that the left-handed seedlings and the plants developing from them were superior in root and nodule characters than right-handed seedlings. Similar differences in the primary root length of left- and right-handed seedlings were reported by Bahadur et al. (1978b) in *Bambusa arundinacea*.

23.2.11 AROIDS

Venkateswarlu (1978) studied the prefoliation (ptyxis) of *Colocasia esculenta* and *Xanthosoma sagittifolium*. In *C. esculenta*, seven collections behaved clockwise and three as counterclockwise throughout their growth period. On the other hand, *X. sagittifolium* showed 10 collections as left-handed and 8 collections as right-handed prefoliation. In both species, a high incidence of left-handers was noticed. He mentioned that in *X. sagittifolium*, right-handed plants recorded a higher yield of tubers than the left-handed plants.

23.2.12 *Lycopersicon esculentum* Mill. (Tomato) and *Capsicum annum* L. (Chili Pepper)

Bible (1976) studied five cultivars of *L. esculentum* and three cultivars of *C. annum* in relation to L/R alternate phyllotaxy and noted equality in the relation of L/R plants. He reported an increase of 20% fruit weight and 16% fruit number in right-handed tomato than left-handed plants. In the chili pepper, however, both fruit weight yield and fruit number were higher by 24% and 21%, respectively, for right-handed than for left-handed plants. In both tomato and chili pepper plants, fruit size was not related to spirality. The results suggest that the regulation of foliar spirality can perhaps be used for increasing productivity of certain crop plants.

23.2.13 *Phoenix dactylifera* L. (Date Palm)

Date palms show either left- or right-handed foliar spirals, but the orientation is apparently not genetically controlled. Palms with a left-spiral and with alternate phyllotaxy yielded more fruits than those with a right-spiral during four years of an experiment (Reuveni 1986). He reported that individual date palms showed either left- or right-oriented foliar spirals. Palms with a left spiral yielded more fruits than those with a right spiral during four years of an experiment. Elhoumaizi et al. (2002) conducted experiments on four Moroccan cultivars of *P. dactylifera* on 20 trees per cultivar and four offshoots per tree. The divergence angle was measured on adult trees, where a phyllotactic direction (clockwise or counterclockwise) was noted for both offshoots and adult trees. They showed that the phyllotaxic direction of the offshoots presents a clockwise or counterclockwise phyllotaxis independently of the mother plant. The divergence angle varied among the cultivars. The authors concluded that this character was interesting criterion for selection of the best adapted cultivars for the marginal date palm growing regions as well as for optimal fruit production.

23.2.14 *Elaeis guineensis* Jacq. (Oil Palm)

According to Arasu (1970) noted in Malaysia, left-handed oil palms (*Elaeis guineensis*) are more frequent (approx. 52%–53 %) than right-handed ones. The direction of the foliar spiral does not appear to be genetically determined or to influence yield.

23.3 DISCUSSION AND CONCLUSIONS

Narayana (1942) was the first to examine the relationship between spirality and yield in coconuts based on observations of 70 trees selected at random, from different yield groups, and concluded that the yield of two groups was not significant. Davis (1972) followed the work further at the Central Coconut Research Station, Kasaragod, Kerala, where in details of yield parameters like number of spathes produced, female flowers produced, setting percentages, yield of nuts per bunch, and copra were investigated over a period of time. On the basis of the above studies, it was concluded that although the left- and right-handed spirality in plants are not genetically controlled, fruit yield is found to be associated with spirality in many plants. Palms in general, show an increase number of fruits in right-spiralled plants than in left-spiralled plants. In his review, Kihara (1972) cites an old Japanese theory of "male and female" is inherited in plants based on vegetative characters (not male and female). According to Sinoto (1972), two principles (yin-yang) of Chinese philosophy prevailed through the 18th and 19th century in Japan. The English translation as given by Kihara is as follows: *Under the heaven all organisms are divided into male and female. If you sow the seed from female plants, you get enormous profit. In rice the yield will be 25% higher than the average of all Japan. Accordingly the excess rice for ten thousand hectares of rice fields is 250,000 koku.* (ca. 125,000 bushels). He further states that two plants belonging to Araceae (*Colocasia* and *Amorphophallus*) are divided into males and females according to the way the overlapping of the leaf stalk margin. The right-handed individuals are "females." It is interesting to note that all leaves on one branch have the same handedness in *Colocasia*. The offspring from the "female" plants should always give a higher yield, and the farmers were recommended to use bulbs of the females.

Kihara cited the comments of Prof. R. D. Preston F.R.S.: "Since the sign of the spiral is not inherited then one is compelled to assume the orientation of the conducting tissue affects the disposal of the materials being conducted." Incidentally, no work has been carried out in this regard even after the suggestion was made!

23.3.1 BIOPHYSICAL ASPECTS

Darwin (1865) was the first to mention that Mikania (Asteraceae) grows clockwise in the Northern Hemisphere while another species in Brazil twine counterclockwise. Similarly, Prof. Van Oye (1926) noted majority of stems of *Castanea sp.* (Fagaceae) twist counterclockwise. Led by these observations of geographic latitude, Davis (1964a) made observations on coconut palms and noted left-spiralled plants to be in excess over the right-spiralled plants in the Northern Hemisphere, and the right-spiralled ones more in Southern Hemisphere, and the differences between the proportions were statistically significant. Minarsky (1998) reevaluated the latitudinal differences based on existing data on coconuts of 42 locations around the world; left-handed coconut palms are in the majority in the Northern Hemisphere, and right-handed ones are in the majority in the Southern Hemisphere. He proposed the **Induced Current Hypothesis**, according to which that latitude-dependent bias in foliar direction may be associated with temporally varying component of the earth's magnetic field. It is known that the changes in the current strength of the ionospheric dynamo induce clockwise earth currents in the Northern Hemisphere and counterclockwise earth currents in the Southern Hemisphere. These earth currents are measureable in trees and bias the diffusion of auxin or auxin transport proteins in young developing embryos, such that left-handed trees are produced in the Northern Hemisphere and right-handed trees in the Southern Hemisphere. Since the genetic bias is non-Mendelian, this hypothesis lends some supports to the R/L handedness phenomenon in coconut palms. Kundu and Sharma (1965) reported the direction of sunlight exposure of jute plants in field plots and suggested that this might either stimulate or inhibit the growth of the opposite type. If the association between spirality and productivity is manifested consistently in some crops, the regulation of direction of foliar spirality by growth inhibitors could be used as advantage (Charlton 1974).

23.3.2 EFFECT OF MAGNETIC FIELD

The response of plants from left-hand and right-hand sugar beet fruits to terrestrial magnetic field orientation and forms of nitrogen fixation add in the discussion (Nikulin and Leisle 1970). Similarly, presoaking magnetic treatment of barley seeds (*Hordeum vulagare*) also result in yield increases in 13 of 19 field trials in Southern Alberta, Canada. Furthermore, treatment of spring and winter wheat seeds (*Triticum aestivum*) resulted in yield in 14 of 23 tests. Pittman (1977), incidentally the author did not mention handedness in his work on barley, wheat and oats, although seedling handedness is well known in these plants (Compton 1912). Similarly, the mechanism of embryo rotation in *Polygala arvensis* described by Krishnamurthy and Senthil Kuma (1992) is also not known. It may be interesting to mention two cases that are relevant for the present discussion.

Itoh et al. (2012) identified a gene that regulates the phyllotactic pattern in rice. Loss-of-function mutants of the DECUSSATE(DEC) gene displayed a phyllotactic conversion from normal distichous pattern to decussate. The DEC mutants had an enlarged shoot apical meristem with enhanced cell division activity. In contrast to the shoot apical meristem, the size of the root apical meristem in the DEC mutants was reduced, and cell division activity was suppressed. This suggested that DEC controls the phyllotactic pattern by affecting cytokinin signaling in rice. Itoh et al. (2012) commented that "this extremely regular pattern is one of the most fascinating enigmas in plant biology."

Such a transformation to decussate opposite condition can have significant effect on photosynthetic potential and yield.

Yet another example of distichous phyllotaxy is maize plants, which normally have alternate phyllotaxy but also show seedling and endosperm handedness (Compton 1912; Weatherwax 1948). Jackson and Hake (1999) and Tan et al. (2011) described a recessive maize mutant, *abphyll*, that initiates leaves in opposite pairs, in a decussate phyllotaxy than an alternate fashion. The decussate shoot meristems are larger than normal throughout development, though the general structure and organization of the meristems are not altered. *abph1* mutants are first distinguished during embryogenesis, prior to true leaf initiation, by a larger shoot meristem and coincident larger expression domain of the homeobox gene *knotted1*. Therefore, the *abph1* gene regulates morphogenesis in the embryo and plays a role in determining the phyllotaxy of the shoot. The mechanism by which these phyllotaxy are generated remains unclear and the authors (Tan et al., 2011) commented this condition as precious material for maize breeding and plant evolutionary studies.

We hope that these interesting findings will be commercially useful for marker-assisted selection breeding of *OP* maize and will also strengthen the basis for cloning of the opposite leaf gene for more productivity. To conclude, the mechanism that controls yield in various plants *vis-à-vis* phyllotaxy, spirality and handedness do not follow a fixed rule and vary considerably from species to species. There is a need for further research in this long-neglected emerging area of modern plant biology.

REFERENCES

Allard, H.A. (1946). Clockwise and counter clockwise spirality in the phyllotaxy of tobacco. *J. Agric. Res.* **75**:237–242.

Arasu, N.T. (1970). Foliar spiral and yield in oil palms (*Elaeis guineensis* Jacq.) *Malays. Agric.* **47**:409–415.

Bahadur, B., Rao, K.L., Rao, M.M., and Chandraiah, M. (1978a). The Physiological basis of handedness in relation to yield in *Cocos nucifera* L. A proposal Incompatibility. *New Lett.* **9**:108–111.

Bahadur, B., Rao, K L., and Rao, M.M. (1980). Asymmetry and yield in pigeonpea. *Cajanus cajan* (L.) Millsp. International Workshop on Pigeonpea, Icrisat, Patancheru, India, **2**:15–19.

Bahadur, B., Rao, K.L., and Rao, M.M. (1978b). Left and right-handedness in the seedlings of *Bambusa arundinacea* Willd. *Curr. Sci.* **47**:584–586.

Bible, B.B. (1976). Non-equivalence of left handed and right handed phyllotaxy in tomato and pepper. *Hort. Sci.* **11**:601–602.

Charlton, W.A. (1974). Studies in the Alismataceae V. Experimental modification of phyllotaxis in pseudostolons in *Echinodorus tenellus* by means of growth inhibitors. *Can. J. Bot.* **51**:1131–1142.

Compton, R.H. (1912). A further contribution to the study of right- and left-handedness. *J. Genet.* **2**:53–70.

Darwin, C. (1865). On the movements and habits of climbing plants. *J. Linn. Soc. London.* **9**:1–118.

Davis, T.A. (1962a). The non-inheritance of asymmetry in *Cocos nucifera*. *J. Genet.* **58**:42–50.

Davis, T.A. (1962b). Asymmetry and yield in *Cocos nucifera* L. *Experientia.* **18**:321–322.

Davis, T.A. (1963). The dependence of yield on asymmetry in coconut palms. *J. Genet.* **58**:186–215.

Davis, T.A. (1964a). Leaf spiral and yield in coconuts. *Nature.* **204**:496–497.

Davis, T.A. (1964b). Aestivation in Malvaceae. *Nature.* **201**:515–516.

Davis, T.A. (1972). Effect of foliar arrangement on fruit production in some tropical crop plants. Tropical Ecology with an emphasis on Organic production, Athens (USA). 147–164.

Dixon, R. (1983). The mathematics and computer graphics of spirals in plants. *Leonardo.* **16**(2):86–90.

Dubrov, A.P. (1978). *Geomagnetic Field and Life: Geomagnetobiology (Translated from Russian by F. A. Sinclair)*. New York, Plenum Press.

Elhoumaizi M.A., Lecoustre, R., and Oihabi, A. (2002). Phyllotaxis and handedness in date palm (*Phœnix dactylifera* L.) *Fruits.* **57**(5–6):297–303.

Ghosh, S.S. and Davis, T.A. (1978). Foliar spirality and aestivation of flowers in *Hibiscus cannabinus* Linn. *Experientia.* **34**:348–349.

Imai, Y. (1927). The right and left handedness of phyllotaxy. *Bot. Mag. Tokyo.* **41**:499–526.

Itoh, J.I., Hibara, K.I., Kojima, M., Sakakibara, H., and Nagato, Y. (2012). Rice decussate controls phyllotaxy by affecting the cytokinin signaling pathway. *Plant J.* **72**:869–881. doi:10.1111/j.1365-313x.2012.05123.

Jackson, D. and Hake, S. (1999). Control of phyllotaxy in maize by the abphyl1 gene. *Development* **126**:315–323.

Kanahama, K., Saito, T., and Qu, Y.-H.. (1989). Right and left handedness of phyllotaxis and flower arrangement and developmental order of flowers in the inflorescence of Solanaceae plants development flowers. *J. Japan Soc. Hort. Sci.* **57**:642–647.

Kihara, H. (1972). Right- and left-handedness in Plants. *Zeiken Ziho.* **23**:1–37.

Krishnamurthy, K.V. and Sentil Kumar, T. (1992). Embryo rotation in *Polygala arvensis* Willd. *Beitr. Biol. Pfllanen.* **67**:55–58.

Kundu, A. (1968). Levo and dextro rotatory situations in plants and their relationship to fruit production. PhD Thesis Calcutta University, India.

Kundu, B.C. and Sharma, M.S. (1965). Direction of leaf spiral in *Corchorus capsularis* L. *Trans. Bose Res. Inst.* **28**:107–112.

Minarsky, P.Y. (1998). Latitudinal differences in coconut foliar spiral directions: A reevaluation and hypothesis. *Ann. Bot.* **82**:133–140.

Narayana, G.V. (1942). Reports of Agric. Stations, Madras Presidency, 414, Government Press, Madras, 1940–1941.

Narsaiah, G. (1984). Handedness in twining plants with reference of *Psophocarpus tetragonolobus* (L.) DC and *Dioscora alata* L. in relation to yield. PhD Thesis. Kakatiya University, Warangal, India.

Nikulin, A.V. and Leisle, V.F (1970). Response of plants from left-hand and right-hand sugar beet fruits to terrestrial magnetic field orientation and forms of nitrogen fixation. *Sov. Plant Physiol.* 17:384–389.

Patel, J.S. (1938). *The Coconut: A Monograph.* pp. 71, Government Press, Madras, India.

Pittman, U.J. (1977). Effects of magnetic treatment on yield of barley, wheat and oats in Sothern Alberta. *Can. J. Plant Sci.* **57**:37–45.

Rao, M.M. (1980). Studies on seedling and corolla handedness in Papilionaceae with special reference to *Vigna* and *Phaseolus* spp. and its possible influence on yield. PhD Thesis, Kakatiya University Warangal.

Rao, M.M. and Bahadur, B. (1982). Note on the relationship between seedling handedness and root nodule characters in Greengram and Blackgram. *J. Agric, Sci. India.* **52**:706–708.

Rao, M.M., Rao, K., and Bahadur, B. (1985). Significance of corolla handedness and stylar hairs in the pollination biology of some Papilionaceae. In: *Pollination Biology: An Analysis.* E.R.P. Kapil, Inter-India Publications, New Delhi, pp. 221–229.

Reber, G. (1960). Reversed bean vines. *Castanea.* **25**:122–124.

Reber, G. (1964). Reversed bean vines. *J. Genet.* **59**:37–40.

Reuveni, O. (1986). Effect of left and right handed phyllotaxy on the date palm. *Acta. Hortic.* **175**:257–258.

Satyabalan, K., Ninan, C.A., and Krishna Marar, M.M. (1964). Foliar spiral and yield in coconuts. *Nature.* **202**:927–928.

Schroder, C.A. (1953). Spirality in Citrus. *Bot. Gaz.* **114**:350–352.

Sinoto, Y. (1972). The theory of male and female individuals in crop plants. (The history of biological sciences in relation to cultivated plants and domesticated animals in Japan Ed: Kihara et al. Yokendo, Japan.

Tan, Y.Q., Xie, C.X, Jiang, H. Y., Ye, H., Xiang, Y., Zhu, S., and Wand Cheng, B.J. (2011). Molecular mapping of genes for opposite leafing in maize using simple-sequence repeat markers. *Genet Mol Res.* **10**(4):3472–3479.

Toar, R.E., Rompas, T.M., and Sudasrip, H. (1979). The non-inheritance of the direction of foliar spiral in coconut. *Experientia.* **35**:1585–1587.

Van Oye, E. (1926). Sur la tordion des frones d' arres. *Bull. Soc. Botanique.* **11**:270–288.

Venkateswarlu, T. (1978). Unichirality in edible aroids. *J. Root Crops.* **4**:15–18.

Weatherwax, P. (1948). Right-hand and left-handed corn embryo. *Ann. Mo. Bot. Garden.* **35**:317–321.

24 Concluding Remarks

*K. V. Krishnamurthy, Bir Bahadur,
Monoranjan Ghose, and S. John Adams*

This edited volume contains 24 chapters, including this last chapter on concluding remarks. Although small in size, this collection of articles successfully connects together different aspects of plant asymmetry. Symmetry/asymmetry is one of the very important features in plant (and also in animal) developmental biology that unifies evo-devo research on form and function that has been so far carried out. This collection of articles also shows how a coherent study on asymmetries yields considerable insights into the evolution of diverse developmental patterns in plants catering to the needs of their different functional adaptations.

The first chapter by Krishnamurthy et al. introduces the concepts of handedness, asymmetry, chirality, spirality, and helicity with particular reference to plants. The confused and overlapping uses of these terms in botanical literature are addressed. Although Palmer (2005) has brought some order into their uses in animal literature, the need for a similar effort in plants is suggested. The authors have suggested that the word "symmetry" may be used to accommodate all the words such as chirality, handedness, spirality, and helicity; the word "chirality" may be restricted to molecular- and cellular-level asymmetry and may be avoided for higher levels of biological organization and integration. Similarly, it was suggested that handedness may be limited to left-right asymmetry at the tissue, organ, and whole organism levels but may not be used to the direction of spirality and helicity, where the use of "dextral" and "sinistral" may be encouraged. Such forced restriction of the use of these different words may avoid the prevailing confusion and the highly overlapping uses of all these words to a very large extent.

The second chapter by Akbar Sha et al. analyzes left-right symmetry of plants and animals in a comparative way. Besides citing similarities between the two groups of organisms, such as the presence of internal and external asymmetries, their role in reproduction, presence of antisymmetry and directional asymmetry, and the role of genetics, environment, and chance in breaking symmetry, this chapter also brings out the main differences that include the greater instances of external asymmetries in the plants (animals show greater instances of internal asymmetry). This chapter also emphasizes that studies on asymmetry in animals have been much greater and more in-depth than in plants. Many more comparative studies on asymmetry between animals and plants are needed in future to highlight the evolutionary similarities and dissimilarities as well as features of parallel evolution between the two groups of organisms.

Krishnamurthy and John Adams discuss at length asymmetry at the cell level in Chapter 3. The most important of the asymmetric events in the plant cell is the presence of asymmetric cell divisions, which decide the subsequent differential fates of the two daughter cells. This is a very vital morphogenetic event at the cell level that affects, many a time, the plant part or even the whole plant. Of the three models to explain cell level asymmetry, the morphogen model and positional information model largely remain theoretical, at least with reference to plant cells. The role of the cytoskeleton (as per the cytoskeleton model) in cell-level asymmetry has been demonstrated. However, much more research is needed to definitely link cell cycle events and polarity with events happening during cell-level asymmetric changes. More detailed studies are also needed to connect cell-level asymmetry with the asymmetry observed at higher levels of plant organization and integration.

One of the characteristic features of the wood of dicotyledons and conifers is the presence of spiral grains, although other types of grains are also known. The presence of these grains is a sort of necessary evil because these grains often affect wood quality (especially when the spiral grain angle crosses a tolerable limit) on the one hand, and, on the other, help in the proper supply of water to all the branches of the tall tree. The direction of the spiral grain is a variable feature within the same tree and may be to the right (Z) or to the left (S) to the theoretical central vertical axis of the tree spiral grain angles; in addition, their direction is largely dependent on the vascular cambium that produces the wood tissue. In Chapter 4, Krishnamurthy et al. describe the behavioral changes in the vascular cambial cell components, particularly the cambial domains that result due to these behavioral changes that produce spiral-grained wood. They further suggest that the usage "spiral grain" is a misnomer and it should be called "helical grain," as they do not make typical logarithmic spirals. The editors would like to record here that more work needs to be done on the dextral-sinistral changes in spiral grains of dicot and conifer woods that are associated with increasing age of the tree.

Robert W. Korn has dealt with broken symmetry and handedness in crytogams in Chapter 5. When the apical cell cuts off derivatives, asymmetry sets in with the first unequal division in a cell lineage. The location of cell plate formation during this division is vital in determining the asymmetry, and later divisions, while generating cycles of three to four merophytic clones, retain this asymmetry. As in other asymmetric divisions in plants (see Chapter 3 of this volume), a cytoplasmic cortical site of the about-to-divide cell determines polarity and asymmetry. This is probably a universal mechanism determining asymmetric divisions through all groups of plants. Although merophytic organization and phytomers are also known in plants with multicellular apical meristems, we do not know very clearly whether an asymmetry associated with unequal cell divisions is also seen in them. Whether the körper-kappe organization reported in multicellular root meristem derivatives is a reflection of asymmetry in merophytic clones needs to be verified, according to the editors of this volume.

Manoharachary and Nagaraj discuss the role of handedness in fungi in the sixth chapter of this volume. This is evident in the asymmetric movement that is often observed in some Mycomycete members, and in dextral and sinistral "spiral" growth in some *Phycomyces* species and cultured filamentous fungi. Asymmetry is also seen in the reproductive/sporulation phase of fungi. This article clearly depicts the lack of detailed work on asymmetry in fungi and the need for much more detailed work on this group of organisms in the future.

The seventh chapter of this volume written by Tennakone highlights some aspects of left-right asymmetries in plants and highlights the fact that these symmetries have received less attention in plants than in animals. This paper also emphasizes the importance of utilizing left-right asymmetry to achieve effective wind dispersal of seeds through autogyration. This paper seems to emphasize the need for more studies on fruit and seed dispersal and its relation to left-right symmetry. For instance, many dry legume pods dehisce, during which the two valves of the fruit spirals to left or right direction; it would be worthwhile to know whether the direction of valve spiraling is a constant feature for any legume taxon or is a variable trait.

In Chapter 8 of this volume, Andrey A. Sinjushin and Bir Bahadur summarize the molecular and genetic aspects of handedness in plants. These authors emphasize that most data were obtained from experimental studies on *Arabidopsis thaliana*, which normally lacks chirality. The observations made on mutants form the basis for the interpretation of asymmetry. They also cite the work on *Medicago truncatula* where a single Mendelian gene controls handedness of pod coiling, as discussed in the earlier paragraph. The editors of this volume feel that molecular studies are almost lacking in plants that show definite handedness, when compared to animals. We also feel whether results obtained in the normally achiral *Arabidopsis* can be extended or applied to cases that are normally handed.

The ninth chapter of this volume concerns embryo rotation written by Krishnamurthy et al. This was first observed by the first and second authors in *Polygala* species. Right from the proembryo stage onwards, the embryo rotates along its longitudinal axis and temporarily stops its rotation at seed maturation. At this time, the mature embryo is oriented either parallel to the two cotyledons or anti- to the cotyledons. Embryo rotation is likely to be related to seedling handedness observed immediately after germination. More studies are needed, using better techniques, to verify the presence of embryo rotation in all plants or whether it is restricted only to plants where seedling handedness is known.

The subject of the 10th chapter, circumnutation, first highlighted by Darwin, is a phenomenon that appears to be restricted to plants, probably because of their stationary habit. It would be worthwhile to check whether it is present in stationary animals like corals. The phenomenon also appears to be seen in woody plant species. So far, four models have been proposed to explain circumnutation, and according to the authors of this chapter, Krishnamurthy et al., some aspects of all four models appear to be involved. The direction of circumnutation seems to be more easily explainable, but changes in the direction are not easily explainable, as also its speed and amplitude. More research is needed on circumnutation with reference to these aspects; also needed is research on circumnutation in woody plants. Is it related to spiral grains, and is it related to reduce the resistance offered to shoot tips by air and to root tips by soil? Is circumnutation a continuation of embryo rotation reported in Chapter 9? These questions need to be addressed by future research.

In Chapter 11, Bir Bahadur et al. discuss seedling handedness in Fabaceae, Poaceae, and Polygonaceae. Seedling handedness relates to the left or right position of the first leaf of the seedling, in an alternate phyllotactic pattern, to the two cotyledons. Invariably, the first leaf primordium encloses the second primordium in the vernation condition of the plumule. Although Compton and Kihara reported seedling handedness, much was not known about this. This chapter highlights this phenomenon. Still many questions remain unanswered or needs critical further examination. The first one relates to finding whether seedling handedness is also known in plants with opposite leaves. The second relates to finding the relationship between embryo rotation discussed in Chapter 9 and seedling handedness. The third relates to the contribution of circumnutation to seedling handedness. The fourth is whether seedling handedness extends until flowering and, if so, whether it is related to phyllotactic spirals.

Robyn J. Burham and his coworkers have made a comparison of chirality patterns in climbing plants, especially lianes, in moist tropical forests of Peru and Brazil in Chapter 12. Climbing involves circumnutation. In contrast to the general assumption that dextral and sinistral climbing directions are equally preponderant, earlier work showed a greater distribution of dextral chirality. Dextral chirality in Brazil is similar in proportion to that observed in Peru. What causes this dextral and sinistral difference in climbing direction? Is it related to the direction of circumnutation? What causes the presence of both directions of climbing in the different individuals of same species? These are some questions according to the editors of the volume that need to be answered by intense future research.

Chapter 13 concerns handedness in tendrils, the most common of all climbing organs of plants. This chapter contributed by Silva et al. discusses the helical handedness of tendrils (technically, not spiral shapes) as well as perversions that are observed in them and which help in changing their directions. The importance of tendril-shape changes and perversions in material sciences is discussed by the authors. The biomechanics of tendrils is one example where a plant structure and its behavior have great potential value in the physical and mathematical fields.

Monoranjan Ghose and Bir Bahadur give a general account on the Fibonacci sequence in Chapter 14. The origin of the name for this sequence and its presence in several spheres of Nature are discussed by these authors. With successively dividing each term in the Fibonacci sequence by the previous term, a ratio appears to be settling down to about a particular value (1.618034), which is known as the golden ratio. The branching rates in plants occur in the Fibonacci pattern, where the first branch level has 1 branching (the trunk), the second has 2 branches, then 3, 5, 8, 13, and so on. Most commonly, the Fibonacci number is associated with phyllotaxy and the arrangement of flowers in a highly condensed inflorescence like that of the sunflower. The Fibonacci sequence-related research has been extensively done in mathematical biology. Such a research, with reference to sunflower spirals, has been described and discussed by Oleh Bodnar in Chapter 15. He has shown that the golden functions can be used to represent the Fibonacci series through specific mathematical functions. A golden function makes it possible to provide an exhaustive mathematical interpretation of phyllotactic growth in plants. Minkowski's geometry is implemented by growth mechanisms of spiral and asymmetrical plant structures.

The 16th chapter of this book by Riichirou Negishi deals with the determination of parastichy numbers and its applications using sunflower inflorescence. The number of spirals exhibited by the flowers of an inflorescence is often called the parastichy number. Since the two adjacent spirals are asymmetric in clockwise and counterclockwise directions and the ratio between them in actual sunflower capitula is equal in his studies, he had used a sunflower model and a pineapple model. He had applied the discrete Fourier transform to the distance between each point in the point distribution, which was obtained by changing in divergence angles to golden angles or arbitrary angles.

Pullaiah and Bir Bahadur have dealt with floral symmetry in Chapter 17 of this volume. They have discussed the evolutionary importance of symmetric versus asymmetric organization of floral parts, particularly petals and styles. They have also discussed the role of genetic factors in controlling the symmetry of flowers. Zygomorphic flowers are considered more evolved than actinomorphic flowers. Several aspects of floral asymmetry need to be subjected to detailed research in the coming years. These relate to the roles of pollinating insects in the evolution of floral asymmetry, particularly that of the petals/perianth lobes, disposition of stamens, vis-à-vis pistil segments (style and stigma), etc.

Chapter 18 discusses the labile handedness of contorted corolla. Karpunia et al. discuss the transference of positional information from bracteoles and calyx lobes to petals with contorted aestivation. The authors have shown that in some eudicots, contorted aestivation is fixed at the species level, while in other dicots it is not fixed. In the latter cases, both the left- and right-handed contorted condition are seen in different flowers, often in the same inflorescence. They further try to explain the possible mechanisms involved in their developmental regulation. In species with unfixed contort direction, they speak of the transference of positional information from sepals to petals, which in turn is determined by the arrangement of bracteoles. The way in which sepals determine corolla handedness is believed by these authors to depend on two possible mechanisms: (1) signal transference by mechanical forces caused by sepals during floral development and particularly during corolla ontogeny, and (2) signal transference taking place much earlier during prepatterning of petals on the floral meristem so as to result in an asymmetric petal initiation site from the beginning. It would be worthwhile to study whether positional information from bracts/bracteoles/sepals controls other types of petal aestivation. It would also be worthwhile to find out whether positional information is involved in spiral arrangement of petals/stamens. It is also important to find out the nature of this positional information.

Chapters 19 through 21 deal with symmetry/asymmetry in flowers caused by the style. Solomon Raju reviews stylar polymorphism in flowering plants and concluded that this feature evolved differently in different bisexual flowers, as an adaption to insect cross-pollination and to ensure reproductive success. In addition to petals, the other floral structure that is responsible for floral symmetry is the style. Heterostyly, enantiostyly, inversostyly, flexistyly, etc. are discussed in this paper. It will be interesting to discuss floral asymmetry in relation to variation in the category of insect visitors to flowers. The editors of this volume feel that since insects co-evolved with the evolution of flowers, particularly with asymmetry in floral organization, more studies are needed to relate floral asymmetry with the insects that visit the flowers. Almeida and Castro have reviewed enantiostyly in angiosperm in Chapter 20. According to these authors, this feature was evolved independently in 11 angiosperms families. Tribe Cassinae of Fabaceae is the dominant plant group that exhibits enantiostyly. They point out that several aspects on enantiostyly require attention in future research; these relate particularly to the genetics and maintenance of monomorphic enantiostyly. The editors of this volume feel that the relation between enantiostyly and type of insect visitor to flowers with that condition need to be addressed in greater length in future studies. The topic of the 21st chapter is inversostyly. The authors of this chapter emphasize that inversostyly is a fairly newly recorded floral polymorphism and has been known in only one species, *Hemimeris racemosa*. This species is characterized by style-down and style-up flowers in different individuals as well as homostylous individuals. The authors speak of two other species with inversostyly where the condition is the result of floral resupination (called dimorphic resupination). Species with inversostyly are zygomorphic and show a specialized pollination mechanism. Again, as stated earlier, detailed studies are needed to specifically address the relationship between inversostyly and the type of pollinating insect(s) that visit these flowers.

In Chapter 22, Amorim et al. have made a very interesting study of asymmetric flowers *vis-à-vis* buzz pollination by hymenoptera. They record that buzz pollen released from anthers rebounds on the floral leaves and then get dusted onto the insect's body to help in the process of pollination. The editors of this volume agree with the authors of this chapter that the ecological role(s), the evolutionary history, and the biophysical aspects of ricochet pollinating mechanisms need more detailed future studies in many other honey-bee pollinated taxa.

Chapter 23 by Bir Bahadur and Monoranjan Ghose have made an interesting review of the relationship between left- and right-handedness and the yield of plants showing them. Citing work on different taxa such as coconut, oil palm, date palm, *Dioscorea* species, some legumes, and a few Solanaceae taxa, they show that yield is often related to handedness. A critical study of their paper, however, shows that either of the two conditions, right or left, may be related to higher or lower yield or vice versa. More studies on a number of plants that show distinct handedness are needed to make definite conclusion on the relationship between bilateral asymmetry and yield.

The papers contained in this volume together raise several pertinent questions that need to be addressed in greater depth in the coming years (see Levin et al., 2016). They are as follows: (1) What are the actual and relative roles of genes, environment, and chance (stochastic factors) in the origin and evolution of asymmetry in plants? Although Palmer (2016) has addressed this question at greater length in animals and to a small extent in plants, there is need for more research in plants. There is also a greater need for bringing together the scattered bits of information available on plants (than has been covered by Palmer, 2016) and to indicate the gaps in our knowledge. A clear and exhaustive list of asymmetric plant features that are controlled by genes, environment, and stochastic factors need to be prepared; (2) Are homologous asymmetric traits in different taxa of plants having a homologous developmental basis, as has already been raised by Schilthuizen and Gravendeel (2012) for animals?; (3) How far have asymmetric or symmetric developmental patterns in plants become immune to disruption during evolutionary history? Some cladistic analyses of asymmetry have been done on various animal groups, but so far, not many such analyses seem to have been done in various plant groups. This is one area where future research on plant asymmetry might have to be focused on; (4) In what respects are asymmetric developmental pathways more (or less) advantageous than symmetric ones to the diverse types of plants, both in vegetative and reproductive development? Asymmetric development in floral construction has obvious advantages in reproductive success of plants, particularly through promotion of out-crossing. However, more supporting studies are needed to validate this point; (5) Have there been cases of reverse changes from asymmetry to symmetry in plants? If we assume that asymmetry was derived from symmetry, that is, retrograde evolution, are changes irreversible? So far, no effort has been taken to address this aspect in plants. If asymmetry is caused by environmental or stochastic factors, this should be possible; (6) A related question is, to what extent are asymmetry changes congenital or ontogenetic, and can we really classify asymmetries reported into either of these categories? There is some information on these aspects found scattered, but so far, no one has attempted to consolidate this information; and (7) Of the three models proposed, explain asymmetry, morphogen, positional information, and cytoskeletal models. The first two largely remain theoretical and conjectural, at least in so far as plants are concerned, while the third has some experimental evidence. The third-mentioned model operates at the cellular level, but whether this can be extended to higher levels of plant organization is still doubtful. More studies on these three models are essential.

Symmetry/asymmetry is an exciting area for more in-depth multidisciplinary research that would greatly enhance our knowledge of evolution and development of form in plants; this in turn will advance our knowledge in the relation between form and function.

REFERENCES

Levin, M., Klar, A.J.S., and Ramsdell, A.F. 2016. Introduction to provocative questions in left-right asymmetry. *Phil. Trans. R. Soc. B.* 371: 20150399.

Palmer, A.R. 2005. Antisymmetry. In: *Variation, A Central Concept of Biology.* Benedikt Hallgrimsson, B & Hall, B.K. (Eds.). Elsevier, New York. pp. 359–397.

Palmer, R.A. 2016. What determines direction of asymmetry: Genes, environment or chance? *Phil. Trans. R. Soc. B.* 371: 20150417.

Schilthuizen, M. and Gravendeel, B. 2012. Left-right asymmetry in plants and animals: A gold mine for research. *Contrib. Zool.* 81: 75–78.

Index

Note: Page numbers in italic and bold refer to figures and tables, respectively.